TCC South

Crowley Learning Center

Roald Hoffmann on the
Philosophy, Art, and
Science of Chemistry

Roald Hoffmann on the Philosophy, Art, and Science of Chemistry

EDITED BY JEFFREY KOVAC

and

MICHAEL WEISBERG

OXFORD
UNIVERSITY PRESS

OXFORD
UNIVERSITY PRESS

Oxford University Press, Inc., publishes works that further
Oxford University's objective of excellence
in research, scholarship, and education.

Oxford New York
Auckland   Cape Town   Dar es   Salaam   Hong Kong   Karachi
Kuala Lumpur   Madrid   Melbourne   Mexico City   Nairobi
New Delhi   Shanghai   Taipei   Toronto

With offices in
Argentina   Austria   Brazil   Chile   Czech Republic   France   Greece
Guatemala   Hungary   Italy   Japan   Poland   Portugal   Singapore
South Korea   Switzerland   Thailand   Turkey   Ukraine   Vietnam

Published by Oxford University Press, Inc.
198 Madison Avenue, New York, New York 10016
www.oup.com

Oxford is a registered trademark of Oxford University Press

Library of Congress Cataloging-in-Publication Data
Hoffmann, Roald.
Roald Hoffmann on the philosophy, art, and science of chemistry / edited by
   Jeffrey Kovac and Michael Weisberg.
   p. cm.
   Includes bibliographical references and index.
   ISBN 978–0–19–975590–5 (hardback)
   1. Chemistry—Philosophy.   2. Communication in science   3. Art and science.
   I. Kovac, Jeffrey.   II. Weisberg, Michael.   III. Title.
   QD6.H64 2012
   540–dc23   2011022118

9 8 7 6 5 4 3 2 1
Printed in the United States of America
on acid-free paper

Acknowledgments

From the Editors

Throughout the process of bringing this book to press, the editors have had the complete cooperation of Roald Hoffmann. To begin, Roald sent us copies of his papers with suggestions of what to include. Once the selection had been made, he revised each of the papers to eliminate unnecessary redundancies and to bring the scientific content and the references up to date. He also provided original figures. He has read many drafts of the introduction, providing suggestions for improvement but allowing us the intellectual freedom to introduce his work as we felt most appropriate. Without his help, this book would not have possible. Working with Roald has been both intellectually stimulating and personally rewarding.

From Jeffrey Kovac

I am grateful to my long-time administrative assistant, Rai-Lynne Broach Alexander, for her help in preparing the final manuscript. As always, the support and encouragement of my wife, Susan Davis Kovac, has been essential to the success of this project.

From Michael Weisberg

I would like to thank Simon Garcia and Deena Skolnick Weisberg for conversations and correspondence that greatly improved this project.

From Roald Hoffmann

It was Jeff Kovac's and Michael Weisberg's idea that it might be useful to bring out a collection of my philosophical, social, artistic, and educational writings, scattered as they were across the literature. That it was necessary to do this is in a way testimony to my lack of ability to integrate these writings, the carving out of my own land twixt philosophy, chemistry, and poetry, into a coherent book. But more than that, the crafting of this assemblage is an act of friendship on the part of two fellow thinkers. We are on the same wavelength, so to say. Or the same wave, at least. I am very grateful to both for their work on the project.

At the end of chapter 1 I thank the people, so many of them, that have encouraged and supported me over the years. I would add here Catherine Kempf for research and help in editing the essays for this volume. Thank you all for helping me reflect.

Contents

Roald Hoffmann on the
Philosophy, Art, and
Science of Chemistry

Introduction

MICHAEL WEISBERG AND JEFFREY KOVAC

Roald Hoffmann has been widely recognized for his fundamental research in chemistry, as well as for his literary career as a poet, playwright, and essayist. But Hoffmann is also a philosopher of science who has made significant contributions to our conceptual, historical, and normative understanding of chemistry. This book collects many of Hoffmann's philosophical essays, which represent a significant and unique understanding of chemistry and its broader context. It is divided thematically into five sections: Chemical Reasoning and Explanation, Writing and Communicating Science, Art and Science, Education, and Ethics.

Hoffmann characterizes his own scientific research as "applied theoretical chemistry." Reflecting on his own work in this area has led Hoffmann to consider the problem of explanation in chemistry. Hoffmann rejects a strictly reductionist framework in which the best explanation of a chemical phenomenon is always its underlying physical processes. While the methods of quantum chemistry are powerful tools for computing molecular properties, he argues that chemists need to provide explanations using chemical concepts. Hoffmann calls this explanatory practice *horizontal reasoning*.

Science is public knowledge and must be communicated both to the scientific community and to the general public. Not only has Hoffmann written extensively for both audiences, he has also made a study of scientific writing, both the form and the language. Perhaps uniquely among the sciences, the language of chemistry is pictorial and has what Hoffmann calls an "iconic vernacular." Hoffmann's reflection on the graphical presentation of chemistry has led him to a wider consideration of the relationship between art and chemistry, a connection made even stronger by the centuries-old craft tradition which is one of the sources of modern chemistry.

One of the ways that chemists communicate is as teachers. Hoffmann himself is a gifted and dedicated teacher who believes, quite correctly, that research and teaching are intertwined. We have included several essays that discuss his insights into chemical education.

The final section of the book is a collection of essays and two unpublished lectures on ethics in science. Although many of Hoffmann's popular essays and his plays discuss ethical themes, we have chosen to include these essays because they take up ethical questions explicitly, focusing on the contemporary problem of environmental degradation. Hoffmann explores the paradoxical position of the chemical industry as a source of significant pollution but also a needed partner in solving environmental problems. He argues that *green chemistry* can and must play a central role in reducing environmental degradation.

Hoffmann is *reflective*, in a way that is relatively rare in chemistry. He thinks and writes about the way chemistry is practiced, how chemistry is embedded in society, and the psychology of its practitioners. He also writes in a direct and expressive manner, perhaps as a consequence of his writing career in poetry and theater. This book is a collection of his thinking about the molecular science.

Chemical Reasoning and Explanation

Throughout his writing on this subject, Hoffmann has argued that chemical reasoning and explanation have a distinct character. This is at odds with the classical tradition in philosophy of science that seeks a uniform logic for explanation and confirmation, largely based on paradigm episodes in the history of physics. Recent developments in philosophy of science have emphasized the extent to which biology deviates from these logical patterns. The essays in this section argue for a similar thesis and develop a theory about the form of chemical explanation across most areas of chemistry.

Hoffmann's account of chemical explanation begins from the recognition that the central chemical activity is the synthesis of molecules. Chemists make new molecules, some for practical applications, others because they are of theoretical interest, or just because they are a challenge. Although chemistry is a diverse field, encompassing many research activities, much of this activity is aimed at understanding molecular structures and transformations between these structures, in other words, chemical synthesis. To use Hoffmann's own expression, creation rather than discovery is the goal of much of the chemical community.

The key to understanding how chemists create molecules is found in the molecules' three-dimensional structure. One account of how scientific explanations work says that to understand this structure we need to show how it can be derived from the laws of physics. This covering law account of scientific explanation is what Hoffmann terms the "vertical" model of chemical explanation. Quantum theory allows the chemist to write down the molecular Hamiltonian and find an approximate solution for the molecule's wavefunction and the energy. This, in turn, allows for the prediction of molecular structure and other properties of the molecule.

With increasingly sophisticated mathematical techniques and ever faster computers, it is now possible to compute the properties of relatively large molecules this way. But do these calculations constitute a chemical explanation of molecular structure? According to the covering law account, with a few emendations to allow for approximations, they are paradigm cases of scientific explanations. However, Hoffmann argues that these are not explanations at all because they convey no chemical understanding, which is the kind of understanding needed by the experimentalist.

What is chemical understanding and why isn't it conveyed by quantum mechanical calculations? Hoffmann argues that chemical understanding arises when explanations deploy purely chemical concepts in a form that can be used across a range of similar situations. He calls such explanations "horizontal," because of the use of chemical, not physical, concepts. When the conformation of a chemical structure is explained by reference to a chemical property like steric hindrance, this gives the chemist an idea about what the structure of a similar, but distinct molecule would be. Further, horizontal explanations are contrastive. They don't just explain why p happened, but why p happened instead of q.

It is easy to see why contrastive explanations are important in chemistry. The key is chemical synthesis. To manipulate matter, one needs to know more than "quantum mechanics says that this bond angle is the most stable." Rather, one needs to be able to anticipate how the bond angle will change as a bulky protective group or a new substituent is added. There is thus a deep connection between identifying trends, contrastive explanations, and chemical synthesis.

Hoffmann's essays also point to another interesting feature of the chemical concepts used in chemical explanations. Such concepts, he argues, are often fuzzy, vaguely defined, and changing through time. Concepts such as acidity, electronegativity, aromaticity, and functional group involve vague predicates. While chemists can give definitions and accounts of them, most attempts to draw rigid boundaries around them have not been fruitful. Moreover, penetrating empirical observations and Gedanken experiments can take what seems like a straightforward definition, acidity as the concentration of H^+ in an aqueous solution, and show that this definition is inadequate. In the case of acidity, what should we say when water isn't the solvent? Or when other acceptors of electrons are present, but H^+ is not? Or when we are dealing with gases or solids instead of liquid solutions?

Chemical explanations and predictions often rely on analogical arguments, narrative, and stories. Why are chemical arguments developed this way instead of via exclusive appeals to mathematics? To answer this question, Hoffmann emphasizes the psychological needs of the intended audiences of chemical argumentation. He argues that since science is made by human beings, it must be understood in human terms, even insofar as it is aiming to tell us objectively about the world. As Hoffmann points out in the essay "Nearly Circular Reasoning," "Science is a curious mixture of the real and the ideal, the material and the spiritual, held together by discourse or argument. The latter is sometimes mathematical, but more often it transpires in the words of some language. [...] Try to imagine a scientific article or seminar without the glue of words or argument."

Hoffmann's essays also address issues of confirmation and testing, in particular the overreliance on heuristics of simplicity. Many scientists accept as an axiom the doctrine known as Ockham's Razor: When two hypotheses explain a phenomenon equally well, one should choose the simpler hypothesis. Medieval theologian and philosopher William of Ockham applied the principle to ontological arguments, but chemists are more likely to appeal to it as a justification for favoring a simple reaction mechanism over a more complex one.

In their essay, "Ockham's Razor and Chemistry," Hoffmann and his collaborators, Vladimir I. Minkin and Barry K. Carpenter, look at the strengths and limitations of the use of the Razor in chemistry, particularly in the analysis of chemical reaction mechanisms. Agreeing with much of the philosophical literature, they point to the role of Ockham's Razor in curve fitting. They point out that even with highly correlated data, where a simple linear fit would be deemed appropriate, there are always many more nonlinear curves that will fit that collection of points. However, this nonlinear curve requires many more parameters, so allowing this possibility opens us up to what they call "real indeterministic chaos—the infinity of hypotheses that fit."

As in curve fitting, chemists are often drawn to the simplest mechanism, even when many other, more complex mechanisms cannot be ruled out. Although this view is often reasonable, Hoffmann, Minkin, and Carpenter argue for caution because nature is sometimes not as simple as we might hope. For example, a reaction can proceed through multiple paths depending on the circumstances. They also point to another consideration in the choice of reaction mechanism: future productivity. The more ornate hypothesis might be richer in its ability to lead to experiments that help us understand the phenomenon. Hoffmann and his collaborators conclude that we ought to look for simplicity, but be wary of it.

The final set of papers in this section concern the relationship between theory and experiment. In a science such as chemistry, where more than 95% of its practitioners are pure experimentalists, what is the proper role of the theorist? Hoffmann's view is that chemical theorists have a dual role in the chemical community: First and foremost, theorists should be explainers and should construct horizontal explanations. Second, Hoffmann believes that theorists should stimulate experiments. One way that they can do this is by imagining new molecules—interesting structures that are likely to be stable but have not yet been made.

Given that more than one hundred million structures have already been characterized by chemists, why would one want to imagine even more molecules? One motivation is to allow or encourage theorists to participate in the creative part of chemistry: synthesis. But a second and more important reason is to test fundamental ideas about molecular structure and properties. In this context, there is much to be learned from what Hoffmann calls "molecules in distress," structures that bend the normal rules. For example, carbon prefers to form four bonds that are oriented in a tetrahedral arrangement. But molecules can be made, or imagined, where the bond angles are significantly distorted. One can learn much about molecular stability and reactivity by studying these structures. These distressed structures also present a significant challenge for the synthetic chemist. Can they actually be made? Finally, Hoffmann points out the sheer beauty and pleasure of this type of work. Humans love to play and inventing new things is one of the most delightful of intellectual games.

The study of distressed molecules raises a more general question about experimentation. Hoffmann believes that standard discussions of scientific practice overemphasize observation and underemphasize the heart of experimental method: intervention. Hoffmann and Pierre Laszlo use the metaphor of Proteus, the Greek sea god who would only foretell the future for those who can seize him to illustrate this point. Proteus avoids detection by changing his form, but Menelaos wrestles with Proteus, pins him down, and makes him talk. Francis Bacon uses the same image, wrestling, to describe the interaction between the scientist and nature. Like Proteus, chemists are also the masters of change, by engineering transformations of molecules. They don't just observe matter, but pin it down, forcing it to speak.

Writing and Communicating Science

As a scientist, Hoffmann understands the role of writing in the creation of scientific knowledge; as a poet, playwright and essayist, he has a deep appreciation of language. Because of his dual perspectives as a scientist and a humanist, Hoffmann has unique insights on scientific communication, particularly the language of chemistry.

The language of chemistry is especially visual. As noted above, a central tenet of chemistry is that the three-dimensional structure of molecules determines properties and reactivity. Chemists have thus developed a language of drawings to represent these structures on the page, what Hoffmann calls chemistry's "iconic vernacular." Typically, substantial portions of chemical articles are occupied by drawings, and even the tables of contents of many chemistry journals are given pictorially. Pictures of structures, reaction schemata, and spectra complement narrative to give a full description of chemical phenomena.

In "Representation in Chemistry," Hoffmann and Pierre Laszlo discuss the intricacies of chemical representation using molecular formulas. Like the characters of any other language, molecular formulas are ultimately symbols, and hence what they denote is intrinsically arbitrary. Their relations of denotation to the structures that they denote are set up according to the conventions of the chemical community. However, the symbols used to denote molecules were not chosen arbitrarily. They were specifically intended to call up a mental picture of the actual structure. At the same time, real molecules are far more complicated than any of their representations. Thus, chemists' structural formulas also have a theoretical role. They are abstractions, designed to emphasize those aspects of the molecule that are important for the argument being developed. Therefore, drawing chemical symbols can be viewed both as a rhetorical device and as a form of art, a symbol used by the chemist to communicate with an audience.

Despite being rich with illustrations, chemical articles are often written impersonally and in the third person. But since scientific articles, as opposed to the phenomena they describe, are human constructions, this style of writing is optional. Have scientific articles always been written this way, and is this a good thing? Hoffmann answers both questions negatively. He argues that while early scientific articles were much more discursive and personal, nineteenth-century German chemists struggled to distinguish their work from the more romantic perspective of the *Naturphilosophen*. In the process, they developed a third person, passive style in which facts were emphasized and the role of the investigator suppressed. This has since become the canonical form of the chemical article.

Although this communication system has worked well for nearly two hundred years, Hoffmann argues that the canonical form of the chemical article ought to be changed. He argues that writing in a dry, passive style removes the human side of science. Scientific discovery is a human activity, motivated by epistemic, practical, and even aesthetic considerations. Although scientific knowledge itself must not depend on the discoverer, suppressing the human side misrepresents the actual process of science. Hoffmann argues that when scientific discovery is presented in a more relaxed, autobiographical, and historical format, chemists can more easily "tell their stories." They can emphasize what is most important and communicate the wonder of scientific discovery.

As evidenced by his own prolific career of popular essay writing, Hoffmann also believes that chemists have an obligation to educate the general public. He urges chemists to communicate with the general public so that the citizenry can better understand the chemical issues at stake in public policy. But he also notes that communication with a wider audience requires especially creative thinking about chemistry. This can sometimes lead to new insights and new ideas for research.

Art and Science

Hoffmann is passionate about the fine and performing arts, and that passion pervades his work. He has said that at the end of his college years, he considered making a career of the history of art instead of chemistry, but ultimately lacked the courage to do so. With time, he has managed to find ways to meld his interest in art with his research in chemistry. In 1993, he collaborated with the visual artist, Vivian Torrence, on the book *Chemistry Imagined*, which integrated prose, poetry, and art to communicate chemistry to the general reader. The book was modeled after emblem books of the Renaissance period. His 1995 book, *The Same and Not the Same,* is similarly a mixture of visual art and chemistry. The three essays in this section explore the connections between art and science, focusing, of course, on chemistry.

Chemists are visualizers; they must be able to imagine and represent the three-dimensional structure of molecules that cannot be seen. Although their thinking about structures is often aided with molecular model kits and, more recently, computer programs, pencil and paper are still the primary tools of the working chemist. One of the skills that an undergraduate must learn is how to draw molecules in ways that show the spatial arrangement of the atoms.

Are these drawings a form of art? Hoffmann argues that they are—not great art, but art nonetheless. Drawings, he tells us, make the molecules come to life. The way that a molecule is drawn emphasizes those aspects of the molecule that the person making the drawing thinks are most important. Sometimes molecular drawings can even be beautiful.

As a science, chemistry owes as much to craft as to natural philosophy. Chemical crafts, such as metallurgy, ceramics, and pharmacy, have a long history, and the skills and techniques of these practical pursuits have found their way into the research laboratory. The synthetic chemist is as much a craftsman as any jeweler; getting a reaction to work can be as delicate a task as throwing and glazing a pot.

In the essay, "Science and Crafts," Hoffmann explores the connections between these two pursuits based on his experiences at the Penland School of Crafts in North Carolina. Most crafts involve some kind of chemistry, usually learned through apprenticeship and practice. Hoffmann asks, "How much chemistry does the craftsman need to know?" Usually, the answer is not much, but a deeper knowledge can be enriching and useful, particularly if the process fails. In that case, a knowledge of chemistry might help in finding and correcting the problem. In chemistry as in craft, hands and minds work together creatively to produce useful, and often beautiful, things.

This leads us to the final question taken up in this section: What, if anything, makes a molecule beautiful? As Hoffmann points out, it is rare to see an aesthetic judgment in a scientific article, but in more unguarded rhetorical circumstances, chemists may remark on the beauty of a particular molecule, just as a physicist or mathematician might say that an equation is beautiful. In "Molecular Beauty," Hoffmann undertakes an investigation of molecular aesthetics.

In some simple cases, molecules are beautiful to the untrained, usually because of their simple but surprising symmetric shapes. Perhaps the most famous molecule of this kind is buckminsterfullerene, the molecular soccer ball. More frequently, substantial chemical background is needed to appreciate chemical beauty. Some of this beauty comes from structure alone, such as the interplay of octahedra in a complicated inorganic crystal or a pair of interlocking rings. In other cases, the beauty comes from function. And in still other cases, the beauty is conceptual, and can be exquisitely concentrated, as when a molecule is created with a single, double, and triple bond all coordinated to the same metal atom.

In considering the criteria for beauty that can be abstracted from these examples, Hoffmann argues that many of these criteria have a counterpoint in classical thinking about esthetics. One major exception he points to involves utility: In most arts, being useful is considered an entirely separate issue from being beautiful. But in chemistry, beauty often comes from molecules and reactions that are especially useful for making other molecules.

Education

Throughout his career, Hoffmann has emphasized the importance of teaching and has been recognized as an outstanding educator, most notably with the George C. Pimentel Award for Chemical Education of the American Chemical Society in 1996. In the late 1980s he took an active part as the presenter and script writer for an influential Annenberg-PBS series of 26 half-hour videos, "The World of Chemistry."

As might be expected, Hoffmann has an unorthodox perspective on teaching and learning that provides the chemical educator with important insights. It is commonplace to say that research improves teaching, but Hoffmann turns that statement around and argues that teaching improves research. In fact, he goes further, by saying that teaching and research are inseparable and symbiotic. However, simply being a good researcher does not automatically make one a good teacher. There are important lessons to be learned both from those who study the teaching and learning process and from successful teachers.

Hoffmann argues that we should view learning not in terms of its location (lecture room, laboratory, etc.) but rather in terms of its audience (undergraduates, graduate students, etc.), noting that audiences often overlap and complement each other. Seen this way, the process of research, which involves developing arguments to convince first oneself and then an expanding series of audiences that something new has been discovered, is essentially the same as teaching. Both involve explaining in words, diagrams, models, and metaphors. Both require clarity of thought and of expression. Both require an understanding of the audience, including what can be assumed and what must be explained.

Those of us who teach undergraduates quickly learn that our students' expectations of the course are different from ours. Hoffmann and Brian Coppola provide us with a nice list of what they call "heretical thoughts on what our students are telling us." Some of the things we consider important, such as providing a broader context, many students find distracting. As we push them to understand principles and see the big picture, students find that memorization and compartmentalization are often effective learning strategies. These lessons from students are important, but as professionals and expert learners, we need to design courses that communicate what is essential and valuable about chemistry.

Along with their philosophical and theoretical perspectives, teachers also need techniques, practical suggestions on ways to help students learn. Based on forty years of experience, Hoffmann, along with Saundra Y. McGuire, provide the working teacher with some down-to-earth advice supported by references to the teaching and learning literature showing why these techniques actually work. Some of the advice seems obvious: Students should take lecture notes by hand, then rework them after class. But there are good reasons, based on cognitive research, why this old idea is still a good idea. The suggested strategies are all ones that promote active learning by taking advantage of a productive interplay of group and individual learning. But technique needs to be supplemented by empathy and empowerment. Teaching is a relational human activity and the human connection is essential.

Ethics in Science

Hoffmann has not written many articles that discuss ethics explicitly, although moral considerations are an important subtext in much of his writing. The four essays in this section, including two previously unpublished lectures, grapple with several important ethical questions involving chemistry, particularly concerns about environmental degradation.

The essence of Hoffmann's ethical stance can be found in the title of his *Chemical and Engineering News* editorial: "Mind the Shade." Humans are constantly presented with ethical decisions, few of which are black and white. Because of the natural tendency to rationalize, compartmentalize, and diffuse responsibility for actions that result in hurt, it is essential that we pay special attention to those moral decisions that are in the gray area. Because chemists create new molecules and new understanding of the natural world, they need to be mindful of the consequences of their discoveries. It is too easy to say that the responsibility lies with those who use our creations, and not with the creator.

Where do we look for guidance in making those difficult decisions in the shade? Hoffmann reminds us of the work of Jacob Bronowski, who pointed out that the practice of science leads to the "habit of truth." Experiment helps to keep us honest as does the reporting of experiment. There are those who believe that science is ethically neutral, but Hoffmann rejects this view. He believes that "the invention or implementation of a tool without consideration of the consequences of its use is deeply incomplete." Much of his ethical writing explores this theme, pointing out that chemists are forced to make moral choices, whether they like it or not.

One of the most important contemporary ethical issues is environmental degradation. The marvelous inventions of modern chemistry have been accompanied by environmental damage that, in Hoffmann's view has moved from just being a mess to what he characterizes as something like a sin. It would be easy to scold the industries responsible for environmental degradation, but Hoffmann chooses instead to make practical suggestions for how to make chemistry more ethical. Industry will adopt green chemistry practices if they make economic sense; the bigger problem is in the academy where green chemistry is often regarded as uninteresting. Making new compounds is sexy; developing a green process to make an industrially useful compound may seem boring. More federal funding will certainly attract interest, but more important is a change in attitude among academic chemists.

A good model for how green chemistry could thrive in academic departments can be found in the history of polymer chemistry. Once considered second-rate, applied chemistry, polymer chemists are now in high demand in chemistry departments and they seem to have no trouble attracting good graduate students to their research groups. Part of this new found legitimacy is due to the work of some first rate academic chemists, such as Robert Grubbs, who found that polymer synthesis presented some fascinating area of research. Green chemistry needs similar role models. It also needs a different pedagogical approach, one that emphasizes the particular. Case studies can provide an entry point for students, showing them the intellectual puzzles that make chemistry so irresistible.

A final ethical contribution is Hoffmann's suggestion of a new way to evaluate green chemical technologies. He suggests the creation of a "transformation index," a new way of measuring efficiency primarily based on thermodynamic considerations, but that also recognizes ethical dimensions.

In the end, the answers to these broad ethical questions depend on our values which in turn reflect our vision of ourselves as human beings. Here Hoffmann turns to Genesis, which contains two creation stories that show the two sides of the human character: the creative transformer of nature and the keeper of the garden. Both sides of the human personality are needed for complete ethics of science.

Conclusion

Roald Hoffmann's perspective on science derives from his unusual career as a successful theoretical chemist who collaborates closely with experimentalists, and his parallel career as a writer of essays, poems, and plays. Hoffmann's research has ranged broadly through organic, inorganic, and solid state chemistry. His writing outside of science encompasses both scientific and nonscientific themes. In a world where scholarship has become increasingly specialized, Hoffmann's interdisciplinary perspective provides unique insights into the philosophy of science in general and the philosophy of chemistry in particular.

Hoffmann's view is that chemistry, at its core, is a creative human activity in which synthesis, making new molecules, plays a central role. Because of the importance of synthesis to chemistry, chemical explanations must employ those sometimes fuzzy, purely chemical concepts that the experimental chemist can readily use to design and control chemical reactions. Chemical communication is likewise a creative human activity that involves both words and a unique iconic vernacular of drawings that represent the three-dimensional molecular structures that determine chemical properties. Through both the drawings that decorate the typical chemical article and the historic connections with the craft tradition, chemistry has surprising connections to art. Finally, because chemists create new substances, they incur a moral responsibility to consider the consequences for society of the use of the materials they make. Hoffmann provides us with a truly human and humane philosophy of science.

1

Trying to Understand, Making Bonds

ROALD HOFFMANN

In 2007, on the occasion of my 70th birthday, Bassam Shakhashiri organized a symposium for me at the Boston meeting of the American Chemical Society. The session was entitled "Roald Hoffmann at 70: A Craftsman of Understanding." I began my talk with thanks to many. That section has been shifted to the end of this chapter.

Beginnings

I was born in a happy young Jewish family in unlucky times, 1937. In that war, most of us perished, 3800 of the 4000 Jews of Złoczów, now Zolochiv in Ukraine. Among those who were killed were my father, three of four grandparents, three aunts, and so on. I just want to show you three photos which relate to that time, one old and two recent. The last 15 months of the war we were hidden by a good Ukrainian man–Mikola Dyuk, the schoolteacher in the small village of Univ. The first year we were in an attic of the schoolhouse, the second year in a storeroom with no windows, maybe 6 x 10 feet, on the ground floor.

Here are two photos from 2006, when my sister, my son, and I visited Univ. Here is the attic in which we were hidden, with its one window (Figure 1-1).

The storeroom, a passageway, another ground floor room are gone, rebuilt into a new classroom of Univ's school.

It's a chemistry classroom (Figure 1-2). Such is fate. Under the plank floor we dug a bunker to sit in if the police came to the house.

I was five and a half when we went in. And nearly seven when we went out. Here's a photo of me, a few months after we came out (Figure 1-3).

We survived. Some of us. Good people helped us, I tell their story. I am also the speaker for the dead—the three million Polish Jews who were killed do not have good stories to tell, or photos to show.

We built a new life, in refugee camps where I read of Marie Curie and George Washington Carver, and then came to America in 1949.

(a)

(b)

Figure 1-1 (a) The attic of the Univ schoolhouse, (b) and the schoolhouse itself, where Roald, his mother, and three others were hidden in 1943.

Then we lived happily ever after! Or, to put it another way—it was wonderful, the only unhappiness in our lives put there by ourselves and no one else. My mother said I must never say anything bad about America.

Moving toward Chemistry

I got to our science sideways, so to speak. By first going toward (under parental pressure) medicine, then edging away. Not having enough courage to go into the humanities—that wonderful opening up of the world to me at Columbia, which you can sense in my mention of my teachers at the end of this chapter. Even in my

Figure 1-2 A present day chemistry classroom, Univ, Ukraine, in the space where Roald and relatives were hidden in 1944.

Figure 1-3 First photograph of Roald after war, in fall 1944, age 7.

first two years in graduate school at Harvard I wasn't sure about chemistry—I sat in on courses in astronomy, public policy, archaeology. Going to the Soviet Union for a year was a way to avoid commitment.

When I came back in 1961, I knew chemistry was my science. I did my Ph.D. work in one productive year.

You know what I've done, with more than a little help from my friends. I want to talk here about some lines that seem to emerge, ever so faintly, from more than 45 years of scientific activity.

The Rhetoric of Pedagogy

So the chairman of the department (Harold Scheraga) said more or less this: "We thought that after teaching physical chemistry lab, you might want to try an introductory chemistry course." Jim Burlitch and I did as we were told, and rebuilt an honors course that had withered. We were lucky; in that course were Cornell's brightest students. And I learned how to explain to them thermodynamics without those partial derivatives.

I've taught introductory chemistry every year since; I finished teaching at Cornell with two such courses. The pedagogical imperative entered my research. I was in the business of explaining anyway—in time I realized that most of the difference between teaching an introductory course and writing my technical papers was in the audience—the strategies of communication were different on the surface, but underneath similar.

As a result of teaching, my research improved. I wanted people to understand—students, researchers. So I tried, using colloquial language, using strategies of optimal redundancy, many pictures, an occasional poke of colorful language to wake people up. No mysteries (beyond those of nature) were to be created. And if one could communicate to the reader (so much easier in a lecture) that I cared that they understand, I was home free. They learned.

I've always written my papers not for my colleagues, but for the 3rd year graduate student.

Calculations and Explanations

My research career is intimately tied to the computer age; there are all those tables in my very first theoretical chemistry paper, with Lipscomb. An IBM 604 diagonalized that matrix for $C_2B_{10}H_{12}$. The sound of the key punch is deeply embedded in my consciousness. As is FORTRAN. But from the beginning, there was this love-hate relationship I had with calculations. For Lipscomb emphasized symmetry and structure, and the work with Woodward taught me the tremendous power of a qualitative argument.

I could sense that the qualitative arguments, based on perturbation theory, were appreciated, used, and cited by people. I think I understood this firmly for the first time (outside of the context of the work I did with RBW) in my initial independent studies at Cornell—on the way a methylene approaches an ethylene, and in the explanation (through bond coupling) of the large splitting of the lone pair orbitals of pyrazine or diazabicyclooctane.

The tension is exacerbated in our times, as simulation crowds out model-building. I've written of this elsewhere.[1]

I was lucky; I became an explainer in an age that needed them. Now how much of explaining is telling just-so stories, you will have to decide. The stories I've told—of orbitals, of two and four electron two-orbital interactions, of through-bond coupling, the isolobal analogy, of a fragment MO analysis, of crystal orbital overlap populations, of how bands run, and bonds and bands in general—are pretty good ones; they have a way of holding on.

I'm very proud of teaching chemists (together with Jeremy Burdett and Mike Whangbo) the connection between the language and phenomena of solid state physics, of extended periodic systems and the molecular orbital theory of discrete molecules. As for motion in the other direction, convincing physicists that there is value in chemical intuition on bonding, well, the best I can say is that I haven't given up trying...

The Graphical Imagination, or Why Orbitals Matter

So organic and inorganic chemists, with no artistic training, became in the 20th century masters of simple 3-dimensional communication. The context was molecular structure; I am thinking of drawing a chair cyclohexane, benzene-$Cr(CO)_3$, or the organometallic molecules shown in Figure 1-4.

All along, people were drawing orbitals and using them. Leslie Orgel's work in inorganic chemistry in the sixties in exemplary, as is Howard Zimmerman's stereo-electronic reasoning in the organic realm. So there was nothing novel in what we did (now that is a recurring theme, and true), only a persistent, pedagogically informed application of orbitals and interaction diagrams to one real chemical case after another, to a multitude of bonding puzzles, to build frameworks of understanding of some generality, in a language accessible to that 3rd year graduate student.

The net result of hundreds of interaction diagrams of the kind you see here (Figure 1-5) is the grafting of electronic information in a graphical code onto an already existing stereochemical imagination. So, just as molecules could be seen as structures made up of a framework of definite three-dimensionality, with functionality (donors, acceptors, chromophores) geometrically dispersed on that framework, the same molecules could be furnished, so to speak, with orbitals, and those would explain and control reactivity, color, etc.

The necessary orbital theory was within reach of an intelligent graduate student; driven by my teaching instincts (and maybe my deflected art historical interests) we created a pictorial perturbation theory that seemed to be psychologically just right for the sophistication of the community. Michael Dewar couldn't understand how people didn't take to his numerical perturbation theory for organic chemistry—after all, it had everything in it. It did. But it was not taught nor used, just because it was not graphic. People like pictures. Chemists live off them.

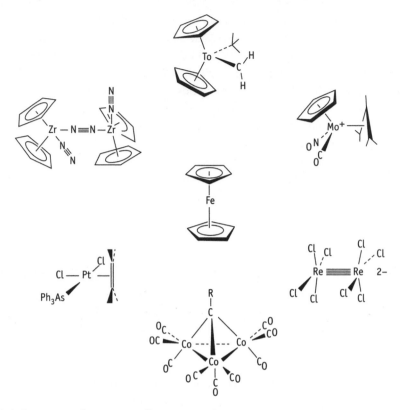

Figure 1-4 Some typical organometallic compounds.

Building Bridges, Making Bonds

I believe leitmotifs emerge, they are not programmed. If someone asked me, "Roald Hilelovich, what is your philosophy of research?" I'd answer—as I answer all your mail, some by hand because my mother, the strong force in my life, said you have to answer mail—I'd answer, "I don't work on important problems, I work on what the world gives me. I am an amoeba." Now this is said in part for effect, I am, after all, a writer. But here is what I mean—the world is complex and particular, the bead of dew on that blade of grass, and no other blade, no other dew, no other observer than me. And the world is connected—the drop's water is my water, the grass blade's biochemistry has much overlap with mine, the chlorophyll is green whether in that blade or in R. B. Woodward's synthesis. There are differences—the weed respires in a way different from me. The differences are telling.

I have patience. And like a figurative omnivorous amoeba I will look at, listen to, taste anything in this world—organic or inorganic, heavy metal or Bach, Korea or Brazil. I will transform the riches of the world not as an amoeba, but as a human

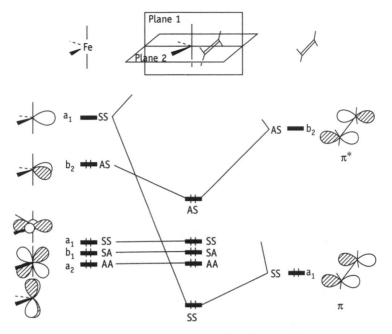

Figure 1-5 An orbital interaction diagram for $(CO)_4Fe(ethylene)$. Note the mix of stereochemical drawings and orbitals.

being. Which means I will try to understand the disparate chemical aspects of a rich universe (and not just the chemistry), try to see relationships between them. My faith—not a religious one, but certainly a spiritual one—is that the connections are there, that our humanity lies in seeking such connections.

I've been lucky, within theoretical chemistry. A method came my way—extended Hückel theory, but more broadly qualitative molecular orbitals and the perturbation theory-based way of following their interactions. That method could be used on any molecule under the sun. The fragment molecular orbital analysis attached to this methodology took molecules apart, and put the pieces together, on paper. This permitted connections to be made—one could see similarities, while at the same time the differences emerged in a way consilient with chemical thinking, with variations in electronegativity and donor and acceptor character.

I used that gift which came my way to make connections. So my Nobel Lecture is not a looking back to the story I am immensely proud of, the complex of orbital symmetry ideas developed with R. B. Woodward. Instead, the lecture sets out the isolobal analogy, under the title of "Building Bridges between Inorganic and Organic Chemistry."

In time I came back to my love for the word and for art, and slowly began to build connections between chemistry and the makers of the spirit—artists, writers, actors, performers, religious thinkers. I could do this only by entering their world, on their terms—the world of struggling to get a poem published, or a play produced. I did this; it remains hard, but tremendously satisfying.

Quiet Subversion, and Fun

By this I mean many things, inter alia my gentle campaign to relax the constraints of the ossified format of a scientific article, or of journals for that matter. I am very proud of having convinced the nonpareil editor of *Angewandte Chemie*, Peter Gölitz, to publish essays. The *Journal of the American Chemical Society* and *Physical Review* wouldn't dream of doing it, right? In an article on silver fluorides, we were allowed to insert a commentary by a sage protagonist, Neil Bartlett, into the text. We could write a discussion on valence bond and MO theory as a dramatic trialogue.

Fun: I think science is fun, as does Bassam, even as it is as serious as a search after reliable knowledge, knowledge that can hurt or heal, must be. The gatekeepers won't let you crack jokes in a paper, but one can be playful in the heart of science. A recent paper on squeezing CC bonds we did traipsed through a jungle gym of organic structures in the service of torturing a nice normal CC bond. The work Roger Alder, Charlie Wilcox, and I once did on square planar carbon was I think, in its own way, compassionate. What could be done, electronically, to give comfort to this unhappy structure?

In the "Entertaining Science" cabaret I run monthly in New York City, we juxtapose science with performance—music, dance, theater, magic. The audience makes every connection we want, and then some. They laugh, with affection, at everything we do. Even I, an awfully serious character (not really), can come off as Harry Gray.

Emotional Engagement

You can't write poetry without it, nor without personal exposure. But I'm speaking of other things as well. It seems to be given to me to be faced with ethical problems. I did not evade the painful consideration of E. J. Corey's claim of having told R. B. Woodward of the MO argument for control of electrocyclic reactions. Nor have I avoided thinking about the actions of Peter Debye in prewar Germany. And in returning to Złoczów, the place where I was born, in 2006, we, the survivors and their children, were drawn into the agony of remembering, and forgiving. There are times I wish I had been spared; but then I reflect, and I am thankful to the world for giving me the opportunity to face them.

I think I've been a good teacher because I've allowed the emotional into what passes between me and my students. So I've been able to communicate to students that the subject is ever new to me, and that I care that they learn, and I care about them as human beings.

We have come full circle. There is a time to thank: For the existential act of living, day by day; the wonder of connections out there to be seen; the microcosm of chemistry; the power of art and the word; the great reach of explaining and teaching with feeling. For being able to try to do these things, ultimately to teach and to retain the desire to be taught, for this I am grateful.

So, at 70, let me give thanks:

1. To the people who saved us in the war: Mikola Dyuk and his wife Maria.
2. To the people who ran the education program in the U. S. Occupation Zone in postwar Germany; aside from doing it well, they sponsored the translation of books, so-called Overseas Editions, meant to set Germans on a democratic path. Two of these inspired a 10-year-old boy in a refugee camp in Wasseralfingen: Eve Curie's hagiography of her mother, and Rackham Holt's life of George Washington Carver. I read these in German.
3. To my teachers at P. S. 93, Queens (especially my cousin, Sally Stewart), at P. S. 16, Brooklyn, Nathan Green and Abe Schwartz, and at Stuyvesant H. S., especially Abe Penzer, a great biology teacher.
4. To Abraham Wajngurt, a landsman who worked in the stockroom at Yeshiva University, and here and there slipped me pieces of chemical glassware.
5. To a great teaching assistant at Columbia, Bob Schneider. And two chemistry teachers there who introduced me to quantum mechanics: George Fraenkel and Ralph Halford.
6. To the teachers at Columbia, who opened up the world to me: Mark Van Doren, Donald Keene, Howard McParlin Davis, and Martin Ostwald.
7. To the people who introduced me to research, in summer jobs at the National Bureau of Standards: Ed Newman, Bob Ferguson, and at Brookhaven National Laboratory: Jim Cumming.
8. To my Ph.D. Advisors: Martin Gouterman and William Lipscomb. For both I was their first graduate student at Harvard.
9. To Per-Olov Löwdin and his Summer School in Sweden, where I learned group theory and met my wife.
10. To the IREX exchange, which brought me to Moscow for a year in the middle of graduate school, against Harvard's and my family's wishes.
11. To the Society of Fellows, for three years of freedom which allowed me to learn organic chemistry. During that time, it was E. J. Corey who, by simply telling me of his work, got me deeply interested in organic chemistry.
12. To R. B. Woodward, intellect incarnate, the patterner of chaos. He allowed a pair of helping hands to be transformed into a collaborator, and, over the years we worked together, ever so subtly pushed me from calculation to explanation.
13. To Cornell, for having faith and offering me a job, when I had sullied a good theoretical reputation by doing organic theory.

14. To Chem. 215, and 216, and 207, and 208, and 103, and 106, and 206 all the introductory chemistry courses I taught.

15. To my Cornell colleagues, for collegiality over four decades.

16. To Cornell again, to two Presidents: Dale Corson and Frank Rhodes, who supported me in so many ways.

17. To places that gave me intellectual shelter on sabbaticals and leaves: the ETH Zurich, the Universities of Stockholm and Lund in Sweden, Cambridge University, Moscow State University, NYU, and Columbia.

18. To my secretaries and assistants over the years, Valerie Watt, Eda Kronman, Joyce Boda, Ellie Stagg, Patricia Giordano, Jenny Cleland, and Catherine Kempf.

19. To Jane Jorgensen, the mistress of the lined orbital, who did my drawings over the years.

20. To my graduate students over 42 years: Chien-Chuen Wan, David Hayes, Jerrald Swenson, James Howell, Maynard M. L. Chen, David L. Thorn, Birgitte E.R. Schilling, Charles N. Wilker, Steve Cain, David M. Hoffman, Kevin Haraki, Timothy R. Hughbanks, Sunil D. Wijeyesekera, Shen-shu Sung, Chong Zheng, Ralph A. Wheeler, Marja C. Zonnevylle, Yat-Ting Wong, Edith A. Chan, Jing Li, Zafiria Nomikou, Kimberly Lawler-Sagarin, Michael Bucknum, Qiang Liu, Hugh Genin, Grigori Vajenine, Gregory Landrum, Abds-Sami Malik, Erika Merschrod, Garegin Papoian, Wingfield Glassey, Ying Wang, Melania Oana, Mihaela Bojin, Daniel Frederickson, Pradeep Gutta, Chinmoy Ranjan, Nicholas Gerovac, Ji Feng, and Anne Poduska.

21. To my postdocs, undergraduates, visitors: Patricia Clark, George Van Dine, Mircea Gheorghiu, Akira Imamura, Rolf Gleiter, Robert Bissell, S. Swaminathan, Donald Boyd, James Williams, Robert Davidson, Brian G. Odell, Charles C. Levin, Lawrence Libit, Victor Neagu, Wolf-Dieter Stohrer, Angelo Rossi, Alfred Anderson, Phillip Mollère, Hiroshi Fujimoto, C. Stephen Kim, Jack C. Thibeault, Notker Rösch, P. Jeffrey Hay, Mihai Elian, Alain Devaquet, Joseph W. Lauher, Richard H. Summerville, Daniel L. DuBois, Thomas A. Albright, Peter Hofmann, Carlo Mealli, Prem Mehrota, Heinz Berke, Alain Dedieu, Myung-Hwan Whangbo, Armel Stockis, E. D. Jemmis, Sason Shaik, Richard Goddard, Odile Eisenstein, Kazuyuki Tatsumi, Pavel Kubacek, Zdenek Havlas, Christian Minot, Miklos Kertesz, Jean-Yves Saillard, Georges Trinquier, Boris Schubert, Richard Dronskowski, Ruslan Minyaev, Karl Anker Jørgensen, Jerome Silvestre, Santiago Alvarez, Douglas Keszler, William Bleam, Wolfgang Tremel, Dragan Vuckovic, Susan Jansen-Varnum, Kenneth Merz, Jr., Jean-François Halet, Roy Johnston, Yitzhak Apeloig, Paul Sherwood, Christoph Janiak, Lalitha Subramanian, Meinolf Kersting, Christian Kollmar, Haibin Deng, Vladimir Pichko, Lars A. Kloo, Thomas R. Ward, William A. Shirley, Joel Bernstein, Andrei L. Tchougreeff, Hans-Jürgen Meyer, Katrin Albert, Pere Alemany, Yuri Slovokhotov, Birgit Schiøtt, Davide Proserpio, Kazunari Yoshizawa, Gerhard Nuspl, F. Matthias Bickelhaupt, Udo Radius, Norman Goldberg, Huang Tang,

Ruedi Rytz, Robert Konecny, Ralf Stowasser, Jesus Ugalde, Dong-Kyun Seo, Nguyen Trong Anh, D. M. P. Mingos, Jeremy Burdett, Paul Dobosh, John Eisch, Clifford McGinn, Victor Tortorelli, Evgeny Shustorovich, Robert Bach, Richard Harcourt, San-Yan Chu, Helga Dunken, Yuansheng Jiang, Debbie Fu-tai Tuan, Daniel Zeroka, Reinhard Nesper, Jingling Huang, Maria José Calhorda, P. A. Cox, John Lowe, Mikhail Basilevsky, Peter Kazmaier, Vladimir Minkin, Gion Calzaferri, Boris Simkin, Maria Matos, Hassan Rabaâ, Donald H. Galván, Sung Hong, Jürgen Evers, Valeria I. Rozenberg, Hoseop Yun, Lan-Feng Yuan, Antonio B. Hernandez, Ray Torralba, Kee Hag Lee, Wojciech Grochala, Andrea Ienco, Pattath Pancharatna, M. M. Balakrishnarajan, Beate Flemmig, Carol Parish, Deborah Huntley, Gabriel Merino, Dean Tantillo, Peter Kroll, Marketa Munzarova, Kelling Donald, Jason D'Acchioli, Edyta Greer, Warren Hehre, Roy R. Gould, Peter Rossky, Karen Goldberg, Robert Weber, Geoffrey Zeiss, Chris Zeiss, Leigh Ann Henderson, Mikhail Velikanov, Lori Rayburn, Stephen Goldberg, Jeffrey R. Long, Aliya Courtney, Pooja Pathak, Georgios Markopoulos, Tomasz Jarón, Tom Cahill, Milan Randić, Ming-Der Su, Sergei Kryutchkov, Mikhalil Grigoryev.

Special thanks to the German, French, and Japanese science research agencies – DAAD, DFG, CNRS, JSPS – which kept a steady flow of postdocs coming from these countries.

22. The National Science Foundation has supported my work for 50 years. Except for a lapse several years ago, they have provided most of the economic base for my free-ranging research. I hope I've repaid their confidence in my science.
23. My literary collaborators – the editors at *American Scientist*; coauthors R. B. Woodward, Vivian Torrence, Shira Leibowitz Schmidt, and Carl Djerassi.
24. And my family; not a large one, but consistently supportive of what I have done over the years: my mother Clara, who died in 2006, who led our family group in the period of survival, my father Hilel Safran (1911–1943), my stepfather Paul Hoffmann (1904–1981), my sister Elinor, my wife Eva, and my children Hillel Jan and Ingrid Helena. Who, between them, took exactly one semester of chemistry.

Note

1. R. Hoffmann, Qualitative Thinking in the Age of Modern Computational Chemistry, or What Lionel Salem Knows. *J. Mol. Str. (Theochem.)* 424 (1998): 1. Chapter 7 in this book.

Part 1

CHEMICAL REASONING
AND EXPLANATION

2

Why Buy That Theory?

ROALD HOFFMANN

The theory of theories goes like this: A theory will be accepted by a scientific community if it explains better (or more of) what is known, fits at its fringes with what is known in other parts of our universe, and makes verifiable, preferably risky, predictions.

Sometimes it does go like that. So the theory that made my name (and added to the already recognized greatness of the man with whom I collaborated, *the* synthetic chemist of the 20th century, R. B. Woodward) did make sense of many disparate and puzzling observations in organic chemistry. And "orbital symmetry control," as our complex of ideas came to be called, made some risky predictions. I remember well the day that Jerry Berson sent us his remarkable experimental results on the stereochemistry of the so-called 1,3-sigmatropic shift. It should proceed in a certain way, he reasoned from our theory—a non-intuitive way. And it did.

But much that goes into the acceptance of theories has little to do with rationalization and prediction. Instead, I will claim, what matters is a heady mix of factors in which psychological attitudes figure prominently.

Simplicity

A simple equation describing a physical phenomenon (better still, many), the molecule shaped like a Platonic solid with regular geometry, the simple mechanism (A→B, in one step)—these have tremendous *aesthetic* appeal, a direct beeline into our soul. They are beautifully simple, and simply beautiful. Theories of this type are awesome in the original sense of the word—who would deny this of the theory of evolution, the Dirac equation or general relativity?

A little caution might be suggested from pondering the fact that political ads patently cater to our psychobiological predilection for simplicity. Is the world simple? Or do we just want it to be such? In the dreams of some, the beauty and simplicity of equations becomes a criterion for their truth. Simple theories seem to validate that idol of science, Ockham's Razor.[1] In preaching the poetic conciseness and generality of orbital explanations, I have succumbed to this, too.

A corrective to the infatuation of scientists with simplicity might come from asking them to think of what they consider beautiful in art, be it music or the visual arts. Is it Bach's Goldberg Variations or a dance tune where the theme plays ten times identically in succession? Is any animal ever *painted* to emphasize its bilateral symmetry?

Still, there's no getting away from it; a theory that is simple, yet explains a lot, is usually accepted in a flash.

Storytelling

What if the world is complex? Here, symmetry is broken; there, the seemingly simplest of chemical reactions, hydrogen burning to water, has a messy mechanism. The means by which one subunit of hemoglobin communicates its oxygenation to a second subunit, an essential task, resembles a Rube Goldberg cartoon. Not to speak of the intricacies of *any* biological response, from the rise of blood pressure or release of adrenalin when a snake lunges at us, to returning a ping-pong serve with backspin. Max Perutz's theory of the cooperativity of oxygen uptake, the way the ribosome functions—these require complicated explanations. And yes, the inherent tinkering of evolution has made them complex. But simpler chemical reactions—a candle burning—are also intricate. As complex as the essential physics of the malleability, brittleness and hardness of metals. Or the geology of hydrothermal vents.

When things are complex yet understandable, human beings weave stories. We do so for several reasons: A→B requires no story. But A→B→C→D, and not A→B→C'→D *is* in itself a story. Second, as Jerome Brunner writes, "For there to be a story, something unforeseen must happen." In science the unforeseen lurks around the next experimental corner. Stories then "domesticate unexpectedness," to use Brunner's phrase.[2]

Storytelling seems to be ingrained in our psyche. I would claim that with our gift of spoken and written language, this is the way we wrest pleasure, psychologically, from a messy world. Scientists are no exception. Part of the story they tell is how they got there—the x-ray films measured over a decade, the blind alleys and false leads of a chemical synthesis. It is never easy, and serendipity substitutes for what in earlier ages would have been called the grace of God. In the end, we overcome. This appeals, and none of it takes away from the ingenuity of the creative act.

In thinking about theories, storytelling has some distinct features. There is always a beginning to a theory—modeling assumptions, perhaps unexpected observations to account for. Then, in a mathematically oriented theory, a kind of development section follows. Something is tried; it leads nowhere, or leaves one dissatisfied. So one essays a variation on what had been a minor theme, and—all of a sudden—it soars. Resolution and coda follow. I think of the surprise that comes from doing a

Fourier transform, or of seeing eigenvalues popping out of nothing but an equation and boundary conditions.

Sadly, in the published accounts of theories, much of the narrative of the struggle for understanding is left out, because of self-censorship and the desire to show us as more rational than we were. That's okay; fortunately one can still see the development sections of a theoretical symphony as one examines an ensemble of theories, created by many people, not just one, groping toward understanding.

The other place where narrative is rife is in the hypothesis-forming stage of doing science. This is where the "reach of imagination" of science, as Jacob Bronowski referred to it, is explicit. Soon you will be brought down to earth by experiment, but here the wild man in you can soar, think up any crazy scheme. And, in the way science works, if you are too blinded by your prejudices to see the faults in your theoretical fantasies, you can be sure others will.

Many theories are popular because they tell a rollicking good story, one that is sage in capturing the way the world works and could be stored away to deal with the next trouble. Stories can be funny; can there be humorous theories?

A Roll-On Suitcase

Theories that seek acceptance had better be *portable*. Oh, people will accept an initiation ritual, a tough-to-follow manual to mastering a theory. But if every application of the theory requires consultation with its originator (that's the goal of commercialization, antithetical to the ethic of science), the theory will soon be abandoned. The most popular theories in fact are those that can be applied by *others* to obtain surprising results. The originator of the theory might have given an eyetooth to have done it earlier, but friends should hold him back—it's better if someone else does it. And cites you.

Relatively uncomplicated models that admit of an analytical solution play a special role in the acceptance and popularity of theories among other theorists. I think of the harmonic oscillator, of the Heisenberg and Hückel Hamiltonians, of the Ising Model, my own orbital interactions. The models become modules in a theoretical *Aufbau*, shuttled into any problem as a first (not last) recourse. In part this is fashion, in part testimony to our predilection for simplicity. But, more significantly, the use of soluble models conveys confidence in the value of metaphor—taking one piece of experience over to another. It's also evidence of an existential desire to try something—let's try this.

Productivity

The best theories are productive, in that they stimulate experiment. Science is a wonderfully interactive way for gaining reliable knowledge. What excitement there is in person A advancing a view of how things work, which is tested by B, used by C to motivate making a molecule that tests the limits of the theory, which leads to D (not C) finding that molecule to be superconducting or an antitumor agent, whereupon a horde of graduate students of E or F are put to making slight modifications! People need reasons for doing things. Theories provide them, surely to test the theories (with greater delight if proved wrong), but also just to have a reason for making the next molecule down the line. Theories that provoke experiment are really valued by a community that in every science, even physics, is primarily experimental.

A "corollary" of the significance of productivity is that theories that are fundamentally untenable or ill-defined can still be immensely productive. So was phlogiston in its day, so in chemistry was the idea of resonance energies, calculated in a Hückel model. People made tremendous efforts to make molecules that would never have been made (and found much fascinating chemistry in the process) on the basis of "resonance energies" that had little connection to stability, thermodynamic or kinetic. Did it matter that Columbus miscalculated in his "research proposal" how far the Indies were?

As Jerry Berson has written, "A lot of science consists of permanent experimental facts established in tests of temporary theories."[3]

Frameworks for Understanding

Stephen G. Brush has recently studied a range of fields and discoveries to see what role predictions play in the acceptance of theories. Here's what he has to say about the new quantum mechanics: "Novel predictions played essentially no role in the acceptance of the most important physical theory of the 20th century, quantum mechanics. Physicists quickly accepted that theory because it provided a coherent deductive account of a large body of known empirical facts...."[4] Many theories predict relatively little (quantum mechanics actually did eventually) yet are accepted because they carry tremendous explanatory power. They do so by classification, providing a framework (for the mind) for ordering an immense amount of observation. This is what I think 20th century theories of acidity and basicity in chemistry (à la Lewis or Brønsted) do. Alternatively, the understanding provided is one of mechanism—this is the strength of the theory of evolution.

It is best to distinguish the concepts of theory, explanation, and understanding. Or to try to do so, for they resist differentiation. Evelyn Fox Keller, who in her brilliant book, *Making Sense of Life*, has many instructive tales of theory acceptance, says this of explanation: "A description or a phenomenon counts as an explanation...if and only if it meets the needs of an individual or a community. The challenge, therefore, is to understand the needs that different kinds of explanations meet. Needs do of course vary, and inevitably so: they vary not only with the state of the science at a particular time, with local technological, social, and economic opportunities, but also with larger cultural preoccupations."[5] As Bas van Fraassen has incisively argued, any explanation is an answer.[6] If we accept that, the nature of the question becomes of essence, and so does our reception of the answer. Both (the reconstructed question of "why?" and our response) are context-dependent and subjective. Understanding, van Fraassen says, "consists in being in a position to explain." And so is equally subjective in a pragmatic universe.

Incidentally, explanations are almost always stories. Indeed, moralistic and deterministic stories. For to be satisfying they don't just say A→B→C→D, but A→B→C→D *because* of such and such propensities of A, B and C. The implicit strong conviction of causality, justified by seemingly irrefutable reason, may be dangerously intoxicating. This is one reason why I wouldn't like scientists and engineers to run this world.

The acceptance of theories depends as much on the psychology of human beings as on the content of the theories. It is human beings who decide, individually and as a community, whether a theory indeed has explanatory power or provides understanding. This is why seemingly "extrascientific" factors such as productivity, portability, storytelling power and aesthetics matter. Sometimes it takes a long time (witness continental drift), but often the acceptance is immediate and intuitive—it fits. Like a nice sweater.

'Tis a Gift

There is something else, even more fundamentally psychological, at work. Every society uses gifts, as altruistic offerings but more importantly as a way of mediating social interactions. In science the gift is both transparent and central. Pure science is as close to a gift economy as we have, as Jeffrey Kovac has argued.[7] Every article in our open literature is a gift to all of us. Every analytical method, every instrument. It's desired that the gift be beautiful (simple gifts are, but also those that bring us a good story with them), to be sure. But that the offering be useful (portable, productive) endows it with special value. The giver will be remembered, every moment, by the one who received the gift.

The purpose of theory, Berson writes, is "... to bring order, clarity, and predictability to a small corner of the world." That suffices. A theory is then a special gift, a gift for the mind in a society (of science, not the world) where thought and understanding is preeminent. A gift from one human being to another, to us all.

Acknowledgment

The author thanks Michael Weisberg for incisive questions and comments.

Notes

1. R. Hoffmann, V. I. Minkin and B. K. Carpenter, Ockham's Razor and Chemistry. *Bulletin de la Societe Chimique de France* 133 (1996): 117.
2. J. Brunner, *Making Stories* (New York: Farrar, Straus and Giroux, 2002).
3. J. A. Berson, *Chemical Creativity: Ideas from the Work of Woodward, Hückel, Meerwein, and Others* (Weinheim: Wiley-VCH, 1999).
4. S. G. Brush, Dynamics of Theory Change: The Role of Predictions. In *Proceedings of the 1994 Biennial Meeting of the Philosophy of Science Association*, vol. 2, ed. by D. Hull et al. (Philosophy of Science Association, 1994), p. 133.
5. E. F. Keller, *Making Sense of Life* (Cambridge, MA: Harvard University Press, 2002).
6. B. C. van Fraassen, *The Scientific Image* (London: Clarendon Press 1980), pp. 132–157.
7. J. Kovac, Gifts and Commodities in Science. *Hyle* 7 (2001): 141.

3

What Might Philosophy of Science Look Like If Chemists Built It?

ROALD HOFFMANN

Implicit in the title might be two presumptions. The first one, that there is (or should be) a single philosophy of science, is not a claim I intend—I do think one should look for a common core, in a way that allows for differences.[1] The second presumption, that philosophy of science, as it is construed today, would be different if it were based on chemistry, is what I wish to examine.

And behind that latter supposition is the notion that philosophers of science, their professionalism and good will not impugned, nevertheless are likely to construct their worldview of science based on the sciences they know best. These are usually the sciences that they studied (a) as a part of their general education, or (b) the science they came from, so to speak, if they made their transition to philosophy at some later point in their career.

I have not made a rigorous examination of the education of philosophers of science. But my anecdotal feeling is that, for those who entered the profession directly, an exposure to mathematical logic is more likely than to geology or chemistry. And, for many of the philosophers of science who came to their field after an initial scientific career, their scientific expertise was likely to be in the first instance physics, after that biology, and rarely chemistry.

I will argue that this matters, for chemistry is different.

There are exceptions. In the English-writing community, the most striking one is Michael Polanyi, a very distinguished physical chemist. In the French philosophical community, Pierre Duhem, Emile Meyerson, Gaston Bachelard, and Hélène Metzger had professional chemical backgrounds. Bernadette Bensaude-Vincent has argued convincingly that this background shaped their philosophical outlook, in contrast with the analytic philosophers of their time.[2]

In recent times the situation may have changed. A subfield of "philosophy of chemistry" has emerged, with annual meetings and two journals (*Foundations of Chemistry, Hyle*). The practitioners of this field are more likely to have had substantive experience in chemistry.

I feel we should put some blame for the situation on the chemists as well, asking why more of them have not wandered into philosophy. Are they so unreflective? If so, why? Is there something in the practice of chemistry that drives them toward becoming kings—à la Margaret Thatcher and Angela Merkel—rather than philosophers? Is the employment situation too good in chemistry?

What Makes Chemistry Different: A Chemical Paper

Chemical Abstracts indexed in 2004 685,796 articles, 5,601 books, and 173,669 patents in chemistry and related fields. Let's look at one such recent paper to get an idea of the science.

But.... is that possible, to glean knowledge of all chemistry from one paper? The science is highly compartmentalized and the historical antecedents, modes of analysis and argument, and goals of a study in physical chemistry differ from those of medicinal chemistry or solid state inorganic chemistry.

As I value the differences that make chemistry and life interesting, I also believe in a poetic perspective on science. Archie Ammons wrote a small poem, "Reflective":[3]

> I found a
> weed
> that had a
> mirror in it
> and that
> mirror
> looked in at
> a mirror
> in
> me that
> had a
> weed in it.

That weed, that dew drop on it, that particular weed and no other dew drop has within it the essence of the universe.

The paper is by a group from Kyoto University of K. Komatsu, M. Murata, and Y. Murata, led by Koichi Komatsu.[4] It is in the field of buckminsterfullerene chemistry, that beautiful, metastable C_{60} allotrope of carbon which has been with us for only 20 years. C_{60} is made by unsporting methods (i.e., not by methodical construction from smaller pieces), by striking a carbon arc in just the right pressure of helium. There is a little room, not much, inside the nanometer-wide ball of buckminsterfullerene. But by equally unsporting (yet reproducible) methods, by brute force of discharge and high temperature covaporization, people have inserted atoms inside C_{60}—He, Sc, N. Nothing bigger, no molecule.

Meanwhile, there has sprung up a vast and ingenious chemistry around C_{60}. People, those manipulative experimenters, have added groups of atoms—few, many, changing in one way or another the buckyball. In 2005, Komatsu's group carved a 13-membered ring orifice into C_{60}. Soon thereafter, they showed that H_2 entered that hole. Now, in this paper, they do something much more interesting, and important, and amusing—yes, all of these, and this is to reclose the hole by a sequence of chemical reactions. All the time the H_2 stays inside. At the end of this "molecular surgery" (a term introduced by Yves Rubin, another worker in the field, who had earlier made molecule **1** of this paper, and had found that some hydrogen goes into it), they have H_2 trapped in the fullerene, $H_2@C_{60}$. The @ sign here stands for a very fixed address, encapsulation.

Let's look at the illustrations in the paper (Figure 3-1a, Figure 3-1b, Figure 3-1c). The drawings are important—the graphic content of the molecular science and its connection to issues of representation, of iconic and symbolic signing, are things I will return to.

The molecule the authors made previously, **2**, is shown in Scheme 1, along with another precursor, $H_2@2$. Then Figure 1 of their paper shows the sequence of

Encapsulation of Molecular Hydrogen in Fullerene C_{60} by Organic Synthesis

Koichi Komatsu,* Michihisa Murata, Yasujiro Murata

In spite of their importance in fundamental and applied studies, the preparation of endohedral fullerenes has relied on difficult-to-control physical methods. We report a four-step organic reaction that completely closes a 14-membered ring orifice of an open-cage fullerene. This process can be used to synthesize a fullerene C_{60} encapsulating molecular hydrogen, which can be isolated as a pure product. This molecular surgical method should make possible the preparation of a series of C_{60} fullerenes, encapsulating either small atoms or molecules, that are not accessible by conventional physical methods.

Endohedral fullerenes, the closed-cage carbon molecules that incorporate atoms or a molecule inside the cage (1–6), are not only of scientific interest but are also expected to be important for their potential use in various fields such as molecular electronics (7), magnetic resonance imaging [as a contrast agent (8)], and nuclear magnetic resonance (NMR) analysis (9, 10). However, development of their applications has been hampered by a severe limitation in their production, which has relied only on physical methods, such as co-vaporization of carbon and metal atoms (2, 3) and high-pressure/high-temperature treatment with gases (9–14), that are difficult to control and yield only milligram quantities of pure product after laborious isolation procedures.

An alternative approach to synthesizing endohedral fullerenes is "molecular surgery,"

Institute for Chemical Research, Kyoto University, Uji, Kyoto 611-0011, Japan.
*To whom correspondence should be addressed. E-mail: komatsu@scl.kyoto-u.ac.jp

in which the cage is opened and then closed in a series of organic reactions (15, 16). For example, an open-cage C_{60} derivative **1** with a 14-membered ring orifice has been synthesized (17), and the insertion of molecular hydrogen into **1** in 5% yield has also been achieved (15). However, the closure of its orifice was not attempted. A C_{60} derivative **2**, which we synthesized recently (18), has a 13-membered ring orifice with a sulfur atom on its rim and a relatively circular shape compared with the elliptic orifice of **1**. This opening has enabled us to insert molecular hydrogen through this orifice in 100% yield (19). When matrix-assisted laser desorption/ionization time-of-flight (MALDI-TOF) mass spectrometry was conducted under enhanced laser power on compound **2** encapsulating hydrogen ($H_2@2$), we observed a molecular ion peak for $H_2@C_{60}$ at a mass-to-charge ratio (m/z) of 722 (19). This result suggested that $H_2@2$ could be a precursor for $H_2@C_{60}$ in an actual chemical transformation. We now report the synthesis of 100% pure $H_2@C_{60}$ from $H_2@2$ (Scheme 1).

Encapsulated H_2 escapes from the cage when the compound $H_2@2$ is heated above 160°C (19), so high temperatures must be avoided if the chemical synthesis of $H_2@C_{60}$ is attempted from $H_2@2$. With such a precaution being taken, we performed a stepwise reduction of the orifice size of $H_2@2$ and completed its closure by a thermal reaction. The application of heat to the last step did not cause a serious loss of H_2, because the orifice size was already reduced sufficiently to prevent such loss.

The first step involved the oxidation of the sulfide unit (-S-) in $H_2@2$ to a sulfoxide unit (>S=O) to give $H_2@3$. The resulting >S=O unit was removed by a photochemical reaction to produce $H_2@4$ (Fig. 1, steps A and B) (20). Both reactions proceeded at room temperature with yields of 99% and 42% (68% for step B based on consumed $H_2@3$), respectively. The MALDI-TOF mass spectrum of $H_2@4$ exhibited the molecular ion peak of $H_2@C_{60}$ as a base peak, indicating its enhanced accessibility from $H_2@4$ as compared to $H_2@2$. The spectrum, however, also showed the presence of empty C_{60} in 20% yield relative to $H_2@C_{60}$ and indicated that further reduction of the orifice size was needed. Thus, in the next step, two carbonyl groups in $H_2@4$ were reductively coupled by the use of Ti(0) (21) at 80°C, to give $H_2@5$ with an eight-membered ring orifice (Fig. 1, step C).

At each process in these three steps, complete retention of encapsulated H_2 was confirmed by observing the characteristically upfield-shifted NMR signal of the incorporated H_2. The integrated signal intensity exactly corresponded to 2.00 ± 0.05 H for the signals at a chemical shift δ of –6.18 parts per million (ppm) in $H_2@3$, at –5.69 ppm in $H_2@4$, and at –2.93 ppm in $H_2@5$, with reference to the 2.00 H signal for two aromatic protons. The gradual downfield shift of the hydrogen signal

Figure 3-1a First page of Komatsu, Murata, and Murata 2005 paper. Reprinted with permission from AAAS.

Fig. 1. Size reduction and closure of the orifice of the open-cage fullerene encapsulating hydrogen, in a four-step process. Percentage values are product yields; that shown in parenthesis is that based on the consumed precursor. m-CPBA, r.t., and o-DCB stand for m-chloroperbenzoic acid, room temperature, and o-dichlorobenzene, respectively.

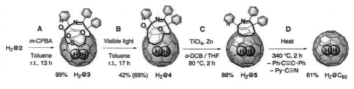

$H_2@2$ → **A** m-CPBA, Toluene, r.t., 13 h → 99% $H_2@3$ → **B** Visible light, Toluene, r.t., 17 h → 42% (68%) $H_2@4$ → **C** $TiCl_4$, Zn, o-DCB / THF, 80 °C, 2 h → 88% $H_2@5$ → **D** Heat, 340 °C, 2 h, – Ph-C≡C-Ph, – Py-C≡N → 61% $H_2@C_{60}$

observed at steps B and C reflects the formation at each step, within the fullerene cage, of a fully π-conjugated pentagon, which exerts a strong deshielding effect through its paramagnetic ring currents (22).

Finally, complete closure of the orifice was achieved by heating powdery $H_2@5$ in a glass tube at 340°C for 2 hours under vacuum (Fig. 1, step D). The desired product $H_2@C_{60}$ (118 mg, contaminated with 9% empty C_{60}) was obtained in 67% yield by passing a carbon disulfide solution of the crude product through a silica-gel column. Similar results were obtained when $H_2@5$ was heated at 300°C for 24 hours, at 320°C for 8 hours, or at 400°C for 2 min. Thus, $H_2@C_{60}$ was synthesized in a total yield of 22% from $H_2@2$, which can be obtained in 40% yield from consumed C_{60} (18, 19).

We presume that the closure of the orifice takes place by way of a thermally allowed [π2s + π2s + π2s] electrocyclization reaction that produces two cyclopropane rings (Fig. 2). Sequential radical cleavage and a retro [σ2s + σ2s + σ2s] reaction produce C_{60} by splitting off 2-cyanopyridine and diphenylacetylene.

The ^{13}C NMR spectrum of the desired product exhibited a signal at δ = 142.844 ppm together with a very small signal at δ = 142.766 ppm (Fig. 3A), the latter corresponding exactly to the signal of empty C_{60}. In an expanded spectrum obtained with 56,576 data points for a 50-ppm spectral width, the integrated peak areas of these signals yield an estimated ratio of $H_2@C_{60}$ and empty C_{60} of 10:1.

We separated $H_2@C_{60}$ from C_{60} through recycling high-performance liquid chromatography on a semipreparative Cosmosil Buckyprep column (two directly connected columns, 25 cm by 10 mm inner diameter, with toluene as a mobile phase; flow rate, 4 ml min^{-1}; retention time, 395 min for C_{60} and 399 min for $H_2@C_{60}$). Isolated $H_2@C_{60}$ was judged to be 100% pure on the basis of a single ^{13}C NMR signal at 142.844 ppm (Fig. 3B), the results of high-resolution fast-atom-bombardment mass spectrometry (calculated molecular weight for $C_{60}H_2$: 722.0157; found: 722.0163), and the agreement of the

Fig. 2. Proposed reaction mechanism for the formation of C_{60} from compound 5 by heating. Only the tops of the molecules are shown. Ph and Py stand for phenyl and 2-pyridyl groups, respectively.

5 [π2s+π2s+π2s] → → – Py-C≡N – Ph-C≡C-Ph → C_{60}

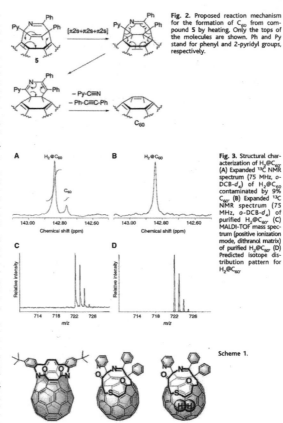

A $H_2@C_{60}$ / C_{60} — 143.00 142.80 142.60 Chemical shift (ppm)

B $H_2@C_{60}$ — 143.00 142.80 142.60 Chemical shift (ppm)

C Relative intensity — 714 718 722 726 m/z

D Relative intensity — 714 718 722 726 m/z

Fig. 3. Structural characterization of $H_2@C_{60}$. (A) Expanded ^{13}C NMR spectrum (75 MHz, o-DCB-d_4) of $H_2@C_{60}$ contaminated by 9% C_{60}. (B) Expanded ^{13}C NMR spectrum (75 MHz, o-DCB-d_4) of purified $H_2@C_{60}$. (C) MALDI-TOF mass spectrum (positive ionization mode, dithranol matrix) of purified $H_2@C_{60}$. (D) Predicted isotope distribution pattern for $H_2@C_{60}$.

Scheme 1.

1   2 $H_2@2$

Figure 3-1b Second page of Komatsu, Murata, and Murata 2005 paper. Reprinted with permission from AAAS.

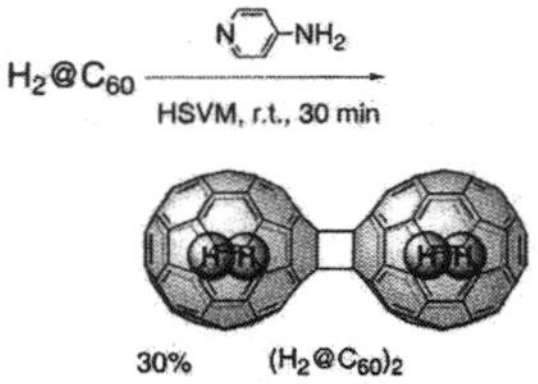

$$H_2@C_{60} \xrightarrow{\text{HSVM, r.t., 30 min}}$$

30% $(H_2@C_{60})_2$

Fig. 4. Mechanochemical solid-state dimerization of $H_2@C_{60}$ by the use of a high-speed vibration milling (HSVM) technique.

observed and predicted isotope distribution patterns in the MALDI-TOF mass spectrum (Fig. 3, C and D), in addition to correct elemental analysis for hydrogen (calculated for $C_{60}H_2$: C, 99.72, and H, 0.28%; found: C, 99.04, and H, 0.24%).

The very small downfield shift (0.078 ppm) observed for the ^{13}C NMR signal of $H_2@C_{60}$ (as compared to empty C_{60}) indicates that the electronic property of the fullerene cage is largely unaffected by the encapsulation of H_2. The ultraviolet-visible and infrared spectra of $H_2@C_{60}$ are also exactly the same as those of empty C_{60}. This situation contrasts with the cases of $Kr@C_{60}$ (13) and $Xe@C_{60}$ (12), in which larger downfield shifts are observed (0.39 ppm and 0.95 ppm, respectively), caused by appreciable electronic and van der Waals interactions between the C_{60} cage and the encapsulated atoms, which are much larger than H_2.

The ^1H NMR signal for the encapsulated hydrogen of $H_2@C_{60}$ in o-dichlorobenzene-d_4 was observed at $\delta = -1.44$ ppm, which is 5.98 ppm upfield-shifted relative to the signal of dissolved free hydrogen. The extent of this upfield shift is comparable to that observed for $^3He@C_{60}$ (6.36 ppm) (9, 10) in ^3He NMR relative to free ^3He. This result shows that the shielding effect of total ring currents of the C_{60} cage is nearly the same, regardless of the paramagnetic species inside the cage.

The irrelevance of the encapsulated H_2 to the electronic character of the outer cage was also demonstrated by cyclic voltammetry (0.5 mM in o-dichlorobenzene with 0.05 M Bu_4NBF_4 for reduction and 0.5 mM in 1,1,2,2-tetrachloroethane with 0.1 M Bu_4NPF_6 for oxidation). The voltammogram of $H_2@C_{60}$ exhibited four reversible reduction waves and one irreversible oxidation peak at the same potentials as C_{60}, within an experimental error of ±0.01 V.

In order to clarify the reactivity of $H_2@C_{60}$, the solid-state mechanochemical [2+2] dimerization reaction (23) was conducted. A mixture of $H_2@C_{60}$ and 1 molar equivalent of 4-aminopyridine as the catalyst

(24) was vigorously shaken by the use of a high-speed vibration mill for 30 min under N_2 according to our previous procedure (23, 24). The ^1H NMR spectrum of the product mixture exhibited a signal at $\delta = -4.04$ ppm of the [2+2] dimer, $(H_2@C_{60})_2$, and a signal of unchanged $H_2@C_{60}$ at $\delta = -1.44$ ppm, in an integrated ratio of 3:7. This result indicates that the dumbbell-shaped dimer of $H_2@C_{60}$ is formed in the same yield as that for the reaction of empty C_{60} (24) (Fig. 4). No effect of the encapsulated H_2 was observed upon reactivity of the C_{60} cage. The extent of the upfield shift of the ^1H NMR signal (2.60 ppm) observed for the dimer $(H_2@C_{60})_2$ was similar to that observed upon the same dimerization reaction in ^3He NMR (2.52 ppm) (24) for ^3He encapsulated in the ratio of ~0.1% in C_{60} (9, 10).

The endohedral fullerene $H_2@C_{60}$ is nearly as stable as C_{60} itself. For example, the encapsulated H_2 does not escape even when heated at 500°C for 10 min. Thus, $H_2@C_{60}$ can be viewed as a stable hydrocarbon molecule that has neither C-H covalent bonds nor C···H interactions. It is likely that our method could be used to synthesize endohedral fullerenes such as $D_2@C_{60}$ and $HD@C_{60}$, as well as the homologous series with C_{70}. Our work here complements the total chemical synthesis of C_{60} recently achieved by Scott and co-workers (25) and implies that organic synthesis can be a powerful means for the production of yet unknown classes of endohedral fullerenes.

References and Notes
1. T. Akasaka, S. Nagase, Eds., *Endofullerenes: A New Family of Carbon Clusters* (Kluwer, Dordrecht, 2002).
2. H. Shinohara, in *Fullerenes: Chemistry, Physics and Technology*, K. M. Kadish, R. S. Ruoff, Eds. (Wiley, New York, 2000), pp. 357–393.

3. H. Shinohara, *Rep. Prog. Phys.* 63, 843 (2000).
4. S. Liu, S. Sun, *J. Organomet. Chem.* 599, 74 (2000).
5. S. Nagase, K. Kobayashi, T. Akasaka, *Bull. Chem. Soc. Jpn.* 69, 2131 (1996).
6. D. S. Bethune et al., *Nature* 366, 123 (1993).
7. S. Kobayashi et al., *J. Am. Chem. Soc.* 125, 8116 (2003).
8. H. Kato et al., *J. Am. Chem. Soc.* 125, 4391 (2003).
9. M. Saunders, R. J. Cross, H. A. Jiménez-Vázquez, R. Shimshi, A. Khong, *Science* 271, 1693 (1996).
10. M. Saunders et al., *Nature* 367, 256 (1994).
11. R. J. Cross, A. Khong, M. Saunders, *J. Org. Chem.* 68, 8281 (2003).
12. M. S. Syamala, R. J. Cross, M. Saunders, *J. Am. Chem. Soc.* 124, 6216 (2002).
13. K. Yamamoto et al., *J. Am. Chem. Soc.* 121, 1591 (1999).
14. M. Saunders et al., *J. Am. Chem. Soc.* 116, 2193 (1994).
15. Y. Rubin et al., *Angew. Chem. Int. Ed. Engl.* 40, 1543 (2001).
16. Y. Rubin, *Top. Curr. Chem.* 199, 67 (1999).
17. G. Schick, T. Jarrosson, Y. Rubin, *Angew. Chem. Int. Ed. Engl.* 38, 2360 (1999).
18. Y. Murata, M. Murata, K. Komatsu, *Chem. Eur. J.* 9, 1600 (2003).
19. Y. Murata, M. Murata, K. Komatsu, *J. Am. Chem. Soc.* 125, 7152 (2003).
20. Materials and methods are available as supporting material on *Science* Online.
21. J. E. McMurry, *Chem. Rev.* 89, 1513 (1989).
22. M. Rüttimann et al., *Chem. Eur. J.* 3, 1071 (1997).
23. G.-W. Wang, K. Komatsu, Y. Murata, M. Shiro, *Nature* 387, 583 (1997).
24. K. Komatsu et al., *J. Org. Chem.* 63, 9358 (1998).
25. L. T. Scott et al., *Science* 295, 1500 (2002).
26. Supported by a Grant-in-Aid for Center of Excellence Research on Elements Science (no. 12CE2005) and by a Grant-in-Aid for Creative Scientific Research Collaboratory on Electron Correlation Toward a New Research Network Between Physics and Chemistry (no. 13NP0201), both from the Ministry of Education, Culture, Sports, Science, and Technology, Japan.

Supporting Online Material
www.sciencemag.org/cgi/content/full/307/5707/238/DC1
Materials and Methods
Figs. S1 to S9
References

8 October 2004; accepted 23 November 2004
10.1126/science.1106185

Corrected Late Triassic Latitudes for Continents Adjacent to the North Atlantic

Dennis V. Kent[1,2*] and Lisa Tauxe[3]

We use a method based on a statistical geomagnetic field model to recognize and correct for inclination error in sedimentary rocks from early Mesozoic rift basins in North America, Greenland, and Europe. The congruence of the corrected sedimentary results and independent data from igneous rocks on a regional scale indicates that a geocentric axial dipole operated in the Late Triassic. The corrected paleolatitudes indicate a faster poleward drift of ~0.6 degrees per million years for this part of Pangea and suggest that the equatorial humid belt in the Late Triassic was about as wide as it is today.

Paleomagnetism is used to determine ancient latitude, but its reliability depends on two assumptions: (i) that the time-averaged geomagnetic field is closely approximated by that of a geocentric axial dipole (GAD), and (ii)

that there is no systematic bias in how the geomagnetic field is imprinted in rocks. Although the GAD hypothesis (1) is supported by paleomagnetic data for the past few million years (2, 3), departures from the GAD model

Figure 3-1c Third page of Komatsu, Murata, and Murata 2005 paper. Reprinted with permission from AAAS.

sewing up the hole after the H_2 enters. This is the subject of the Komatsu et al. paper before us. The last step of the reaction, a pyrolysis, kicks out many atoms. Figure 2 of the paper shows a proposed "mechanism" for this step, the presumed sequence of elementary reaction happening.

Figure 3 is the only illustration in the Komatsu paper which shows experimental measurements—nuclear magnetic resonance measurements (giving information on the magnetic environment of the carbons) and mass spectra (effectively weighing the molecule) of $H_2@C_{60}$. Figure 4 shows a first reaction (a dimerization) of the new molecule, $H_2@C_{60}$.

Does this paper look like a philosopher of science's typical paper? I know, I am about to take the easy road, set up a straw-man. Please forgive me, it allows me to enter a discussion on just how different this chemical enterprise is.

In my opinion, the Komatsu et al. paper bears no resemblance to a paradigmatic hypothetico-deductive, or Carnap's deductive-nomological, or inductive, or Popperian conjecture/refutation process. There are observations in the $H_2@C_{60}$ paper, of course. But they are not passive, with the experimenters perplexed by what nature shows them. The measurements are willed, to achieve a transformation. There are no alternative hypotheses, just one developing narrative. There are further experiments, indeed (the nmr and mass spectra), but they are more in the context of building a coherent story rather than falsification.

The various mid-20th century logical-positivist models of the scientific process by and large reduce experiment to mere observation. They privilege theory construction and to me reveal indifference to the affective-cognitive link forged in the performance of experiment. A rotation in a laboratory would have been good for these philosophers.

What theories are being tested in this beautiful paper? None, really, except that this molecule can be constructed. The theory building in that is about as informative as the statement that Archie Ammons' poem tests a theory that the English language can be used to construct novel and perceptive insights into the way the world and our minds interact. The power of that tiny poem, the cleverness of the molecular surgery that Komatsu et al. perform in creating a new molecule, just sashay around any analytical theory-testing concerns.

Michael Weisberg, who read a draft of this paper, remarks that I should anticipate attempts to reconstruct out of my presentation and the Komatsu paper some sort of hypothesis (Weisberg, personal communication). He suggests I say:

> This paper gives us evidence that a certain theoretical statement— namely the procedure for making $H_2@C_{60}$—is true. This could be given hypothesis-like form by putting it into a conditional: if you do such and such reactions, you get $H_2@C_{60}$. So yes, we get evidence for a hypothesis or theoretical statement. However, H-D and falsificationism also suggest a certain PROCEDURE for doing science. And this is what I am claiming is not present.[6]

I agree.

In a carefully reasoned book, replete with detailed examples from 19th and 20th century chemistry, Jerome A. Berson makes much the same point as I do:

> *It seems clear that much of the activity of synthesis seems to fall outside of the doctrine of conjectures and refutations. That set of guidelines envisions no important role for the largely confirmative aspect of an activity that has proven central to the concerns of chemists. Whether one wants to make a certain compound to confirm its structure, or to make a medicinally significant molecule, or to test a theoretical existence issue, or just because "it is there," the point remains that the actual synthesis itself frequently has an inescapably confirmative purpose, not a refutative one.*
>
> *In trying to adapt falsificationist doctrine to the field of chemistry, we seem to have uncovered something arrestingly curious. If, as Popper says, confirmations count only if they are intended to test a theory, and if, again as he says, a valid test of theory must be refutative, we are led to a dilemma: if a confirmative synthesis succeeds, it doesn't count, but as any chemist knows, if it [RH: the synthesis] fails, it doesn't refute.[7,8]*

What Is Chemistry?

Let's move beyond this paper. Institutionally the enterprise is gigantic: the American Chemical Society has 158,000 members; I have already mentioned the estimate of $\sim 7 \times 10^5$ articles published per year. Underlying the cadres and their production is economic value: ~70% of the close to 2,000 US Ph.D.'s in chemistry find industrial employment.

Chemistry always was the art, craft, and business of substances and their transformation. I say "always," because I want to emphasize that the protochemistries in food preparation, cosmetics, metallurgy, glass and enamel making, textile preparation, dyes and pigments, fermentation, medicinal preparations, and many others *preceded* the formal development of the science. Chemistry remains this—the study (and utilization) of macroscopic matter and its changes. The transformed matter must be macroscopic if it has commercial value, but on the research plane, the actual amount manipulated may be tiny (think of the knowledge built on DNA gels).

With time, we've learned to look inside the innards of the beast, and reasoned out that in the macroscopic matter, static and undergoing transformation, there are atoms, and, much more interesting, persistent groupings of atoms which are molecules. So chemistry is also the art, craft, business and science of molecules and their transformations. It is, symbolically, and essentially

$$A + B \rightarrow C + D$$

I will later add the arrow that makes this an equilibrium, and its consequences.

The reliable knowledge (a phrase I've borrowed from John Ziman,[9] more than van Fraassen's "empirically adequate") gained of the molecular world came from the hot and cool work of our hands and mind combined. Sensory data, yes, but we did not wait for scanning tunneling microscopes to show us molecules; we gleaned their presence, their stoichiometry, the connectivity of the atoms in them, and eventually their metrics, shape and dynamics, by indirect experiments. The reasoning in these was rarely decisive, sometimes productively quasicircular,[10] and often in the nature of extending a story. Amazing that one could design the reality of physiologically active pharmaceuticals and billion-ton industrial production on such seemingly flimsy knowledge, isn't it?

Given this briefest sketch of chemistry, let me look at some leitmotifs of the last century of philosophy of science in the chemical context.

Realism, Reinforced by Transformation

You are not likely to find a doubter of realism, a skeptic on any level, in the chemical community. Yet chemists realize the symbolic essence of their molecular representations and the multiplicity of ways of looking at a molecule. Chemists also are well aware of the ephemeral nature of theories. But underlying their productive activities is a solid realism. Whence the intuitive realism of the chemist?

I think it derives from the reinforcement of sensual perception by the transformation inherent in chemistry. The sensual perception I'm talking about is at one level just that, of some green crystals that precipitate in a flask and are filtered out in a separate process. The feeling that matter is matter is reinforced by a chemical appreciation of differences in chemical and biological behavior. Perhaps 65 million of the 70 million or so well-characterized chemical substances are white crystalline solids. Salt, sugar, penicillin, and tetrodotoxin (the poisonous principle of the pufferfish, fugu) are such. The differences in their biological behavior (or in the nmr spectrum) are a continuous lesson in realism. Taste them, take their spectra.

By itself, the gestalt of sensual perception, though deceivable in its parts, makes a very big difference. No amount of exposure to text and images with sexual content diminishes the nervous wonder of a first sexual experience.

Transformation provides great, great reinforcement to a naïve realist perspective. It is one thing to make molecule A. You may be uncertain about its structure, always more uncertain if it is a molecule that you wanted to make and it was made by someone else. But when you take compound A, with the supporting chemical and physical evidence for its structure, and you transform it into compound B, and the spectroscopic clues for its structure check out, and then B is changed into C, and C into D, and it all makes sense—boy, do you then believe in A!

Ian Hacking says it well:

> *Experimental work provides the strongest evidence for scientific realism. This is not because we test hypotheses about entities. It is because entities that in principle cannot be 'observed' are regularly manipulated to produce a new phenomena and to investigate other aspects of nature. They are tools, instruments not for thinking but for doing.*[11]

Hacking's outlook, reminding philosophers to pay more attention to experimental intervention, rings true.

The constructed realist confidence is perhaps more in the molecular, microscopic structure; there was never much doubt about the reality of those powders.

Reductionism

What's the use of flogging a dead concept, you might think. I agree. But simplistic reductionism, while retaining little standing in the philosophical community, is alive and well within science. Aside from the general good feeling that comes from being at the bottom of a reductionist chain (and for chemists that may derive from providing explanations for biology), there are theoretical chemists who really think that all the concepts of chemistry would be clarified by rigorous definitions, of course based on physics.[12]

My minor tirade against reductionism, stressing its unreality and futility, is elsewhere.[13] More positively, here's what I believe: In dealing with the complexity of this world, human beings adduce explanations in two modes (and then mix them up; more on this below). Vertical understanding is the classical reductionist kind. Horizontal understanding is expressed in the concepts, definitions, and symbolic structures at the same level of complexity as the object to be understood. Horizontal explanations, like dictionary definitions, but richer, are quasi-circular.[14] And none the worse for it. A poem is horizontally understood at the level of the language in which it is written and the psychology of the writer and reader; it is vertically (and impoverishingly) explained by the sequence of firing of neurons (and the biochemical actions behind them) in the author's or reader's mind.

My stronger claim is that most concepts in chemistry that have proven productive in the design of new experiments are not reducible to physics. By this I mean that they cannot be redefined adequately using only the language of physics. To be specific, I'm thinking of aromaticity, the acid-base concept, the idea of a functional group, a substituent effect, and that of a chromophore. And let us not forget the chemical bond.

Few, if any, new molecules would have been made if we had waited around for rigorous definitions of these productive thought constructions. Our icon, achieving that deserved status through its sheer utility, Mendeleyev's Periodic table, would not have been drawn up for 56 years. The mechanism of the deceptively simple dimerization reaction that Komatsu runs on $H_2@C_{60}$ at the end of his paper is quite unclear. That doesn't stop him from making the molecule.

It might seem as if I am advocating a special logic for chemistry, or no logic at all. It's as if I were back in the period of 1700–1750, in which Georg Ernst Stahl, Guillaume-François Rouelle, and Gabriel-François Venel (who wrote the *Encyclopédie* article on chemistry) set out a special philosophical basis for chemistry. Each in his own way privileged the actions of chemistry, the distinctive properties of compounds, the irreducibility of chemistry to physics.[15]

Perhaps I am. But I think the recognition that understanding is as much horizontal as it is vertical, in *any* science (as it is in the social sciences) gives me a way to express the distinctiveness of chemistry.

Let me say it another way: Reductionists claim that they labor to reach the roots. And once the elementary particles are understood, and all the forces too, then one could move upscale, and with the workings of those building blocks and forces, a bit of the aleatory thrown in, the macroscopic would emerge. At any level. I'm skeptical; I've hardly ever seen a soul move upscale.

I think chemistry does move upscale, climb ladders of complexity, creates new molecules and emergent phenomena. It does so by alternating small riffs of reductionist analysis with lots of intuitive thinking on the horizontal level. Making up a story, while making molecules.

Incommensurability

In chemistry, revolutions may be less frequent than the grafting of a new way of thinking onto an old stock. Though some people think that neologisms are the way (*pace* the Condillacian side of Lavoisier), in the long run the way to capture minds may be to introduce the new pretending to be the old.

My model here is the two-step evolution of the nature of the chemical bond. The first conflation is of the simple 19th century line, denoting association, with a shared electron pair in a Lewis structure. This was followed by Pauling's skillful association of the covalent wave function of the new quantum mechanics with Lewis' shared pairs, and through that with the 19th century bond. Meanwhile, other signatures of bonding—length, energy, vibrations—reified the chemical bond.

Another instance, a very recent one in my community of theoretical chemistry, is of using the supposedly unneeded (if not unreal) orbitals of density functional theory in the same ways as the orbitals of a so-called one electron molecular orbital approach to electronic structure. The latter is a poorer theory, with greater explanatory power, and in it my favorite molecular orbitals play the central role.

Still another grafting is that of explanations of electrostatics onto a quantum mechanical calculation which from the start has electrostatics built into it. This is going on, with a vengeance, right now.

Kuhn saw incommensurability as being the consequence of two competing paradigms, and he distinguished incommensurabilities of language, and of standards of evidence. I think incommensurability is no problem whatsoever to chemists. Differences in language are there, the result of different paradigms, but more so of history, and of education. Yet people, eager to make things, with no handwringing on how problematic it all is, graft one way of understanding onto another. So, to return to that electrostatic/quantum mechanical conundrum, a couple of decades down the line there is one language (of opposite charges attracting, *and* of quantum mechanical explanations) which even though it is deeply inconsistent at its core, is rich enough to provide productive extrapolations.

There is a relation here to Peter Galison's concept of the utility of pidgin languages,[16] with the addition I would make that the trading zone forms in the mind of every individual.

A not unrelated phenomenon in chemistry (and maybe not just chemistry) is the fecundity of flawed if not downright wrong theories.[17,18]

Analysis and Synthesis, Creation and Discovery

In a unique 1806 novel, *Elective Affinities*, constructed on the metaphor of a theory already on the way out, in a time when analysis was at the center of chemistry, Goethe has this scene between Eduard and Charlotte:

> *"The affinities become interesting only when they bring about divorces."*
>
> *"Does that doleful word, which one unhappily hears so often in society these days, also occur in natural science?"*
>
> *"To be sure," Eduard replied. "It even used to be a title of honour to chemists to call them artists in divorcing one thing from another."*
>
> *"Then it is not so any longer," Charlotte said, "and a very good thing too. Uniting is a greater art and a greater merit. An artist in unification in any subject would be welcomed the world over."[19]*

The $H_2@C_{60}$ paper I chose details the sequential transformation of molecules. Is that typical? And, if typical, what consequences do such papers have for the philosophy of science?

I would claim the Komatsu et al. paper, beautiful as it is in its individuality, is indeed also typical of organic, polymer, inorganic, organometallic, and solid state chemistry. But not of physical and analytical chemistry. In the formative years of modern chemistry, the primeval analytical question "What is it?" answered at the atomic level, was the dominant one. That query remains important, for Komatsu and his coworkers interrogate every molecule they make as to its identity. They must analyze their reagents, and the intermediate structures in their synthesis. And they most certainly must prove the consistency of their spectra with the structure they suggest for their heroic achievement, $H_2@C_{60}$.

But, as you can just see from the way the work is presented, the analytical enterprise is, if not taken for granted, subjugated. Most of the evidence is in the Supplementary Material to Komatsu's paper. The emphasis is on the new molecule, on synthesis. Making it matters.

The concern may return that I have picked a certain kind of paper out of the multitude. Joachim Schummer has come to similar conclusions.[20] He has also taken a sample of 300 synthetic papers and analyzed them in detail, asking for the papers' *aims*, i.e., why chemists do experiments. And how they accomplish them. Figure 3-2 shows the outcome:

Schummer finds the emphasis on synthesis I mention, as well as what one might call "propagation"—the making of new molecules "in order to improve the abilities to produce more new substances."

When did this change, when did synthesis begin to dominate chemistry? One locus is found in the in the second half of the 19th century, in Perkin's opening to aniline dyes, and the subsequent development of German dye and medicinal chemistry. Another is found in the twin explosions of activity in organometallic and solid

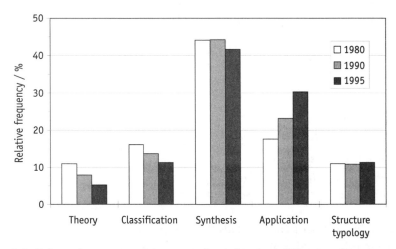

Figure 3-2 Relative frequencies of aims in synthetic chemistry, 300 papers from *Angewandte Chemie*, 1980-1995.[20]

state chemistry in the second half of the 20th century. Certainly an intellectual high turning point was provided by the synthetic tour-de-forces of R. B. Woodward's mid-20th century syntheses.

What Could Be Different?

I believe the dominant interpretations of the nature of science by mid-20th century philosophers come out of a focus on science as discovery. And if those interpretations were instead based on the heart of chemistry, the creation explicit in synthesis, they would be different. As I said above, and as Berson argues convincingly,[21] the extended philosophical discussion of falsification, the prevalent concern with testing theory, has little to with chemical preparative practice.

I fear that the even the chemistry-educated philosophers of note—from Freudenheim to Polanyi—have also come from a certain side of the tracks. This was physical chemistry, which shares much of its research ethos, though not all, with physics. Even as they do fascinating things, by and large, physical chemists don't make molecules.

Emily Grosholz reminded me that I should acknowledge a change in philosophical winds in the 1980s. Ian Hacking, Nancy Cartwright, Bas van Fraassen, Margaret Morrison, as well as others emphasized experiment, construction, and intervention. Their efforts, paralleled by covergent activity in the history of science, have created a philosophical purview sympathetic to the story I tell. Grosholz remarks that these philosophers ".... might have developed their positions in even more interesting ways if they'd drawn their case studies from chemistry as well as physics." Indeed.

So what could have been different? Chemists make compounds/molecules, the objects of their own contemplation. That puts them close to art and artists. Lest we get too romantic about that (only someone who has not tried to make a living as an artist would), the centrality of creation also puts us close to engineers. The molecule is made, and often it is made for a purpose—a desired physical or biological property.

The connection of this one science to art—high and low—could have led philosophers to apply aesthetic theories to science. This would have certainly helped people to be more at peace with the obvious influence (not all for the good, mind you) of aesthetic factors—symmetry, order, telling a good story—in the acceptance of theories. Not just prediction.

The connection to engineering makes utility an important criterion to chemistry. Sulfuric acid, no. 1 on the chemical hit list for 75 years, is beautiful because all 100 billion lbs of it made in the U.S. last year are *used*. And utility immediately creates a problem for my hypothetical new philosophy of science that incorporates aesthetics. Disinterested contemplation has been a central feature of aesthetic theories—vide Kant's *Zweckmässigkeit ohne Zweck*. I personally think this points to a failure of classical aesthetics, in not allowing utility to be a valid determinant in defining what constitutes art and beauty.[22]

One more point: Chemistry is the truly anthropic science—our molecules can heal, and they can hurt, for they are on the scale of the molecules in our bodies. A consequence is that chemists must enter the arena of public ethical debates. Many do not do so willingly.

Semiotics, Too

The number of drawings in the Komatsu et al. paper is probably less than if this paper were published in a chemical journal—*Science* limits the number of illustrations. But it is high enough for you to see centrality of graphical representation in chemistry. And that means pictures of molecules more than curves of measurements.

Though describing compounds and their transformations requires representation (see the writing around the arrows in Figure 1), it is the microscopic worldview, of molecules, which makes for a flowering of representation. Curious representations they are too, polyhedra floating in some neutral space, situated somewhere between the compact messaging of caricature, the conventions of "primitive" art, and cubist perspectives.[23]

And, all the time, trying desperately to evoke in the mind of the chemist reader who is privy to the primitive graphic codes used (for chemists, to whom three-dimensional representation of shape may be a matter of life or death, are not trained in art), the geometry of the molecule.

There is lots going on there, in those visual representations. Perhaps most important is the mixing of symbolic and iconic representations. The names of the compounds, and the elements of which they are comprised, are all symbolic. Even though these are real, tangible substances! The molecules, which are *not* seen directly, are given a more iconic representation—you see that hole in the buckyball, don't you? Though there are awful doses of symbolism hidden in those bond lines.

It seems like a holy mess, this conflation of representations. But it works. Elsewhere, Emily Grosholz and I have argued that the world bridging involved in dealing with a real chemical problem requires both symbolic and iconic representations.[24] We write:

> *How does the iconic form of the chemical structure expressed as a diagram that displays atom connectivities and suggests the three-dimensionality of the molecule, bridge the two worlds of the chemist? The most obvious answer is that it makes the invisible visible, and does so, within limits, reliably. But there is a deeper answer. It seems at first as if the chemical structure diagram refers only to the level of the microscopic, since after all it depicts a molecule.*

But in conjunction with symbolic formulae, the diagram takes on an inherent ambiguity that gives it an important bridging function. In its display of unified existence, it stands for a single particular molecule. Yet we understand molecules of the same composition and structure to be equivalent to each other, internally indistinguishable. . . .

Thus, the icon (hexagonal benzene ring) also stands for all possible benzene rings, or for all the benzene rings (moles or millimoles of them!) in the experiment, depending on the way in which it is associated with the symbolic formula for benzene. The logical positivist in search of univocality might call this obfuscating ambiguity, a degeneracy in what ought to be a precise scientific language that carries with it undesirable ontological baggage. And yet, the iconic language is powerfully efficient and fertile in the hands of the chemist.

The iconic and symbolic languages are each incomplete—they cannot represent all. But that is not a failing; they are "productively ambiguous," serving an essential bridging function.

Violating Categories

Another characteristic of modern chemistry is the inseparable mixing up of the microscopic and the macroscopic perspectives in doing chemistry. Look at the Komatsu et al. paper in Figure 3-1. It contains four figures of molecular drawings, and not a single photo of the material(s) undergoing transformation. The physicality of the processes is hinted at in the telegraphic (sorry, text-messaging) abbreviated reaction conditions around the arrows in Figures 1 and 4 ("toluene, r.t. [= room temperature], 13 hours"). The details are in the "Supporting Online Material"; the only amount of material mentioned in the text is the mass of $H_2@C_{60}$ obtained in the final step–118 mg, contaminated with some C_{60}.

The point here is not how far away (and why) we are from the format of scientific reportage of say, 18th century England. Rather it is that the practicing (and excellent) chemists inextricably mix macroscopic and microscopic viewpoints of substances and molecules in the productive work of their science.

I think there are philosophical implications in the violation of categories. I do not dispute that the philosopher's job is to define the modes and levels of thought and action, and that upon such delineation one can superimpose the messy actions of often illogical human beings. This was Lavoisier's program too, of really defining the elements. And it made modern chemistry possible. For instance, could we ever define impurity in chemical compounds except by getting compounds as pure as they can be (defying entropy in the process) and then comparing made-to-order mixtures with naturally impure substances?

I'm torn about this. I started out in chemistry perturbed by the mixing of categories around me, drunk on logic, mathematics, and symmetry. I was looking, as Primo Levi once was, for the theorems of chemistry. Eventually I came to peace with the multivalency of piecewise understanding around me. And I saw that partially irrational reasoning (oh, prettified for publication) led to stunning molecules and reactions.

My perception of human beings, not just chemists, is

(a) that in the service of either creation or utility, they will naturally and deliberately violate all categorizations (here chemists inextricably mixing up molecules and compounds), and

(b) that the process of creation of the new depends essentially on the transgression of categorization.

Point (a) is weak, and ultimately unimportant: people are people. Point (b) is stronger, with implications for philosophy: I want to claim that people are unlikely to make the new (art, science, religion, new people) without violating categories.[25]

I am here beyond philosophical holism, beyond intellectual bricolage, close to Feyerabend's prescription for "epistemological anarchism." Must be what too much poetry does.

One could imagine the last, strong variant as a restatement of Kuhn's revolutionary program; if so it represents a wholesale democratization of revolutions—they must be there in every creative act.

Change

So, Heraclitus was right. While equilibrium creates a stable, seemingly dormant middle (even as that middle is seen at the molecular level to be all flux), the chemical norm is change. Products are different from reactants. We can, and do, perturb equilibrium.

Could there be philosophical consequences of continuing, ever-present change? I'm not sure, but I wonder if cognizance of chemistry's emphasis on the arrow in $A + B \rightarrow C + D$ could have led people to spend more effort in constructing workable process philosophies. In doing so, perhaps one could learn something from the chemical analysis of complex reaction systems (the atmosphere, metabolism, oscillating reactions, surface catalysis) in building a philosophical frameworks that respect equally things and the changes between them. Some efforts in this direction have been made.[26,27]

We are drifting in that direction in molecular biology, as the relative paucity of the genome does not easily explain the richer proteome and how the environment influences an organism. Evelyn Fox Keller has spoken interestingly of "verbing biology."[28] I think we need thought from the philosophical community (with something to learn from artists) on how best to represent change.

Change is desired and resisted by individuals, and mostly resisted by our societies. No wonder then that society has an ambivalent view of the science at whose core is change. Even so, we are fascinated by change—be it a fire burning, a new dance tune, a child growing. As Mircea Eliade said (I paraphrase), chemists, screaming to high heaven that they had nothing to do with alchemy, accomplished everything the alchemists desired—to heal, to turn base things into gold (that's what Merck, Sharp, and Dohme did).[29] At the same time chemists—well, molecular biologists— transplanting genes have entered the transgressive region of free change that bothered the Church about alchemy. Bernadette Bensaude-Vincent writes well of this, of our continually transgressive science.[30]

And yet, as I said, we are fascinated by change. How to capitalize on that fascination, now that is a problem for today's chemistry. We desperately need the magic of alchemy, its natural hold on the imagination.

Philosophy of science changes too. To say that every decade a new "ism" is in, misses the point as much as the current rage for nano-this or nano-that in chemistry. Much more interesting is the wonder of something new made or said. In chemistry, all of a sudden there is this new molecule, C_{60}, or H_2 in it—how come we didn't think of it, did not have it before? So into the library (I'm getting old) comes this new book, as Kuhn's did one day, and it shapes a new world, even as it has antecedents and ambiguities.

I think that philosophy has something to learn from my truly anthropic science, the one with synthesis at its center. That science is in the middle, but the middle—be it for human beings or molecules—is never static. Oh, it may seem quiet, but it is tensed, a perturbable balance of polarities: of natural/unnatural, macro/micro, pure/impure, harm/benefit, the same/not the same, of trust/doubt. Things change. Quiescent, they can be willed to change. They, we must change.

Acknowledgments

I am grateful to Bernadette Bensaude-Vincent, Emily Grosholz, and Michael Weisberg for their reading and commentary on my text.

Notes

1. Indeed, the French name of the discipline explicitly carries the plural "sciences."
2. B. Bensaude-Vincent, Chemistry in the French Tradition of Philosophy of Science: Duhem, Myerson, Metzger, and Bachelard. *Studies in the History and Philosophy of Science* 36 (2005): 627.
3. A. R. Ammons, *The Really Short Poems of A.R. Ammons* (New York: Norton, 1990), p. 16.

4. K. Komatsu, M. Murata, and Y. Murata, Encapsulation of Molecular Hydrogen in Fullerene C_{60} by Organic Synthesis. *Science* 307 (2005): 238.
5. For what I think is a good account of these views, see P. Godfrey-Smith, *Theory and Reality* (Chicago: University of Chicago Press, 2003).
6. M. Weisberg, letter to R. Hoffmann.
7. J. A. Berson, *Chemical Discovery and the Logicians' Program* (Weinheim: Wiley-VCH, 2003).
8. See also B. Bensaude-Vincent, Lessons in the History of Science. *Configurations* 8(2) (2000): 203.
9. J. Ziman, *Reliable Knowledge* (Cambridge: Cambridge University Press, 1978).
10. R. Hoffmann, Nearly Circular Reasoning. *American Scientist* 76 (1988): 182. Chapter 5 in this book.
11. I. Hacking, *Representing and Intervening* (Cambridge: Cambridge University Press, 1983), p. 262.
12. R. F. W. Bader, *Atoms in Molecules: A Quantum Theory* (Oxford: Oxford University Press, 1990).
13. R. Hoffmann, *The Same and Not the Same* (New York: Columbia University Press, 1995).
14. R. Hoffmann, Nearly Circular Reasoning. *American Scientist* 76 (1988): 182.
15. For a succinct review of the period, see J. Golinski, Chemistry, in *The Cambridge History of Science, Vol. 4. Eighteenth-Century Science*, ed. by R. Porter (Cambridge: Cambridge University Press, 2003), p. 377.
16. P. Galison, *Image and Logic: A Material Culture of Microphysics* (Chicago: University of Chicago Press, 1997).
17. Berson, *Chemical Discovery and the Logicians' Program* (Weinheim: Wiley-VCH, 2003), Ch. 8.
18. R. Hoffmann, Why Buy That Theory? *American Scientist* 91 (2003): 9.
19. J. W. v. Goethe, *Elective Affinities*, trans. by R. J. Hollingdale (New York: Penguin, 1971), p. 53.
20. J. Schummer, Scientometric Studies on Chemistry I: The Exponential Growth of Chemical Substances, 1800-1995. *Scientometrics* 39 (1997): 107; J. Schummer, Scientometric Studies on Chemistry II: Aims and Methods of Producing New Chemical Substances. *Scientometrics* 39 (1997): 125; J. Schummer, Why Do Chemists Perform Experiments? in *Chemistry in the Philosophical Melting Pot*, ed. by D. Sobczyńska, P. Zeidler, and E. Zielonacka-Lis (New York: Peter Lang, 2004), pp. 395–410.
21. J. A. Berson, *Chemical Discovery and the Logicians' Program* (Weinheim: Wiley-VCH, 2003).
22. R. Hoffmann, Molecular Beauty. *J. Aesthetics and Art Criticism* 48(3) (1990): 191.
23. R. Hoffmann and P. Laszlo, Representation in Chemistry. *Angewandte Chemie, Int. Ed. Engl.,* 30 (1991): 1 (Chapter 14 in this book); P. Laszlo, *La Parole des Choses* (Paris: Hermann, 1993).
24. E. R. Grosholz and R. Hoffmann, How Symbolic and Iconic Languages Bridge the Two Worlds of the Chemist: A Case Study from Contemporary Bioorganic Chemistry, in *Of Minds and Molecules: New Philosophical Perspectives on Chemistry*, ed. by N. Bhushan and S. Rosenfeld (New York: Oxford University Press, 2000). pp. 230–247, Chapter 16 in this book.
25. P. Feyerabend, *Against Method* (New York: Verso, 1975).
26. J. E. Earley, Self-Organization and Agency in Chemistry and in Process Philosophy. *Process Studies* 11 (1981): 242; J. E. Earley, Modes of Chemical Becoming. *Hyle* 4 (1998): 105.
27. R. L. Stein, Towards a Process Philosophy of Chemistry. *Hyle* 10 (2004): 5.
28. E. F. Keller, The Century Beyond the Gene. *J. Biosc.* 30(1) (2005): 3.
29. M. Mircea Eliade, *The Forge and the Crucible*. Trans. by S. Corrin (London: Rider, 1962).
30. B. Bensaude-Vincent, *Faut-il Avoir Peur de la Chimie* (Paris: Seuil, 2005).

4

Unstable

ROALD HOFFMANN

Words are our enemies, words are our friends. In science, we think that words are just an expedient for describing some inner truth, one that is perhaps ideally represented by a mathematical equation. Oh, the words matter, but they are not essential for science. We might admit there is a real question as to whether a poem is translatable, but we argue that it is irrelevant whether the directions for the synthesis of a molecule are in Japanese or Arabic or English—if the synthesis is described in sufficient detail, the same molecule will come out of the pot in any laboratory in the world.

Yet words are all we have, and all our precious ideas must be described in these history- and value-laden signifiers. Furthermore, most productive discussion in science takes place on the colloquial level, in simple conversation. Even if we know that a concept signaled by a word has a carefully defined and circumscribed meaning, we may still use that word colloquially. In fact, the more important the argument is to us, the more we want to be convincing, the more likely we are to use simple words. Those words, even more than technical terms, are unconsciously shaped by our experience—which may not be the experience of others.

I was led to reflect on this by the reaction of a friend of mine, a physicist, to my use of the word "stable." I had said that an as yet unmade form of carbon was unstable with respect to diamond or graphite by some large amount of energy. Still, I thought it could be made. My friend said, "Why bother thinking about it at all, if it's unstable?" I said, "Why not?," and there we were off arguing. Perhaps we should have pondered why the simple English word "stable" has different meanings for a physicist and a chemist.

First, a little background. Diamond (Figure 4-1 *left*) and graphite (Figure 4-1, *right*) are the two well-known modifications, or allotropes, of carbon.

The carbon atoms are linked up in very different ways: in diamond each carbon atom is tetrahedrally surrounded by four neighbors, whereas in graphite a layer structure is apparent. Each layer is composed of "trigonal" carbons, three bonds going off each carbon at 120° angles. The graphite layers are held weakly, and not by real chemical bonds. They slip easily by each other, which is why graphite serves as

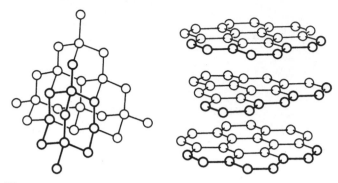

Figure 4-1 The structures of diamond and graphite.

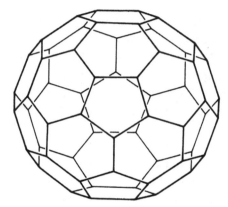

Figure 4-2 Buckminsterfullerene structure.

a lubricant. Isn't it nice that black graphite is more stable (thermodynamically, more on this in a moment) than pellucid, hard diamond? Not by much, but so it is for carbon on the surface of the earth. Under high pressure, however, the stability sequence, which is determined by a combination of energy and entropy, changes; the denser diamond becomes more stable. This is what happens deep within the earth; it is also the basis of a commercial process for making industrial-grade diamonds.

There are other forms of carbon.[1] The random and systematic incendiary activities of men and women have led to a multitude of pyrogenic materials, and most seem to be some form of graphite. A couple of rare, but well-established, allotropes of carbon are related to the diamond and graphite structures, and the existence of some others is disputed.[2] Carbon also turns up in short chains in the tails of comets and in flames, and recently hefty clusters of between two and 100 carbon atoms have been produced in the gas phase. A most abundant cluster is one with 60 atoms, first detected by Kroto, Heath, O'Brien, Curl, and Smalley at Rice University. They suggested the structure of a soccer ball for this remarkable molecule, and named it "buckminsterfullerene"(Figure 4.2).[3,4]

One day, I was trying to think up some alternatives to diamond and graphite. Why? For one thing, it was fun; for another, people have been squeezing elements which are not metals, trying to make them metallic. When you apply a megabar or so of pressure to almost anything, the atoms are forced so close together that their electron clouds overlap, and the material becomes a metal. Some of my friends at Cornell do this routinely—using diamond anvils! Xenon and iodine have been made metallic in this way,[5] and there's an argument whether hydrogen has been so transformed.[6]

The interesting thing about both diamond and graphite is that they are, so to speak, full of nothing. They're not dense at all; a close-packed structure such as that of a typical metal would be much denser. Of course, there is a good reason that the density of the known carbon allotropes is so low: carbon atoms form bonds, and there is a lot of energy to be gained by forming those bonds only directionally, trigonally, or tetrahedrally. Carbon, with its four valence electrons, has better things to do than to try to shuffle its bonding among its 12 or 14 different nearest neighbors, as it might be forced to do in a close-packed structure.

Could there be carbon networks filling space more densely than diamond or graphite, yet forming bonds along trahedral or trigonal directions? If such a structure could in principle exist, applying pressure to one of the known allotropes might be a way to produce it. To sum up a long story, many hypothetical alternative space-filling structures have been designed.[7] But we haven't yet found one denser than diamond.

Peter Bird and I thought up an allotrope that is intermediate in density between diamond and graphite and is quite special. It fills space with perfect trigonal carbon atoms. In the jargon of our trade, these form polyacetylene chains, needles of conjugation, running in two dimensions, and no conjugation at all in the third. The most remarkable thing about this structure (Figure 4-3), something which emerged from the calculations of Tim Hughbanks, is that it should be metallic—as it is, with no pressure applied to it.[8]

Here then is a prediction of a metallic allotrope of carbon. If only one knew how to make it!

Proceeding from structural reveries to matters of stability, the relatively unreliable calculations at our disposal indicate our hypothetical substance to be *unstable*

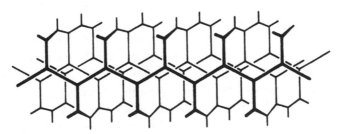

Figure 4-3 A hypothetical metallic allotrope of carbon.

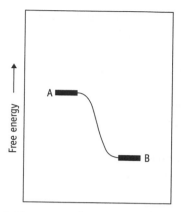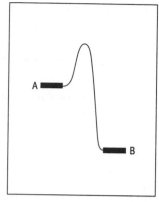

Figure 4-4 Two potential energy surfaces for an exothermic transformation of molecule A into B.

relative to graphite by a whopping 0.7 electron volt per carbon atom, or 17 kilocalories per mole of carbons. This is what made my physicist friend say, "Why bother thinking about this substance?" But this degree of instability didn't bother me at all.

Why the different reactions? Because the common English words "stable" and "unstable" had different meanings for the two of us!

To get at the source of our misunderstanding, let me go back to a scientific definition on which both my friend and I could agree. Real stability has to do both with thermodynamics, the science of energy and entropy relationships, and with kinetics, the rates or speeds of imagined processes by which a system might be stabilized or destabilized. In chemistry we distinguish between thermodynamic and kinetic stability. Suppose we have two molecules, A and B, which have the potential of changing one into the other (see Figure 4-4).

Their relative thermodynamic stability is gauged by a marvelous function called the Gibbs free energy, which contains in it terms for enthalpy (something very much like energy, but with specific conditions placed on it) and entropy. The natural, spontaneous direction in which matter moves is toward lower enthalpy and higher entropy, which means, in turn, greater disorder. A molecule which has the lower free energy is more stable, and a molecule with higher free energy will transform spontaneously into one with lower free energy. To be specific: should it be that B has the lower free energy, then the spontaneous reaction will be A→B. We can represent this in a graph in which the vertical axis is the free energy.

But life is not so simple. Thermodynamics says what *must* happen, but not how fast it *will* happen. To go from molecule A (say, my hypothetical metallic carbon) to molecule B (graphite) is no idle molecular promenade. Bonds have to break, many of them, and then to reform. Before poor A knows all the happiness waiting for it in those lovely rings of B, it's got to suffer a lot of electronic indignity in the form of broken bonds. It resists. In general, molecules have barriers to their transformation. The situation is typically not as shown at left in Figure 4.4, but as at right.

There's a hill in the way. It's like a book that wants to fall under the force of gravity, but has a shelf under it. We might say that A is metastable, or that A is a local minimum on some energy surface. Suddenly it is no longer a question of falling, but of climbing hills!

Will the reaction proceed after all? Yes, if you wait long enough. It depends on the size of the hill and on the temperature. Molecules don't sit still. In a gas or in solution they are bouncing around at great speed, buffeted randomly by collisions with the 10^{20} other molecules in a typical flask. It's a crowded dance floor there. Some of the molecules acquire enough energy through collisions (this is where the temperature comes in, for the higher the temperature, the faster the molecules move) to pass over the hill. Others don't. If the hill is higher than about 30 kilocalories per mole, then at room temperature A will remain A. Unless you wait a thousand years, for it is only then you might be able to see a little B.

A chemist would say A is thermodynamically unstable and kinetically stable or persistent, whereas a physicist might call A metastable. These concepts are quite familiar to chemists and to physicists. So where is the problem? The difficulty is that our everyday discourse is perforce colloquial. We say "stable" and not "thermodynamically and/or kinetically stable." Some may label the colloquial characterization sloppy and say it should be more precise. I say that we wouldn't be human (and therefore have the potential of doing great science) unless we were often imprecise in just this way.

But now comes the crux of the matter. Into that word "stable" goes the history of what we are or have done. When a chemist says "stable," I think he or she means 90% kinetic and 10% thermodynamic. But a physicist, I would hazard a guess, means (not in the sense of making a rational choice, but unconsciously) just about the converse: 90% thermodynamic, 10% kinetic.

From the beginning of one's life in chemistry the importance of kinetic stability and the relative unimportance of thermodynamic stability are highlighted. Every organic molecule in the presence of air (a typical situation in the laboratory and real life) is thermodynamically unstable with respect to CO_2 and H_2O. Think of methane (CH_4 natural gas), the essence of stability, having survived unchanged under the earth for thousands of years. Every time you light a gas stove, you demonstrate methane's thermodynamic instability in the presence of oxygen. But it takes the complicated autocatalytic reaction set off by a match to take those CH_4 and O_2 molecules and get them over the hill, giving off light and heat along the way. Otherwise, methane is stable as a rock. Speaking of rocks, modern air pollution shows that they are not particularly stable when strong acids come around.

One amusing way to define synthetic chemistry, the making of molecules that is at the intellectual and economic center of chemistry, is that it is the local defeat of entropy, the construction of complex thermodynamically unstable molecules. In chemistry, a molecule that is strained, or otherwise thermodynamically disfavored by 1 electron volt per molecule relative to another molecule, is not thought of as an occasion to throw up one's hands. It's a challenge to be made, ingeniously.

Thermodynamic stability is set more firmly in the physicist's mind, for a number of reasons. First, a typical course in elementary physics concentrates on mechanics, dynamics, and electromagnetism in the absence of barriers or obstacles. Motion in the presence of barriers is too difficult to solve explicitly, so such problems are not mentioned. No one ever puts a shelf of variable permeability under that falling weight in Physics 100. Barrier penetration is probably first encountered in quantum mechanics courses.

Second, in thinking about the transformation of matter, physicists most often begin with motions governed by central forces, masses, or charges moving around without hooks or directional valences. Entering the study of matter from the starting point of gases or close-packed metals, one encounters few activated processes, only collisions, or balls sliding frictionlessly past balls, to reach the thermodynamically most stable point. Friction, barriers, and the evolution in time of real systems are just as important in the end for physicists as they are for chemists. But the subtle weighting of concepts which shapes the colloquial language of science is fixed in scientific infancy. The early experiences matter; this is why I think the words "stable" and "unstable" mean different things to chemists and physicists.[9]

Meanwhile, our metallic carbon allotrope is still waiting to be synthesized. I think it will be pretty stable – sorry, enduring – when it is made. If it is made.

Notes

1. J. Donohue, *The Structure of the Elements* (New York: Wiley, 1974).
2. J. Donohue, *The Structure of the Elements* (New York: Wiley, 1974).
3. H. W. Kroto, J. R. Heath, S. C. O'Brien, R. F. Curl, and R. E. Smalley, C_{60}: Buckminsterfullerene. *Nature* 318 (1985): 162.
4. This was written in 1987. The evidence for the buckminsterfullerene structure was at that time still fragmentary.
5. Xenon: D. A. Nelson and A. L. Ruoff, Metallic Xenon at Static Pressures. *Phys. Rev. Lett.* 42 (1979): 383. Iodine: A. S. Balchan and H. G. Drickamer, Effect of Pressure on the Resistance of Iodine and Selenium. *Chem. Phys.* 34 (1961): 1948. Also see K. Syassen, K. Takemura, H. Tups, and A. Otto, in *Physics of Solids under High Pressure*, ed. by J. S. Schilling and R. N. Shelton (Amsterdam: North Holland, 1986), p. 125.
6. Subsequent to the writing of this article, hydrogen was metalized in shock wave experiments.
7. I. V. Stankevich, M. V. Nikerov, and D. A. Bochvar, *Russ. The Structural Chemistry of Crystalline Carbon: Geometry, Stability, and Electronic Spectrum. Chem. Rev.* 53 (1984): 640. Many more hypothetical carbon allotropes have been suggested since then.
8. R. Hoffmann, T. Hughbanks, M. Kertesz, and P. H. Bird, Hypothetical Metallic Allotrope of Carbon. *J. Am. Chem. Soc.* 105 (1983): 4831.
9. For some fascinating observations on the language of physics see C. F. von Weizsäcker, *Die Einheit der Natür* (DTV, 1974), p. 61.

5

Nearly Circular Reasoning

ROALD HOFFMANN

Scientific argument is supposed to be logical. But do scientists study logic? Probably not. Were they asked about the advisability of learning formal or applied logic, most would likely say, "Logic, as studied by philosophers, is just a systemization or description of what we, as scientists, do naturally. So we don't need to study it."

The chain of reasoning that I've ascribed here to a straw-man scientist is, on analysis, full of the fallacies described by Aristotle in *Sophistical Refutations (De sophisticis elenchis)* more than 2,300 years ago. The argument suffers from circular reasoning, the fallacy of false cause, the argument *ad populum* (the *populus* here being scientists, as opposed to philosophers), and more.[1] But actually I do not want to berate here the logically unsophisticated scientist (myself), nor to urge that scientists need study philosophy. Rather, I'd like to examine the curious role of logic in science. Good logical thinking is absolutely necessary to both everyday and revolutionary science. But I will argue that at the same time, reasoning in all science, paradigmatic or ground-breaking, on close scrutiny often turns out to be in part illogical. There is nothing new in this—we see readily the fallacies in the work of others, especially when they disagree with us, don't we? I will try to make a case, however, that there may be a real advantage implicit in occasionally faulty reasoning, especially a mode which I will call nearly circular reasoning.

Science is a curious mixture of the real and the ideal, the material and the spiritual, held together by discourse or argument. The latter is sometimes mathematical, but more often it transpires in the words of some language. The real is the material, say, a vial of a chemical, or its measured spectrum, the relative amount of light a solution of that chemical absorbs. The ideal may be a proposal on the mechanism of formation of the molecule, or a theory that interprets that spectrum as necessarily indicating the molecule contains a carbon-hydrogen bond. The discourse consists of the exposition of several arguments, several alternative models explaining the observable, and a choice between them. Try to imagine a scientific article or a seminar without the glue of words or argument!

The use of language, an absolute necessity in scientific discourse, is one source of possible circularity; the nature of argument is another. A cursory tracing of chains of definitions in any dictionary reveals how quickly such chains become circular. Yet we easily use language to explain and communicate. As Klever says, "Natural language is a complex network in which circular argumentation is not only unavoidable, but even the only means of explication."[2]

Argument has perforce a psychological and rhetorical component—it is an attempt to convince, first oneself, then others, of the validity of a certain conclusion. The dialogue may be an inner one, shaping ideas or prompting one to do the next experiment. It may be an outer one, with imagined audiences as one writes a paper, with real ones at a scientific conference. It is in these rhetorical settings that the natural argument of scientists becomes quasicyclic. Yet, by being connected on one hand to the reality of substances and measurements, on the other hand tied to the inner psychological forces that move us, that possibly illogical argument advances science.

Productive argument in science is mostly about new things. Thirty-five years ago, high-temperature superconductors of two kinds were discovered, $La_{1.85}Ba_{0.15}CuO_4$ and $YBa_2Cu_3O_{-7}$. If one takes the oxidation states (a tremendously useful fiction) of lanthanum and yttrium as 3+, oxygen as 2–, and barium as 2–, then in both cases one comes to the conclusion that one is working with copper atoms in oxidation states between 3+ and 2+. Perhaps some coppers are 3+, some are 2+, just enough to make that balance right. One can begin to build theories which link the extraordinary conductivity to some precarious balance of the energetics of copper ions in different oxidation states.[3]

Meanwhile, several groups have come up with evidence, from a measurement involving x-ray spectroscopy, that in these ceramics there is another type of oxygen ion, in addition to the normal "2–" one.[4] If it's there, it's important, because if you follow through the charge balance, the coppers don't have to bear so much positive charge if some of the oxygens are O instead of O^{2-}. In fact, instead of Cu^{3+} and Cu^{2+}, the discussion shifts to Cu^{2+} and Cu^{1+}. Which, for various reasons, would make chemists happier.

Since oxidation states are a convenient fiction, we mustn't take O^{2-} and O^{-} too seriously.[5] A realization of "O^{-}" would be peroxide ions: O_2^{2-}, diatomic entities. There is an argument building as to whether, in these superconductors, some of the oxygen atoms might have moved off their idealized lattice sites, where they are too far apart to bond, to form peroxide ions. The argument will soon draw in the structural chemists and physicists who determined the positions of the atoms in these substances. Because there is some disorder in these molecules as a result of the oxygen nonstoichiometry, these structures—the best that can be done—are not as accurately known as other stoichiometric solids. Whatever the crystallographers did, and I repeat that they did it very well, and in nine laboratories (or is it seventeen?), it was assumed that the oxygens always sit at certain lattice sites, with no O_2^{2-} or peroxide species present. Well, they're going to go back and think about it.

Someday, there will be definitive proof of the absence or presence of O_2^{2-} units. But then the problem will be solved, dull, and uninteresting. Today there is no proof, the problem is actual—there are just hints, the merest trace of a shoulder in a complex spectrum, less than that. Right now intuition, a jump of the imagination, a nondeductive argument, the following of a hunch can matter. And the published literature, still more the oral presentations at seminars and meetings, and still much more the informal opinions voiced in research group meetings, are full of suppressed or explicit opinions and categorical statements that such-and-such a measurement or theory is nonsense. There is rash judgment, there may be prejudice, there certainly is a lot of disagreement. In the debate that ensues, logic is likely to play a significant role. Nevertheless, because it is a debate, and human beings rather than machines are debating, it is likely that the full spectrum of rhetorical tricks and fallacies that Aristotle saw so clearly will be involved—nearly as much in this discussion as in any presidential campaign.

But what people will do to win an argument or convince others is really not what I want to address. I want to examine some of the ways in which fallacies or illogical thought may be useful in science.

Constructing an explanation or rationalization, then claiming it as a prediction, validating the theory so constructed, seems patently illogical. But I will claim that the process has definite value. Since I get into mighty trouble if I use the work of living or recently deceased colleagues, I'm left with illustrations from my own work or that of people long dead.

Irontetracarbonyl-ethylene, $Fe(CO)_4(C_2H_4)$, is an interesting molecule, known for over thirty years, and quite typical of modern organometallic compounds. Its shape can be described as an iron-centered trigonal bipyramid, or two tetrahedra sharing a face, with an iron in the middle. Structure 1 in Figure 5-1 illustrates the geometry; in it there are distinct axial (a) and equatorial (e) sites.

To a chemist the obvious question is whether the ethylene occupies an axial site, as shown in structure **2**, or an equatorial site, as in **3**. The experimental answer, coming from x-ray crystallography, is that shown in structure **3**, ethylene equatorial.[6]

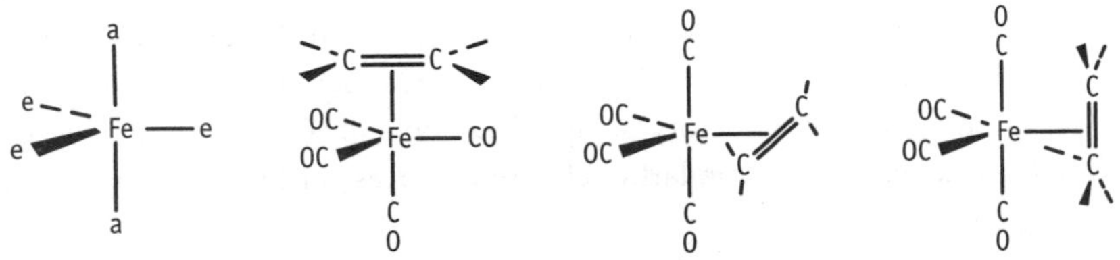

Figure 5-1 **1** Axial and equatorial positions in FeL_5; **2** and **3**: ethylene in axial and equatorial positions in $(CO)_4Fe(ethylene)$; **4**: another geometry for ethylene in the equatorial position.

With that established, a further question may be posed: is the carbon-carbon bond of the ethylene oriented in the equatorial plane (as in **3**), or perpendicular to that plane (i.e., parallel to the vertical axis, as shown in structure **4**), or somewhere in between?

When my co-workers and I constructed a theory of the geometry of this molecule and other organometallic compounds we knew of several experimentally determined structures, all of which showed the ethylene in an orientation near to structure 3. In our explanation we constructed **2**, **3**, and **4** from the orbitals of an $Fe(CO)_4$ fragment and ethylene, and showed that the favored arrangement was ethylene equatorial and that there was a strong preference for the specific geometry illustrated in **3**.

When we found our argument for the observed geometry of structure **3**, we didn't say that we merely "rationalized" the known preference. Neither did we dare say that we "predicted" it, for that would clearly have been too much, a number of experimental structures already being known at the time. In typically ambiguous (or sloppy) language we said that a certain quantum-mechanical interaction "will cause a marked preference for the coordinated ethylene to be in the equatorial plane." Note the subtle appeal to strong causality, a hint that this is the way things must be.

Actually, I think that our argument, whose details are quantum-mechanical, was not a case of circular reasoning. But in the explanation we were admittedly reaching for the status of a prediction. Why? Because, of course, a true explanation must have consequences of a predictive nature. The subtle conversion of a rationalization into a prediction was probably even stronger in our minds than we allowed ourselves to commit to paper. Was this bad?

Not really, I think, and here I will leave my own failing and generalize that much of this, very much, goes on in the work of others. What we gained as a result of that bit of almost circular reasoning is confidence. We were not engaged in a mathematical proof, we were doing chemical theory. Theoretical chemistry, at the level we were practicing it, is happy if it is right 85% of the time on the geometry of a molecule. It is soft theory. So is most theory in science, although the ideology of science tends to single out hard theories—those capable of being disproved by a single experiment—as being emblematic of all theory.[7] This is a romantic fantasy, and science, not only chemistry, would have gotten nowhere if it had waited for such strong theories. The reasoning used by Bednorz and Müller in their discovery of the high-temperature superconductors is a good example of this.[8] They came to their remarkable discovery by a wonderful, chemical process of hints, analogies, facts, and intuition.

When one is in the act of building a theory, a framework for understanding, one needs all the psychological support one can get. When it's all done, it's stuff for textbooks on the scientific method or for sanitized memoirs. But when it's being done, one grasps at straws. One wants a theory to "work." And so when it does work one is happy, encouraged to put in the terrible labor that is often involved in pushing through to the next calculation, the next prediction.

There is another imperative for circularity, inherent in the natural interplay of theory and experiment. One has curious observations, begging for explanation. So one explains them, constructing a trial model, a theory. To see if it works one compares the theory's results with experiment. Note the forward and reverse motion that shapes the circle. If the theory doesn't quite explain all the anomalous observations, then one thinks harder, refines the model. The circle tightens. The way out is a prediction that is new, unexpected.

Fallacious reasoning can play a useful role in experiment as well as in theory. One needs a reason for doing something—it doesn't matter what the reason is. If this is an existentialist argument, so be it. A clever person, with a will to experiment and the proper tools at his or her command, the most important tool being an open and inquisitive mind, can find interesting science in a study undertaken, in retrospect, for the weirdest reason.

In the late 1940s there was a remarkable validation of Hückel's rules for the stability of conjugated electron systems.[9] Molecules predicted to have large resonance energies were suddenly being made in the laboratory (see Figure 5-2): the cyclopropenyl cation shown in structure **5**, cycloheptatrienyl cation in structure **6**, and cyclooctatetraenyl dianion, structure **7**.

These aromatic rings conformed readily to the far-reaching $4n + 2$ rule (the number of π-electrons in these molecules). At the same time it became easy to calculate a simple quantity called the "delocalization energy," which was found to be high for these molecules. It was also high for other molecules that had not yet been made, such as pentalene (structure **8**), fulvalene (structure **9**), and many heterocycles, of the type shown in structure **10**. Organic chemists around the world were spurred on to make these molecules.

But the theoretical reasoning behind these extensions of Hückel's method was weak; the concept of a resonance energy, simplistically defined as it was then,

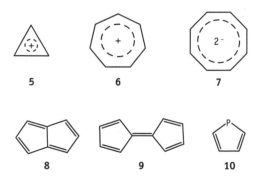

Figure 5-2 Top line: some aromatic organic molecules; bottom line: three organic molecules for which resonance energies (incorrectly) predicted stability.

invalid. The molecules that were synthesized were often kinetically unstable. Was this a wasted effort? Should the experimentalists have waited for a better theory? They would be waiting still. Fascinating molecules, some of pharmacological utility and some leading to novel conducting systems, were made, all chasing a wrong theory.

There's more to it than doing the right thing for the wrong reason.[10] Let's take a typical planned organic synthesis of some complex natural product. In a sequence of 20 to 30 steps, actually hundreds of physical operations, a complicated molecular architecture is constructed. The process is all the more incredible for being a hands-off kind of building. The reagents in each step are mixed, energy is put in in some way, and, voilà, 10^{20} molecules do what we would like them to have done, while the entropy of the universe increases. Or so it appears in the published paper.

But along the way things were different. The reaction steps remain marvelous. In proceeding through each step the designers and executors of the synthesis certainly begin with a plan. A bond must be formed, and carbon A should attack carbon B just on this side of a plane, and not on the other. The professor remembers a paper in which such a reaction took place under certain conditions. He advises a graduate student to try it (this is called "precedent," but in the absence of further reasoning it may shade over into the fallacy of appealing *ad antiquam*. Antiquity in science is usually two years old.) It almost never works on the first try, and often doesn't on the tenth try as well. The graduate student, unwilling to tell the professor that the latter's magnificent plan has failed, asks a postdoctoral fellow working in the same group if he has seen any other way to effect the reaction. The postdoc suggests that perhaps a method he saw in an obscure paper might work. The reasoning proceeds often by crudest analogy, or by primitive ideas that no one wants to admit to in print. This suggestion fails on the first try, too, but then the graduate student changes the order of mixing the reagents, and the synthesis works! He then optimizes the yield, because he knows that his professor knows that the community views a high yield as a sign of elegance and control. There is a happy ending in this for all: the student, the professor, and the whole field of chemistry admiring the ingenious synthesis. How clever of them to apply that reaction (the one that worked) in the most difficult step in the synthesis!

Curious—the true ingenuity and originality that was exhibited in real life, in exploiting the magnificent aleatory process that is much of science, is masked because of some hidden insecurities as to the "illogical" way in which key choices were made.

One might argue against my thesis of a real role for fallacy by pointing out that science could not advance if all arguments were circular or suffering from other failures of logic. I agree that much of paradigmatic science follows logical straight lines. The failures of logic do not impede the progress of science but seem to provide a mechanism for including random deviation as a productive shock setting the system off the beaten track. The awesome motive force of individual psychological conviction then propels one to true innovation. Science as a whole progresses because it is not derailed by isolated failures of logic. But in analyses of the way science works one always chooses simplified cases, unrealistic ones. In these one can only see that circular reasoning is bad. One can't conceive of nearly circular reasoning.

Let me give an analogy. It has been shown that one cannot have a true phase transition, a cooperative phenomenon in one dimension, but can have one in two or three.[11] In one dimension a chance mistake in spin alignment ruins everything—there is no way of communicating information, some favored alignment, past that mistake. In more dimensions than one the linkage is essentially tighter; there is a way, many ways for information to pass "around" the mistake. Well I think science is like that, an intellectual enterprise in a space of immense dimensionality—at least the product of the number of practitioners and the number of objects of their study. Looking at a piece of circular reasoning isolated out of the context of thousands of people reading of the work of others, using it, testing experiments, is like trivializing many dimensions to one.

The closed circle is nonproductive, even if it served Kekulé. My metaphor is the nearly closed circle. One end moves, thinks it is free. It doesn't know how near it is to forming a true circle. It knows it should avoid it, yet the forces that shape its motions keep pressing it near its origin, onto itself. In the space between the end and the beginning, in that little space, it uses the tension of resisting junction to carve out marvels, a universe.[12]

Notes

1. For a hilarious introduction to fallacies see M. Pirie, *The Book of the Fallacy* (London: Routledge and Kegan Paul, 1985).
2. The reader's attention is directed to a remarkable article (in Dutch, with English summary) addressing many of the points made in my column. It is by W. N. A. Klever, *Tijdschrift v. filosofie* 44 (1982): 603. The article, entitled "Circular Argumentation," was brought to my attention by Sven Ove Hansson; I thank Marja Zonnevylle for her translation. For a perceptive discussion of related problems, see E. W. Beth and J. Piaget, *Epistémologie mathématique et psychologie* (Paris: Presses Universitaires de France, 1961), p. 294.
3. See, for instance, A. W. Sleight in *Chemistry of High Temperature Superconductors*, ed. by D. L. Nelson, M. S. Whittingham, and T. F. George (Washington, D.C.: American Chemical Society, Symposium Series 351, 1987), p. 2.
4. Some leading references, as they say, are: D. D. Sarma and C. N. R. Rao, Evidence for Peroxide and Cu^{1+} Species in $La_{1.8}Sr_{0.2}CuO_4$ from Photo-Emission Studies. *J. Phys. C.: Solid State Phys.* 20 (1987): L659; R. A. de Groot, H. Guttfreund, and M. Weger, Valence Fluctuations in $La_{2-x}Sr_xCuO_4$. *Solid State Comm.* 63 (1987): 451; and B. Dauth, T. Kachel, P. Sen, K. Fischer, and M. Campagna, Valence Fluctuations and Oxygen Dimerization in High-Temperature Superconductors from X-Ray Photoemission Spectroscopy. *Z. Phys. B.—Cond. Matter* 68 (1987): 407.
5. See the discussion in endnote number 3. Attitudes toward formal oxidation states are another illustration of the differences in language between chemistry and physics that I discussed in Unstable. *Am. Sci.* 75 (1987): 619.
6. See ref. 17 in T. A. Albright, R. Hoffmann, J. C. Thibeault, and D. L. Thorn, Ethylene Complexes: Bonding, Rotational Barriers, and Conformational Preferences. *J. Am. Chem. Soc.* 101 (1979): 3801.

7. The characterization of theories here as soft or hard connects up to the idea of strong inference, as described by J. R. Platt, Strong Inference. *Science* 146 (1964): 347. Obviously I don't agree with this paper on many matters.

8. K. A. Müller and J. G. Bednorz, The Discovery of a Class of High-Temperature Superconductors. *Science* 237 (1987): 1133.

9. See, for instance, chapter 2 in J. March, *Advanced Organic Chemistry*, 3rd ed. (New York: Wiley, 1985).

10. T. S. Elliot, *Murder in the Cathedral* (San Diego: Harcourt Brace Jovanovich, 1935), p. 44. For an apposite citation, see J. Meinwald, A. Eckell, and K. L. Erickson, The Photoisomerization of α-Phellandrene: A Structural Reassignment. *J. Am. Chem. Soc.* 87 (1965): 3532.

11. L. van Hove, Sur l' integrale de Configuration pour les Systemes de Particules a Une Dimension. *Physica* 16 (1950): 137; R. Peierls, On Ising's Model of Ferromagentism. *Math. Proc. Cambridge Phil. Soc.* 32 (1936): 477; R. B. Griffiths, Peierls Proof of Spontaneous Magnetization in a Two-Dimensional Ising Ferromagnet. *Phys. Rev.* A 136 (1964): 437.

12. In the writing of this paper I benefited from discussions with Sven Ove Hansson, Barry Carpenter, Paul Houston, and Ben Widom.

6

Ockham's Razor and Chemistry

ROALD HOFFMANN, VLADIMIR I. MINKIN, AND BARRY K. CARPENTER

Scientists think they are born with logic; God forbid they should study this discipline with a history of more than two and a half millenia. Isn't it curious that some of our competitors and critics, pretty good scientists (except when they review our papers), seem to be strangely deficient in logic!

While scientists think they can do without philosophy, occasionally principles of logic or philosophy do enter scientific discourse explicitly. One of these philosophic notions is Ockham's Razor, generally taken to mean that one should not complicate explanations when simple ones will suffice.

The context in which Ockham's Razor is used in science is either that of argumentation (trying to distinguish between the quality of hypotheses) or of rhetoric (deprecating the argument of someone else). Either way, we think that today appeal to the venerable Razor has a bit of a feeling of showing off, of erudition adduced for the rhetorical purposes. This attitude reveals a double ambiguity. The first is toward learning—today's science, no longer elitist, does not depend on men steeped in classical learning. And appeal to Ockham's Razor also points to a certain ambiguity in the relationship of science to philosophy.

Let's learn something of the man whose name the principle bears, and its various meanings. Then we give a personal account of the use of Ockham's Razor in chemistry, with specific reference to the analysis of reaction mechanisms. And we enter a dialogue on the validity and limitations of this device.

William of Ockham

Neither today's scientists nor medieval theologians and philosophers, one of whom was William of Ockham (or Occam),[1] can avoid the politics of their times. We know precious little of William of Ockham's early life. He was born in the village of Ockham in Surrey near London, probably within five years of 1285. The first certain date we have in his life is February 26, 1306 when he was ordained subdeacon of Southwork. William entered the Franciscan order, tremendously popular at that time, at an early age. He is likely to have studied at Oxford from 1309 to 1315, and continued his philosophical work there and in London from 1315 to 1324.

Despite the tremendous quantity and quality of William of Ockham's scholarly work in this period (the definitive edition of his work, published by the Franciscan Institute at St. Bonaventure, NY, and by Manchester University Press, runs to ten volumes of theology, seven of philosophy and four of political writings, one volume is still to come[2]) he never held a chair at Oxford. This was due to the enmity of the Chancellor of Oxford at the time, John Lutterell, who has been characterized as "an overzealous Thomist,"[3] a nasty character by all accounts.

In 1324, William's life changes. The politics of the various orders of the Catholic Church, and the interplay of secular and religious power in this period are most intricate. Perhaps reading the section of Umberto Eco's *The Name of the Rose* that most readers skipped might help.[4] The papacy in this time is buffeted by secular power struggles, and resides in exile in Avignon, France (from 1309–1377). John Lutterell travels there in 1323, to Pope John XXII, accusing William of Ockham of 56 instances of teaching dangerous doctrine. Ockham is summoned to Avignon in 1324, and a commission is appointed to examine his teaching. Essentially this was a trial for heresy. It dragged on for three years, and never reached a formal conclusion as other events overtook it.

The Franciscans were at this time involved in a dispute with the Pope, an argument with the usual mix of theological and financial overtones. The theoretical side concerned the question whether Christ and the Apostles possessed property in private or in common. Behind this discussion lay the issue of the ideal of poverty, favored by some orders, and opposed by others. It was a matter of great economic and political concern to the Church whether the Church, the Pope, or the Franciscans were bound to follow literally the path of Christ and its faithful imitation by St. Francis of Assisi.

The General of the Franciscan order, Michael of Cesena, asks William of Ockham to study the issue. William's intellectual honesty and depth of logic leads him past simple disagreement with the Pope on this issue. He finds many of John XXII's statements contradicting earlier authority, and he says so. Eventually, in 1328, William joins his General and two other Franciscans in defying the Pope. They flee to Pisa, and there obtain the protection of the German Emperor, Ludwig (Louis) of Bavaria. Ludwig had his own political agenda; he had installed an Anti-Pope in Rome, and had himself crowned as Emperor of the Holy Roman Empire.

So began the period of the rival papacies and 20 years of mostly political activity for William of Ockham. Excommunicated by the Avignon Pope, the rebellious Franciscans settled at the court of Ludwig in Munich. Upon the death of their protector in 1347, their position became untenable. A document of submission was drawn up. It was never signed; William died, unrepentant, in the same year Ludwig did.[5]

The Theologian and Philosopher

William of Ockham may be known to scientists as the man whose name is associated with Ockham's Razor. To his peers and to the world of theology he was
and is a leading "scholastic" philosopher. This is the end of the Middle Ages;
the wisdom of the Greeks is reintroduced into Europe through Al Andalus,
Islamic Spain. It is a time of great minds in the religions; the time of the Rabbis
Moses ben Maimon (Maimonides) in Cordova and Egypt, Moses ben Nachman
(Nachmanides) in Gerona, Shlomo Yitzhaki (Rashi) in Troyes. It is the time, or
shortly after the time, of St. Thomas Aquinas, of Roger Bacon, of Duns Scotus.
The philosophy of Aristotle, with its far-reaching rationality, finds a resonance in
the agile minds of Catholic theologians. The glory of God merges in their work
with the path of reason; there is no disjunction between faith and rationality for
these men.

Brian Tierney aptly characterizes Ockham's philosophy as "nominalist,
emphasizing the irreducible individuality of external entities, and voluntarist,
emphasizing the primacy of will over intellect, above all the absolute, unfathom-
able will of God."[6] A basic principle of William of Ockham's theology is that all
things are possible for God, save such as involve contradiction. So we may learn
more of our (His) religion by probing its logical depths. In some ways this is an
early statement of the philosophical rationale that produced (much later) the
religious scholar-students of nature, especially the great scientists of the Jesuit
order.

It becomes important (for William and his scholastic contemporaries) to seek
out contradictions, to probe causes, to seek the reason for all but the First Cause.
That search may seem to us abstruse. As in this typical passage:

> ...*Sudden change is not a thing (res) distinct from permanent things
> and destroyed after the first instant at which the subject is suddenly
> changed... Rather for the subject to change suddenly is only for the subject to
> have a form that it did not have earlier or lack a form that it had earlier—
> nevertheless, not part by part in such a way that it has one part of the form
> before the other; nor does it lack one part before it lacks another. But it receives
> the whole form simultaneously or loses the whole simultaneously.*[7]

Perhaps a brake on calling a passage such as this "abstruse"[8] might be the reflection on how a typical paragraph from one of our papers might sound to a scholastic
philosopher, or for that matter to any intelligent human being who is not a chemist. Interestingly, an astute observer, Mary Reppy, remarks that the last part of this
passage sounds awfully like an attempt to define a concerted reaction.[9] Of which
more, anon.

The Razor

William of Ockham was not only a theologian, but a great logician. A case has been made for his awareness of many of the principles of mathematical logic that were not mathematicized until 600 years later.[10] One of the tools he used routinely in his reasoning is what is known in philosophy as the principle of parsimony, and popularly as Ockham's Razor.

Just as for the Golden Rule, there are many ways of stating Ockham's Razor. Here are four that William of Ockham used in his works:

(A) It is futile to do with more what can be done with fewer. [*"Frustra fit per plura quod potest fieri per pauciora."*]
(B) When a proposition comes out true for things, if two things suffice for its truth, it is superfluous to assume a third. [*"Quando propositio verificatur pro rebus, si duae res sufficiunt ad eius veritatem, superfluum est ponere tertiam."*]
(C) Plurality should not be assumed without necessity. [*"Pluralitas non est ponenda sine necessitate."*]
(D) No plurality should be assumed unless it can be proved (a) by reason, or (b) by experience, or (c) by some infallible authority. [*"Nulla pluralitas est ponenda nisi per rationem vel experientiam vel auctoritatem illius, qui non potest falli nec errare, potest convinci."*]

Philosophers and historians are generally puzzled as to why the principle of parsimony should be called Ockham's Razor. The principle is not original to William of Ockham. Versions of it are to be found in Aristotle, and nearly *verbatim* variants occur in the work of most scholastic philosophers.[12,13] Though Ockham used it repeatedly and judiciously, "he clearly does not regard it as his principal weapon in the fight against ontological proliferation."[14]

We suspect that the association is due to the strength of the razor metaphor rather than anything else. Scholastic and theological arguments were complex; to cut through them, to reach the remaining core of truth quickly, was desperately desirable. Whoever rechristened the principle of parsimony as Ockham's Razor (the earliest reference appears to be by Etienne Bonnot de Condillac in 1746[15]) was creating an easily imagined image. Metaphor reaches right into the soul.

The last, most extensive formulation of Ockham's Razor, (D) above, is intriguing. Note the "religious exclusion" in it. It refers to the Bible, the Saints and certain pronouncements of the Church. This testimony to the faith of William did not stop him from questioning the infallibility of Pope John XXII, when the Pope's writings came in conflict with earlier church authority. In the context of science, especially interesting is part (b) of version D of the Razor, that experience (*"experientiam"*) can serve to justify plurality. There is no reason not think of "experience" here as "experiment," even though the idea of a scientific experiment lies centuries in the future. William of Ockham's method (and that of Aristotle) empowers the human senses as arbiters. His method accepts what we now call science.[16]

Reaction Mechanisms

Six and a half centuries is a lot of time; it is also very little time. In the Middle Ages one had protochemistries—fermentation, metallurgy, ceramics, alchemy, dyeing. People have always transformed matter in ingenious ways. The Renaissance came, then the Industrial and Scientific Revolutions. Now there is chemistry, a true science, an industrial empire, a profession. Beautiful molecules are made, seventy million of them unknown to Nature. People ask questions such as "How does this reaction run?" "What is the mechanism (a very Newtonian clock-work type of question) of that reaction?" And remarkably, 650 years after he died, they invoke William of Ockham's restatement of the principle of parsimony, that old Ockham's Razor, to help them reason out what happens.

Let us first define what is to be meant by the term "reaction mechanism." The notion of the mechanism of a chemical reaction consists of a description of all "elementary" steps in the transformation of reactants into products. On the molecular level the mechanism includes, in principle, knowledge of the geometry and relative energy of all structures involved, including isolable or potentially isolable intermediates and transition states, the latter representing the turning points along the minimal energy paths connecting all interconverting species. Following another line of thinking, the reaction mechanism traces the evolution of a chemical system along the reaction trajectory, i.e., the line linking reactant and product molecules in the space of all nuclear coordinates. The concept of a potential energy surface (PES), with all its attendant limitations, is essential to this definition.

Rube Goldberg, who had some chemistry at UC Berkeley, captured something about reaction mechanisms in his cartoons of two generations ago. One is shown in Figure 6-1.[17]

Minimal Action, Least Motion

Given the definition of a reaction mechanism, the drawing of an analogy with the mechanical description of moving particles is obvious. A predictable consequence was the early application of the principles and methods developed so successfully in classical mechanics to the treatment of mechanisms of chemical reactions. Before the idea of a molecule ever took hold, there had been developed the *principle of minimal action*, first introduced by Pierre Louis Moreau de Maupertuis and universally applied by Leonhard Euler in ballistics, central force motion, etc. According to this principle, spontaneous movements are always associated with minimal changes in

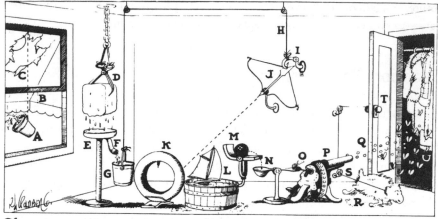

THE PROFESSOR EMERGES FROM THE GOOFY BOOTH WITH A DEVICE FOR THE EXTERMINATION OF MOTHS. START SINGING. LADY UPSTAIRS, WHEN SUFFICIENTLY ANNOYED, THROWS FLOWER POT(A) THROUGH AWNING(B). HOLE(C) ALLOWS SUN TO COME THROUGH AND MELT CAKE OF ICE(D). WATER DRIPS INTO PAN(E) RUNNING THROUGH PIPE(F) INTO PAIL(G). WEIGHT OF PAIL CAUSES CORD(H) TO RELEASE HOOK(I) AND ALLOW ARROW(J) TO SHOOT INTO TIRE (K). ESCAPING AIR BLOWS AGAINST TOY SAILBOAT(L) DRIVING IT AGAINST LEVER(M) AND CAUSING BALL TO ROLL INTO SPOON(N) AND PULL STRING(O) WHICH SETS OFF MACHINE GUN(P) DISCHARGING CAMPHOR BALLS(Q). REPORT OF GUN FRIGHTENS LAMB(R) WHICH RUNS AND PULLS CORD(S), OPENING CLOSET DOOR(T). AS MOTHS(U) FLY OUT TO EAT WOOL FROM LAMB'S BACK THEY ARE KILLED BY THE BARRAGE OF MOTH BALLS. IF ANY OF THE MOTHS ESCAPE AND THERE IS DANGER OF THEIR RETURNING, YOU CAN FOOL THEM BY MOVING.

Figure 6-1 One take on a reaction mechanism, by Rube Goldberg. Reprinted by permission of Rube Goldberg, Inc.

the quantity of "action," the latter a well-defined physical variable. Reporting in 1744 to the Academie des Sciences of Paris on the principle of minimal action, de Maupertuis stressed, in particular, that light chooses neither the shortest line, nor does it follow the fastest path. Instead, light takes the path which gives real *economy* (cf. the law of parsimony), i.e., where the quantity of action is minimal.[18] Minimal action is itself a beautiful, economic way to get at the heart of physical motion. And it found a place in the new quantum mechanics, most elegantly in the work of de Broglie, Schwinger, and Feynman.[19]

It is thus hardly surprising that when in the 1930s studies of mechanisms of chemical reactions had grown in importance, indeed to become the intellectual focus of the rapidly developing area of physical organic chemistry, the key generalizations relevant to reaction mechanisms were made in the spirit and in the terminology of mechanics. Perhaps, the first step in this direction had been taken even earlier, when A. Muller in 1886, i.e., at a time when molecular theory was still young, introduced *the rule of least molecular deformation* in the course of chemical transformation.[20] The idea was appealing, and found its place in a number of textbooks as *the principle of minimal structural change*.[21] In its most general terms it was formulated by F. Rice and E. Teller, who in 1938 proposed the principle of least motion (PLM) according to which "Those elementary reactions will be favored that involve the least change in atomic position and electronic configuration."[22] In the context of the orbital symmetry rules that were to come into organic chemistry 27 years later, the inclusion of electronic configurations in the Rice and Teller formulation is noteworthy.

To apply the PLM to a certain reaction, the constituent atoms of the molecules of reactant and product must be displaced with respect to one another so that their nuclear motions (usually measured by their squares) are minimized. Indeed, a good number of organic reactions of the rearrangement, decomposition, and elimination type have been shown to follow those reaction pathways that do obey the requirements of the PLM. The extreme simplicity of the relevant computational technique and, more importantly, the clarity of the underlying idea, assured broad application of the PLM treatment of reaction mechanisms, particularly where a choice between several conceivable pathways was needed.[23]

It was always perfectly well understood that PLM represents a very, very simplified theoretical model of the actual motion of nuclei and electrons in the course of chemical reaction. That motion is properly described by the equations of quantum mechanics. None doubted that quantization of electronic, vibrational and rotational states mattered. And that one has to take a dynamic view, describing the real reaction by the totality of the myriad trajectories followed by an ensemble of real molecules in phase space. Still, PLM met a desire for simplicity. Given that it was simplistic, deviations from, or even incompatibility with, the PLM predictions, met in a number of applications of the principle, were never regarded, we think, as final indictments of a mechanistic hypothesis.

A Personal Experience

In contrast to this forgiving attitude toward deviations of a simple theory, the chemical community turns out to be not so tolerant when important, accepted ideas seem to be threatened. Let us give an example, drawing on personal experience.

In 1982 one of the authors (V.I.M.) published a preliminary account of the experimental observation of inversion of stereochemical configuration at a tetrahedral boron center.[24] Several possible reaction pathways that might, in principle, connect the interconverting steroisomers were enumerated. These included (Figure 6-2): (a) intramolecular (dissociative) and (b) intermolecular (associative) routes, both involving bond-breaking processes at the tetrahedral boron, as well as (c) intramolecular inversion occurring through an intermediate tetracoordinate planar boron species, in which all four bonds to boron are retained (although their strength changes drastically).

Whereas the intermolecular variant of the bond-breaking mechanism was ruled out on the strength of the experimental evidence then available, no unequivocal choice could be made at the time between the two remaining possibilities, (a) and (c).

The Rostov-on-Don authors could not abstain from the temptation of giving preference to the more exciting non-bond-breaking alternative mechanism (c). This choice turned out to be in error, as detailed experimental study later revealed.[25]

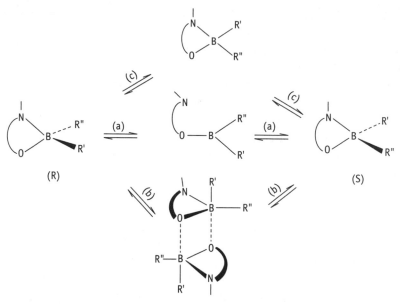

Figure 6-2 Three reaction mechanisms for inversion of stereochemical coordination at a tetrahedral main-group center.

But even before convincing evidence in favor of a bond-breaking mechanism was presented, the uncommon interpretation of the "square-planar boron" mechanism of inversion elicited a quick response. Researchers from the University of East Anglia pointed to the fact that the rate of the inversion process was comparable to that of bond-breaking processes in compounds structurally similar to those studied by the Rostov-on-Don group.[26] On this basis they concluded that the inversion reaction follows the dissociative bond-breaking route, a mechanism with a venerable history going all the way back to the classic 1912 work by Alfred Werner on stereoisomerization of cobalt complexes.

While this was indeed a weighty argument in favor of the bond-breaking pathway, the reasoning of the English researchers was by and of itself not yet conclusive. Perhaps this was why they in turn were seduced by a crumb of philosophy, supporting their argument by the statement that following the dissociative pathway in preference to the bond-conserving inversion "is also a natural result of the application of Occam's chemical razor principle: mechanisms should not needlessly be multiplied."

Ockham's Razor and Reaction Mechanisms

East Anglia and Rostov-on-Don are hardly enemies; the chemistry got sorted out in the end. Nevertheless, it is interesting to reflect on why appeal to such a general modality of reasoning as Ockham's Razor seemed to be quite appropriate in tackling such a specific problem as the mechanism of a certain chemical reaction? The answer is to be found, we think, in the nature of the theoretical construction which the reaction mechanism represents.

In general, the mechanism of a reaction can neither be directly observed, nor can it be deduced with absolute certainty on purely experimental grounds. It would be nice if the world were that simple. But it isn't. We are not convinced either that femtosecond spectroscopy, an incredibly fast and beautiful way of observing nature, will give the requisite mechanistic answers. The mechanism of a reaction is a logical construction based on a perforce limited set of experimental facts, which are then interpreted by human beings in the framework of current, fashionable and ephemeral theoretical models. And it is logic, with its laws and rules, that makes it possible to arrange observations in harmony with relevant concepts and hypotheses. Ockham's Razor belongs to the category of logical rules which indicate how to process experimental facts. It shows the way to the best fit of observables to the least complicated possible interpretation. It is, therefore, by no means accidental that in many textbooks concerned with the problem of reaction mechanisms, from introductory to advanced ones,[27,28] Ockham's Razor is mentioned among the significant criteria to be met when determining a mechanism.

The utility of Ockham's Razor in the selection and classification of reaction mechanisms has proven itself in chemistry, just as it has in various other areas of natural science.[29] Ockham's Razor must indubitably be counted among the tried and useful principles of thinking about the facts of this beautiful and terrible world and their underlying causative links.

Take That, You Naïve Chemist!

In the preceding section we recited the scientist's catechism, of the great importance and utility of Ockham's Razor. It may come as a surprise to our colleagues that not everyone agrees. For instance, in a remarkably perceptive article, N. Oreskes, K. Shrader-Frechette, and K. Belitz write:

> *Ockham's razor is perhaps the most widely accepted example of an extraevidential consideration. Many scientists accept and apply the principle in their work, even though it is an entirely metaphysical assumption. There is scant empirical evidence that the world is actually simple or that simple accounts are more likely than complex ones to be true. Our commitment to simplicity is largely an inheritance of 17th-century theology.[30]*

Now that puts us right into our place, in the company of ancient priests!

Though this quote cuts to the heart of the problem, we'd prefer to approach the difficulties with Ockham's Razor gently, through several chemical examples. And since this is a dialogue, with epistemological intent if not expertise on the part of its authors, we will wend our way back eventually to a balanced view of this principle.

Multiple Reaction Paths

Continuation of the story of the mechanism of inversion of configuration at tetrahedral boron provides the first example. When, in due time, a sufficient body of experimental and computational data had been accumulated concerning the intrinsic mechanisms governing inversion of configuration at a variety of tetrahedral main group metal centers,[31] unequivocal evidence was presented for the simultaneous operation of at least three of the forementioned mechanisms, including the one rejected ostensibly on the basis of Ockham's Razor. Each mechanism has precisely the same net outcome, namely inversion of stereochemistry at the main group metal center. The relative contribution (or energetic preference) of a given mechanism depends on the metal. Structural factors influence the mechanism as well, and may be deliberately manipulated. In some cases (e.g., complexes of zinc and cadmium) all three mechanisms are virtually equivalent in their energetic demands.

Such a diversity of reaction paths for one and the same chemical transformation is by no means a unique occurrence. With rapidly developing experimental and computational techniques for studying reaction mechanisms, a good number of important chemical reactions have been found to follow several competing reaction channels, their relative significance sometimes critically dependent on most subtle variation of structure and reaction conditions. This relatively new development may be illustrated by just a few examples.

Consider first a classic so-called pericyclic reaction,[32] the Cope rearrangement (3,3-sigmatropic shift, Figure 6-3). Here, even rather tiny structural tuning of the parent hydrocarbon, 1,5-hexadiene, appears to lead to a switch from the most typical pathway (a) with its "aromatic" transition state structure (in two isomeric forms), to pathways (b) or (c), which feature, respectively, a biradical-like transition state or a long-lived "intermediate."[33] We will return below to the current state of affairs in this mechanism.

As a second example, let's look at a challenging current mechanistic problem, that of unraveling the mechanism of formation of fullerenes, the polyhedral products of graphite vaporization at plasma temperatures of over 3000°C. Contrary to an "entropic" expectation of the existence at these conditions of structurally little-organized forms of matter, specific, highly symmetric polyhedral C_{2n} molecules, their structure reminiscent of the geodesic domes exploited in architecture by

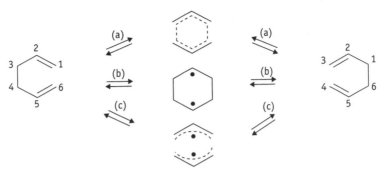

Figure 6-3 Three mechanisms for the Cope Rearrangement.

R. Buckminster Fuller, are created in carbon vapor. C_{60}, possessing the truncated icosahedral geometry of a soccer ball, has attracted special attention because of the perfection of its polyhedral structure, its relative stability, and the horizons opened up with the discovery of a new allotrope of carbon.

How does this thermodynamically unstable molecular soccer ball assemble? Considerable effort has been expended on detailed study of the mechanistic aspects of fullerene formation following graphite vaporization.[34] Several ingenious suggestions for the growth process that generates the C_{60} have been forwarded.[35] Yet a tiny deviation from optimal reaction conditions found in the famous pulse laser vaporization experiment of Smalley, Curl, Kroto and coworkers appears to result in a drastic decrease of the yield of C_{60}, and in alteration of the mechanism of self-assembly of carbon atoms as well. R. Smalley, one of the discoverers of fullerenes says: "Of course, there must be hundreds of mechanisms whereby a fullerene like C_{60} can form."[36] Smalley's statement, with which we agree, by no means signifies a repudiation of attempts to gain insight into the detailed mechanism and the driving forces of the spontaneous self-assembly of carbon atoms. The statement merely emphasizes the great complexity of the problem, and the terrible incompleteness of our knowledge.

The greater the insight gained into the origin of chemical transformation, the more justified seems the view that reaction pathways are inherently manifold. As we said, one usually thinks of a chemical reaction as a geometric rearrangement of the relative positions of the nuclei which make up the interacting molecules, i.e., motion along a path on the potential energy surface (PES), bisected by ridges that form the reaction barriers. Such a picture of a PES reminds one of a hilly landscape; the metaphor continues with the successfully transformed molecule likened to the motion of a mountaineer moving from the valley of reactants to that of products by surmounting one of the lowest possible passes.

But the real hilly landscapes of this world (or those calculated) are not so monotonous as to feature a *unique* pass between valleys. Thus branching of reactive trajectories might be a rather common occurrence. The number of trajectories grows rapidly when reactants are supplied with an additional increment of kinetic energy. The requirement of passing through a single saddle point is then relaxed. Moreover, when the nuclear displacements in the course of rearrangement of reactants to products are sufficiently small, the reaction may proceed by a kind of trickling through (under) the energy barrier, i.e., by quantum mechanical tunnelling.[37]

Ornate Hypotheses May Be Richer

Let us continue our fault-finding with Ockham's Razor:

Supposing there are two explanations for a phenomenon or an observable. Let's symbolize one as

$$\Pi = A, \tag{1}$$

where A is the determining factor. The other explanation can be written symbolically as

$$\Pi = c_a A + c_b B, \tag{2}$$

i.e., Π is viewed as being caused by two factors, A and B, in some admixture.

Now it may be that for a single physical observable the "simple" explanation (1) made good enough sense of the available data, and by Ockham's Razor would be preferred to (2). But the universe is likely to have in it not one phenomenon or observable Π but several, $\Pi_1, \Pi_2, \Pi_3 \ldots$ Adducing the more complex explanation (2), even when only one of these phenomena is known, may lead to the eventual realization that there is some related one, Π_2. The more complex explanation is *productive*, it leads one to think about alternative experiments.

Such an approach may be thought of as one formalization of the epistemologic method of multiple hypotheses that had been advanced at the beginning of this century by Chicago's geologist T. C. Chamberlain and later used by J. Platt (a one-time physicist and chemist) as the basis for the "method of rigorous conclusions."[38] These methods, in a way ramifications of F. Bacon's seminal method of induction, point to the fact that to achieve the right conclusion, simultaneous testing is needed of several hypotheses, each endowed with its own means of uncovering the truth. The summary result of the application of various means and approaches must be richer (and more complete) than the relentless pursuit of any single hypothesis. Do we need to rehearse the myriad examples the history of chemistry (or our colleagues) provides of the sterility of hypotheses held too strongly, too single-mindedly, by individuals?

Complex Nature, Simple Minds

To finish the argument against the trivial application of Ockham's Razor:

Time and time again the process of discovery in science reveals that what was thought simple is really wondrously complicated. If one can make any generalization about the human mind, it is that it craves simple answers. This is true in politics as in science. So we have a President of the USA (pick any recent one) saying that if we control the flow of drugs across our borders, then we will diminish greatly the terrible social problem of drug addiction. Or, just to take something from across the political spectrum, someone (no President would dare) asserting that if we distribute condoms in the schools that such action will reduce significantly the spread of AIDS.

The ideology of the simple reigns in science as well, whereas every real fact argues to the contrary. So we have the romantic dreams of theoreticians (e.g., Dirac) preferring simple and/or beautiful equations. The intricacy of any biological or chemical process elucidated in detail points clearly in the opposite direction.

Let us be specific here, with a chemical and biological vignette, the story of the sex pheromone of the cabbage leaf looper moth, *Trichoplusia ni*. When the pheromone was first discovered in 1966, it was thought to be a simple molecule, (Z)-7-dodecenyl acetate. A few years later a second active ingredient was found, and more recently some clever biosynthetic reasoning by Biostad, Linn, Du, and Roelofs led to the discovery that a blend of six molecules was needed for full biological activity.[39] There is a relationship between the concoction of a new perfume and insect chemistry.

It's not that every physical, chemical, or biological observable needs to have a complicated cause. But we would argue that in the complex dance of ingenuity that is modern science, in the gaining of reliable knowledge, one should beware of the inherent weaknesses of the beautiful human mind. The most prominent shortcoming is not weak logic, but prejudice, preferring simple solutions. Uncritical application of Ockham's Razor plays to that weakness. What's worse it dresses up that weakness in the pretense of logical erudition.

We've fleshed out the argument against the use of Ockham's Razor in science. But now it is time to reverse gears, and argue the other way.

Complex Modules, Simple Molecules

In our guise as critics of Ockham's Razor, we are, perhaps, guilty of pulling off a philosophical sleight of hand. We (and other critics) imply a necessary relationship between the preference for a simple model and the belief in a simple universe. We then go on to argue that the universe is hardly simple, and thereby appear to invalidate the application of Ockham's Razor in scientific investigation. But does it really follow that one must believe in a simple universe in order to be philosophically honest when invoking Ockham's Razor? Is it not inherent in any analytical epistemology, that one attempt to find simple intellectual bricks from which the wonderfully complex architecture of Nature could be reconstructed? And isn't it really the case that Ockham's Razor properly applies to the identification of these individual modules, rather than to the entire *Weltanschauung* that one builds from them? The principle of parsimony is not a metaphysical statement about the way the universe is. Everyone knows it is wondrously complex; Ockham's Razor is a prescription for unraveling and comprehending—piece-wise, never completely—its marvelous complexity. In this pragmatic point of view, Ockham's Razor serves as an operational principle, *not* a rule or a Law of Nature.

In the so-called "scientific method," we seek to devise experimental tests that can falsify our hypotheses. The excommunication of ideas that takes place when a model "fails" one of these trials is taken to be rigorous and irreversible, provided that the experimental tests meet criteria of both intellectual validity and competence of execution, therefore reproducibility.

In the pragmatic interpretation of Ockham's Razor, one would not use such irrevocable language. One might say that the choice between two otherwise equally valid models should be made in favor of the simpler, *but that the rejection of the more complex is only conditional.* The idea that has been set aside could be reconsidered at a later date if the currently favored hypothesis fails some future test. If one adopts such a view, it follows that the temporarily discarded model should not be said to be "ruled out" by or to have "violated" Ockham's Razor, since this language belongs in the domain of the more rigorous exclusionary tests.[40]

But even this liberal prescription for the use of Ockham's Razor begs the underlying question of "why?" Why should we lean in favor of the simpler of two otherwise equally satisfactory models? We can advance several arguments, no one of which has logical rigor beyond an appeal to reasonableness.

1. The simpler model is likely to be more *vulnerable* to future falsification, because with fewer adjustable parameters it will have less flexibility. If, as Popper suggests, a good scientific hypothesis is one that is falsifiable, then perhaps the better of two competing models is the one that is somehow *more* falsifiable. To be vulnerable is not a weakness, in science or human relationships.

2. Or one could say that the simpler model provides a clearer and more readily *comprehensible* description. This view would admit the human difficulty with handling complexity, and relate simplicity to comprehensibility. It is important to *understand*, and the breaking of a complex reality into comprehensible bits is not only the Cartesian method, but a teaching strategy.

3. A third rationale relies on an *assessment of the probability of future success of any model.* Suppose, in some experiment, we made a series of measurements of a property **y** in its response to adjustment of a factor **x**, with results depicted in Figure 6-4.

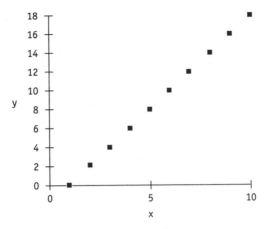

Figure 6-4 Some experimental measurements of a property y in response to variation of a factor x

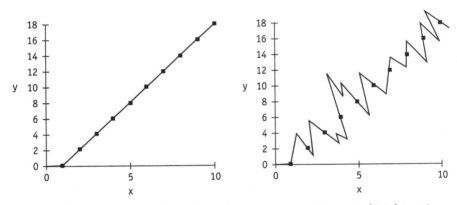

Figure 6-5 (left) A straight line fitted through the data points of Figure 4. (right) Another fit of the same data points.

If one wanted to try to describe **y** as some mathematical function of **x**, one would probably choose a straight-line relationship (Figure 6-5 left) in preference to a more complex functional form such as that shown in Figure 6-5 at right.

But, aside from some intuitive sense that it just seems right, why would one prefer the straight-line model? An answer can come from looking at the degrees of freedom of the fits. In statistics, the number of degrees of freedom of a model is the difference between the number of independent experimental observations and the number of adjustable parameters in the mathematical function that seeks to describe the relationship between **y** and **x**.[41] It is axiomatic that any function with a number of adjustable parameters equal to or greater than the number of observations can be made to pass exactly through all of the (**x,y**) points on the graph. However, it is not necessarily true that a function with fewer adjustable parameters than the number of observations will pass through all of the points. If it turns out that it does, then the function—our model—has already had some success in describing one or more events that we have measured experimentally.

The number of degrees of freedom of a model can be thought of as the number of points whose positions were correctly described by the model, without any algebraic requirement that it come out that way. The world is not static. One measurement will be, must be, followed by another. Models that predict are valued. Since we are presumably seeking a mathematical relationship between **y** and **x** in order to predict future points on the graph, we are naturally more inclined to choose the model that has already had the greater success in "predicting" the measurements we have made so far. This will be the model with the larger number of degrees of freedom, or the smaller number of adjustable parameters—i.e., the simpler model.[42,43]

4. The graphical representation of the **y** *vs.* **x** relationship serves to illustrate a fourth, and here the last, reason for applying Ockham's Razor as an operational principle. The number of equally satisfactory models in a given class is generally related to the complexity of the class. For example, there is one and only one straight line that will pass through all of the (**x**, **y**) points in the graph described above. We do not have to ask which straight line to choose in order to best represent the **x**, **y** relationship. On the other hand, since the number of parameters required to describe the jagged line in the illustration of our more complex model exceeds the number of observations, there exists an infinity of jagged lines, all passing exactly through the points. With the observations made so far, we have no logically defensible way to choose one from this infinity.

To put it another way, if you think Ockham's Razor gets you into trouble by limiting the number of hypotheses, thereby diminishing the imaginative world, then relaxing from Ockham's Razor opens up real, indeterministic, chaos—the infinity of hypotheses that fit.

Those of us who have mystically inclined, nonscientist friends may have used arguments like this last one in our discussions of the lack of general scientific acceptance for extra-sensory perception, UFOs, homeopathic medicine, or astrology. The nonscientist might ask: "Do you scientists think you understand everything about how the universe works?" When we modestly profess our woeful lack of understanding, we might hear in return: "Well then how can you rule out the possibility of...?"

Of course the answer is that we cannot, but in order to make any kind of sense of the world, we must have some procedure for selecting among the plethora of ideas that the collective action of creative human minds has spawned. If we had to operate under an equal opportunity clause for every concept that was ever espoused, we would have such an impossibly complex and self-contradictory description of Nature, that we could never feel that we were making progress in understanding or utilizing our environment.

Why should we make progress? Have we progressed? We are painfully aware of all the ambiguities of the 19th century idea of Progress, in which science flourished. And of the deep mistrust of such progress by thoughtful people in our time. While we're actually ready to do battle for progress, not without internal doubts, this is not the place for that confrontation.

A Statistical Interlude: Principal Component and Bayesian Analysis

The need to have operating principles just to make progress at all in sifting through the complexity of Nature shows up most clearly in the procedure called Principal Component Analysis (PCA).[44,45,46] Many of the observables of nature are multivariate, i.e., each property or phenomenon analyzed yields a series of numbers. Examples are spectra or chromatograms, yielding a datum for each wavelength or retention time. PCA allows one to correlate the data available by deriving a set of orthogonal basis vectors, principal components, so that the first such component represents the best linear relationship, the one showing the greatest variation, exhibited by the data. Each successive principal component explains the maximum variance not accounted for by the previous ones. Identifying the number of significant components enables one to determine the number of real sources of variation within the data. The most important applications of PCA are those related to: (a) classification of objects into groups by quantifying their similarity on the basis of the Principal Component scores; (b) interpretation of observables in terms of Principal Components or their combination; (c) prediction of properties for unknown samples. These are exactly the objectives pursued by any logical analysis, and the Principal Components may be thought of as the true independent variables or distinct hypotheses.

One example of the application of PCA in chemistry may be found in the recent statistical analysis of the concept of aromaticity by A. R. Katritzky et al.[47] Widely applied for the characterization of specific features of conjugated cyclic molecular systems, the notion of aromaticity lacks a secure physical basis. Not that this has stopped aromaticity from being a *wonderful* source of creative activity in chemistry.[48] We can think of no other concept that has led to so much exciting chemistry! Yet, although numerous indices of aromaticity have been designed, based on energetic, geometrical and magnetic criteria, no single property exists whose measurement could be taken as a direct, unequivocal measure of aromaticity.

The PCA analysis of the interrelationship of 12 proposed indices for 9 representative compounds by Katritzky indicated that there exist at least two distinct types of aromaticity. "Classical aromaticity" is well described by certain interrelated structural and energetic indices, whereas the second type of aromaticity, the so-called "magnetic aromaticity," is best measured by anisotropies in the molar magnetic susceptibility. It seems that the concept of aromaticity should be analyzed in terms of ornate hypotheses, a multiplicity of measures.[49] (Katritzky's conclusion has been recently questioned by P. v. R. Schleyer and coworkers.) But notice that the ornate description is reducible to simple components. The universe is not simple, but the models used to describe it can be made of simple pieces.

Several further examples of the power of intelligent PCA may be found in the recent chemical literature. So Murray-Rust and Motherwell have looked at the molecular deformations of 99 β-1′-aminofuranosides, and have shown a very pretty strong correlation with two Principal Components, just those expected to define the so-called "pseudorotation" of the five-membered sugar ring.[50] An analysis of distortions in five-coordinate complexes by Auf der Heyde and Bürgi showed beautifully the relationship of various reaction modes such as the Berry pseudorotation, an S_N2-type mode and an addition/elimination path.[51] And Basu, Gō, and coworkers use a Principal Component analysis of molecular dynamics simulations to trace the path of a $3_{10}/\alpha$-helix transformation in an oligopeptide.[52]

Is there an equivalence between a Principal Component and a physically meaningful factor which, coupled with strong logic, could provide what we usually mean by "an explanation"? In general not. Yet, as Michael Fisher has pointed out to us, an identification of the Principal Components "can, and often does, lead to deeper theoretical insights and constructs."[53] Fisher points, for example, to the Fourier analysis of the tides, in which Lord Kelvin played a principal role, and which led to an understanding of the contributory factors beyond the gravitational pull of the moon.

Incidentally, there is nothing special about chemistry's problems in identifying causes and fundamentals here—the complexity of this task is illustrated just as well by the difficulties arising in the quantitative description of the perception of quality in food. While from the deterministic standpoint, the quality of a steak or a Bordeaux wine may be decomposed into attributes or components, sensory analysis points to simple words (factors) with a world of meanings used by real people to characterize foods. For instance, "toughness-tenderness" and "juiciness" separated as important Principal Components, in an analysis of 69 beef roasts by J. M. Harris, D. N. Rhodes, and B. B. Chrystall.[54]

H. K. Sivertsen and E. Risvik have carried out a detailed multivariate study of French wine profiles. They had eight panelists rate thirty wines for 17 specific and integrated sensory attributes. Two Principal Components accounted for 65% of the variation. The first of these was a fruity aroma axis, "going from berry aromas to vegetative aromas and attributes representing a more ripened aroma, like 'animal' and 'vanilla.' PC2 could be described as a mouth feel and color axis, going from 'suppleness' to 'astringency,' 'color intensity,' 'potential,' and 'color tone.'" Sivertsen and Risvik point to an amusing mapping of the wine varieties separating along these Principal Components onto the geography of France's wine-producing regions.[55]

The science of statistics incorporates Ockham's Razor in its framework in a number of explicit and implicit ways. A particularly useful methodology for fitting

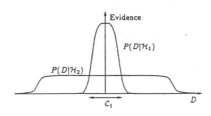

Figure 2: Why Bayes embodies Occam's razor. This figure gives the basic intuition for why complex models are penalized. The horizontal axis represents the space of possible data sets D. Bayes rule rewards models in proportion to how much they *predicted* the data that occurred. These predictions are quantified by a normalized probability distribution on D. In this paper, this probability of the data given model $\mathcal{H}_i$, $P(D \mid \mathcal{H}_i)$, is called the evidence for $\mathcal{H}_i$. A simple model $\mathcal{H}_1$ makes only a limited range of predictions, shown by $P(D \mid \mathcal{H}_1)$; a more powerful model $\mathcal{H}_2$, that has, for example, more free parameters than $\mathcal{H}_1$, is able to predict a greater variety of data sets. This means however that $\mathcal{H}_2$ does not predict the data sets in region C_1 as strongly as $\mathcal{H}_1$. Assume that equal prior probabilities have been assigned to the two models. Then if the data set falls in region C_1, the *less powerful* model $\mathcal{H}_1$ will be the *more probable* model.

Figure 6-6 A figure with its caption (from a paper by D. J. C. McKay [Ref. 57]) describing how Ockham's Razor influences the choice of models in a Bayesian analysis. Reprinted by permission of MIT Press.

models to data and assigning preferences to alternative models is Bayesian inference, introduced by Harold Jeffreys.[56] We reproduce here a figure (Figure 6-6) with its full caption from an important article on Bayesian interpolation by D. J. C. MacKay,[57] which succinctly indicates how Ockham's Razor enters the choice of models in this methodology. A further exposition to the method may be found in the very clear article by W. H. Jefferys and J. O. Berger, entitled "Ockham's Razor and Bayesian Analysis."[58]

Our dialogue is not over; we return to question the arguments made in favor of an operational valuation of Ockham's Razor.

World View or Operating Manual

If we distance ourselves from philosophical implications by treating Ockham's Razor as just an operating principle, aren't we really displaying intellectual cowardice? Take that straight-line graph. If we made the measurements leading to the $(\mathbf{x}, \mathbf{y})$ points already shown, wouldn't we really *believe* that the "proper" value of $\mathbf{y}$ at some new value of $\mathbf{x}$ within the range would be the one that fit on our straight line? Indeed, if we didn't obtain such a result wouldn't we suspect that we had made a mistake in our experiment? And isn't such an expectation really a belief in a simple universe?

In the processing of models we must be especially cautious of the human weakness to think that models can be verified or validated. Especially one's own. The Oreskes, Shrader-Frechette, and Belitz article from which we drew that provocative quote makes this point most convincingly. The main tactical problem in modeling the course of chemical reactions, be they ozone depletion or a pericyclic reaction under new conditions, is to find a reasonable balance between completeness of description of an object or phenomenon under study, and the simplicity of the models applied. The balance is really, really delicate and the razor (Ockham's Razor!) is best wielded by a really skillful barber (experienced chemist) to warrant that essential but hidden features of the object under study were not lost upon modeling its properties and behavior. In the United States, at least, there are not too many barbers left who can give you a razor shave.

Smoothness and Simplicity

The dialogue is not finished. When one infers a linear relationship from empirical observation, be it a linear free energy relationship in physical organic chemistry, or a Hooke's Law relationship in physics, one would indeed be surprised if some of the measurements, made within the range of all the others, failed to fit the model. But that surprise derives not from belief in a simple universe, but rather from belief in a smoothly changing one. With the important and fascinating exception of systems on the threshold of chaotic behavior,[59] or those near phase transitions, our experience suggests to us that the universe is much more a system of smooth curves than jagged edges. It is not often that small changes in some control factor cause wild and unpredictable swings in the response of the system under study. We understand now the importance of bifurcation points in chaotic systems, and know that complex assemblies are subject to chaotic behavior. But most of chemistry is a science of smooth trends.

Take that Cope rearrangement again. For a while it looked like the compromise between the "aromatic" and "biradical" camps was to say that both were right, and that the system flipped from one mechanism to another in response to changes in substituent, as we have described. Such a flip-flop would not be easily described by any linear or smoothly curved function. However, the latest, highest-level *ab initio* calculations have returned us to a smoother description.[60] The multiplicity of reaction channels has disappeared again, and we are now in a situation where the best model seems to be one in which the geometry of the transition structure moves smoothly and continuously from "aromatic" to "biradical" in response to substituent changes.

Even the duality of "concerted" *vs.* "stepwise" mechanisms may be falling to a smoother description. The forced choice between such descriptions is, at least in some cases, a consequence of drawing a potential energy profile in which there is only a single dimension assigned to the reaction coordinate. One then has only two options: one includes a little dip in the curve to imply the existence of an intermediate along the reaction coordinate (stepwise), or one does not (concerted). But of course, for a nonlinear, N-atom molecule there are 3N-6 dimensions to the reaction coordinate. In this space, there is no need to place a local minimum in the potential energy surface on an obligatory path between reactant and product. If such a local minimum exists, and if it is energetically accessible without intervening barriers, then should it be called an intermediate or not? Is the reaction concerted or stepwise? The two descriptions merge smoothly together.[61]

Some barbers will use Ockham's Razor to give you a smooth shave.

Models, Paradigms, and Revolutions

Three final comments in this discussion, neither pro nor con…

1. The gap between the complexity of an object under study and comprehension of its origin is bridged (shaky constructions, to be sure…) through elaboration of suitable models devised to describe the underlying features of the object under study in terms of previously understood phenomena. Every model is, by definition, incomplete.[62] It is thus hardly surprising that a set of complementary models, each of them valid over a certain range of application, is generally needed to describe adequately an object as a whole.

We forward a tentative notion that in the evaluation of models, different criteria may be applied whether one seeks understanding or predictability. We enter an epistemological battleground here (deep trenches recently dug on the field of artificial intelligence…) in positing that there is a difference between human *understanding*, perforce qualitative, and that dream of dreams, a computational model that predicts everything accurately.[63]

Real chemical systems, be they the body, the atmosphere, or a reaction flask, are complicated. There will be alternative models for these, of varying complexity. We suggest that if understanding is sought, simpler models, not necessarily the best in predicting all observables in detail, will have value. Such models may highlight the important causes and channels. If predictability is sought at all cost—and realities of the marketplace and judgments of the future of humanity may demand this— then simplicity may be irrelevant. And impossible, for, as we said, any real problem is complex and will force a complex model. Whatever number of equations or parameters it takes, that's fine. As long as it works.

2. Ockham's Razor is a *conservative* tool. It cuts out crazy, complicated constructions and assures that hypotheses be grounded in the science of the day. So the tool is certain to lead to "normal" science, the paradigmatic explanation. Revolutions in science, to follow Thomas Kuhn's fruitful construction, do not grow from such soil.

Perhaps that is an oversimplification. At the critical turning point when a revolution is about to break loose, Ockham's Razor can turn a conservative into a reluctant revolutionary. We're thinking of Max Planck, interpolating between the Wien and Jeans radiation laws, and following the logic, an Ockham's Razor logic, to the quantum hypothesis. And, it seems, resisting that hypothesis even as the world and he found it necessary.[64]

3. Still another perspective, one which should make a scientist really stop and think, comes from a sensitive reader of this paper, sympathetic to science, Hillel J. Hoffmann.[65] He remarked that to most nonscientists, the very idea that Ockham's Razor is part of the scientific method seems *strange*. This is because to many, science is not about simplicity, but about complexity. Our enterprise seems difficult and obscure to people, even as they use the fruits of that greater knowledge of the world.

Telling Stories, Telling It Straight, Writing Poetry

There was spoken language before writing, before science. And around the fire, when men and women sat and talked of the things of this world, even then there were different ways of telling the story of a failed hunt, of an insect from which one could make a red dye, or what needed to be done to a certain rock to win from it a hard metal. The stories could be embellished, and gods pulled in as causes. No one suffered from these tales, in fact they provided a spiritual matrix for the material world.

Then there came science, and the ritual way of reporting it, the scientific article. To gain reliable, repeatable knowledge, to deal out of the game prejudiced "Nature-Philosophers" (*Naturphilosophen*), the narrative in the standard article tightened. But if you think that scientific articles tell the facts and nothing but the facts, please look again. The facts by themselves are indigestible. They are, and must be, encased in language, connected to frameworks of understanding (theories). Try writing an article with just the facts, and see how many people read it! The narrative may be suppressed (which actually, as suppression usually does, only raises the tension lurking beneath the surface[66]) but the impulse to tell a story remains.

With no nostalgia for those days around the fire, the wielding of Ockham's Razor attacks something most fundamentally human, the love for narrative. There are times when the story has to be told simply, the fire engine sent the shortest route to the fire. But a world without stories is inhuman. It is a world where nothing is *imagined*. Could a chemist be creative in such a world?

Let us put it another way. There is a human tendency to tell elaborate stories if not tall tales. Even scientists succumb to it. And there is a logical emphasis on the succinct, the unembellished, which has certainly been part of the successful method of science. There is danger in going astray, following the person who tells a wild story well. And there is danger—we think perhaps greater—in telling too few stories, in building fewer scenarios which present or future facts may demolish. Or uphold.

There is another very human literary activity. This is to write poetry, to tell essences intensely, in words. The cult of mathematical simplicity as beauty is a reaching for essences that parallels the compact truth-telling of poetry. This is what Dalton, Dirac, and Einstein aspired to. And this perspective has led to "the majesty, subtlety, and grace of science, and her deepest insights and discoveries," as Michael Fisher so aptly puts it.[67] We agree. But poetry is more. Not a stripping to a common nakedness, it aspires to singularly adorned simplicity.

Ockham's Razor and the Struggle for Understanding

The search for true understanding might be compared with the crafting of an endless, absorbing mosaic picture. The pieces already in place, lustrous and dull, have been laboriously and joyously shaped in the creative work of thousands of years of protoscience and a few hundred of "real" Western science. They furnish us with some clues as to the nature of the beast. If simplicity of interpretation (in other words, "beauty of equations," according to P. A. M. Dirac, or "lucidity complementary to truth," according to Niels Bohr) be a desirable quality, the interpretation must be constructed out of simple components.[68] The principle of parsimony is then just what we need as we labor, discover, and create.

If the *desideratum* be a human science open to change and the unexpected, then maybe there are occasions when Ockham's Razor should be sheathed. Or we should remind ourselves ceaselessly of the *conditional* interpretation of a conclusion based on Ockham's Razor reasoning. Cognizance of the complexity that so beautifully contends with simplicity in this evolving world, cognizance of the creative foment of intuition without proof within science, lead us to think so.

William of Ockham himself recognized the conditional nature of all human knowledge. As Brian Tierney has pointed out to us, this is implicit in Ockham's statement that "plurality should not be assumed without necessity." He knew that there would be a time and place for pluralities and complexities.[69] Our fathers, our teachers knew this well too: So Newton wrote "We are to admit no more causes of natural things than such as are both true and sufficient to explain their appearances."[70] And Einstein supposedly said "Everything should be made as simple as possible, but not simpler."[71]

Intuition figures prominently in the strong pull on us toward the simple, the logical, and the beautiful. Plato had that right. And the same concept, intuition, serves us as we argue for a certain sterility of William of Ockham's sharp principle. "Intuitive" is, probably, the best characterization of the law of parsimony, Ockham's Razor. It is also intuition that sometimes leads to the oh so many blind alleys, if not mistakes, of our sciences. And it is precisely human intuition that provided and provides for the disclosure of those mysterious and wondrous ways of Nature, and the creation of so much new. The mosaic grows.

Notes

1. The authoritative work on Ockham's life remains L. Baudry, *Guillaume d'Occam. Sa vie, ses oeuvres, ses idées sociales et politiques* (Paris: Librairie Philosophique Vrin, 1950). There is a substantial literature on his philosophy—here we cite only four leading references: H. Junghaus, *Ockham im Lichte der neueren Forschung* (Berlin: Lutherisches Verlagshaus, 1968); G. Leff, *William of Ockham: The Metamorphosis of Scholastic Discourse* (Manchester: Manchester University Press, 1975); M. M. Adams, *William Ockham*, 2 vols. (South Bend: University of Notre Dame Press, 1987); W. Vossenkuhl and R. Schönberger (eds.), *Die Gegenwart Ockhams* (Weinheim: VCH Acta Humaniora, 1990).

2. The definitive edition of Ockham's work is published by the Franciscan Institute at St. Bonaventure, NY, 1967–1988 (*Opera philosophica et theologica*); and by Manchester University Press, 1940ff (*Opera politica*). An English translation of his philosophical work is provided by P. Boehner (ed.), *Ockham. Philosophical Writings* (Edinburgh: Nelson, 1957).

3. P. Boehner (ed.), *Ockham. Philosophical Writings* (Edinburgh: Nelson, 1957) (Note 2), xii.

4. U. Eco, *The Name of the Rose* (Boston: Harcourt Brace Jovanovich, 1983).

5. G. Gál, William of Ockham Died "Impenitent" in April 1347. *Franciscan Studies* 42 (1982): 90. I am indebted to Brian Tierney for this reference.

6. B. Tierney, personal communication. There is a substantial literature on William of Ockham's philosophy – here we cite only two leading references: G. Leff, *William of Ockham: The Metamorphosis of Scholastic Discourse* (Manchester: Manchester University Press, 1975), and W. Vossenkuhl and R. Schoenberger, *Die Gegenwart Ockhams* (Weinheim: VCH Acta Humaniora, 1990).

7. In *Tractatus de succcessivis*, 36–37, translated by M. M. Adams, *William Ockham*, 2 vols. (South Bend: University of Notre Dame Press, 1987),p. 821.

8. Brian Tierney begins a paper on Ockham with the words, "William of Ockham was a difficult thinker."

9. Mary Reppy, personal communication.

10. Boehner, P. (ed.), *Ockham. Philosophical Writings* (Edinburgh: Nelson, 1957), pp. xxx–xxxix.

11. These are quoted from M. M. Adams, *William Ockham*, 2 vols. (South Bend: University of Notre Dame Press, 1987), (Note 2), vol. 1, pp. 156–157, who provides the citations from Ockham where the formulations quoted may be found.

12. For instance, Odo Rigaldus, *Commentatorium super sententias*, MS Bruges 208, fol. 150a, has "Frustra fit per plura quod potest fieri per unum." (P. Boehner [ed.], *Ockham. Philosophical Writings* (Edinburgh: Nelson, 1957), [Note 1], p. xx). Odo Rigaldus was Archbishop of Rouen from 1248 to 1275. Aristotle in *De Caelo* writes: "Obviously then it would be better to assume a finite number of principles. They should, in fact, be as few as possible, consistently with proving what has to be proved. This is the common demand of the mathematicians who assume as principles things finite either in kind or in number." (*The Works of Aristotle Translated into English, vol. II, De Caelo*, trans. by J. L. Stocks, ed. by W. D. Ross [London: Clarendon, 1930], III, c. 4, 302b). And Thomas Aquinas (1225–1274), in *Summa Contra Gentiles*: "If a thing can be done adequately by means of one, it is superfluous to do it by means of several; for we observe that nature does not employ two instruments where one suffices." (*Basic Writings of St. Thomas Aquinas*, trans. A. C. Pegis, [New York: Random House, 1945], p. 129.). We owe these last two references to R. Ariew, Did Ockham Use His Razor? *Franciscan Studies* 37 (1977): 5. See also G. Leff, *William of Ockham* (Manchester: Manchester University Press, 1975), p. 35, note 141; and the spirited account of W. M. Thorburn, The Myth of Occam's Razor. *Mind* (NS) 27 (1918): 345.

13. See the spirited account of W. M. Thorburn, The Myth of Occam's Razor. *Mind* (NS) 27 (1918): 345.
14. M. M. Adams, *William Ockham*, 2 vols. (South Bend: University of Notre Dame Press, 1987) (Note 1), vol. 1, p. 157.
15. *Rasoir des Nominaux*: E. B. de Condillac, *Essai sur l'Origine des Connaissances Humaines*, 1746, p. 214. The English variant (*Occam's Razor*) apparently appears first in W. Hamilton, *Discussions on Philosophy and Literature, Education and University Reform*, 3rd ed. (Edinburgh: Blackwood, 1852), p. 590. We owe these citations to Thorburn, The Myth of Occam's Razor. *Mind* (NS) 27 (1918) (Note 13). But we may need a Mertonian analysis of the origins of this term.
16. For further discussions of Ockham's Razor, its utilization by Ockham and others, and its utility in philosophy, see: C. K. Brampton, Nominalism and the Law of Parsimony. *Modern Schoolman* 41 (1964): 273; A. A. Maurer, Method in Ockham's Nominalism. *Monist* 61 (1978): 426; J. Boler, Ockham's Cleaver in William of Ockham (1285-1347) Commemorative Issue – Part II. *Franciscan Studies* 45 (1985): 119; K. Menger, A Counterpart of Occam's Razor in Pure and Applied Mathematics: Semantic Uses. *Synthese* 13 (1961): 331; G. O'Hara, Ockham's Razor Today. *Philosophical Studies* 12 (1963): 125 (particularly good in discussing Bertrand Russell's use of Ockham's Razor).
17. Image is from P. C. Marzio, *Rube Goldberg: His Life, and Work* (New York: Harper & Row, 1973), p. 132; P. Garner, *Rube Goldberg: A Retrospective* (New York: Delilah, 1983).
18. P. L. M. de Maupertuis, Accord des Différents Lois de la Nature qui Avoient jusqu'ici para Incompatibles. *Mémoires de l'Académie Royale* 423 (1744).
19. L. de Broglie, Recherches sur la Théorie des Quanta. *Annales de Physique* 3 (1925): 22; R. P. Feynman, R. B. Leighton, and M. Sands, Angular Momentum, in *The Feynman Lectures on Physics*, vol. II (Reading, MA: Addison-Wesley, 1964), Chap. 19.
20. A. Muller, Sur la Decomposition Pyrogenée des Amines de la Série Grasse. *Bulletin de Société Chimique, Paris* 45 (1886): 438.
21. W. Hückel, *Theoretische Grundlagen der Organischen Chemie* (Mannheim: Akademische Verlagsgesellschaft, 1934). See also G. W. Wheland, *Advanced Organic Chemistry* (New York: Wiley, 1960).
22. F. Rice and E. Teller, The Role of Free Radicals in Elementary Organic Reactions. *Journal of Chemical Physics* 6 (1938): 489.
23. See the comprehensive reviews of J. Hine, The Principle of Least Nuclear Motion. *Advances in Physical Organic Chemistry* 15 (1977): 1; and M. L. Sinnott, The Principle of Least Nuclear Motion and the Theory of Stereoelectronic Control. *Advances in Physical Organic Chemistry* 24 (1986): 113. See as well the comment, in the context of concerted reactions, by J. A. Berson in *Rearrangements in Ground and Excited States*, ed. by P. de Mayo, vol. 1 (New York: Academic Press, 1980), pp. 375–376.
24. M. S. Korobov et al., Energy Barriers for the Enantiotopomerization of Tetrahedral Boron Chelates. *Journal of the Chemical Society - Chemical Communications* (1982): 169.
25. M. S. Korobov et al., Kinetics and Mechanism of the Enantiomerization of Tetracoordinated Boron Chelate Complexes. *Journal of Molecular Structure – Theochem.* 200 (1989): 61.
26. A. J. Boulton and C. S. Prado, Energy Barriers to the Enantiotopomerisation of Tetrahedral Boron Chelates: A Rearrangement of Some Cyclic Boron Compounds. *Journal of the Chemical Society - Chemical Communications* (1982): 1008.
27. R. A. Jackson, *Mechanism: An Introduction to the Study of Organic Reactions* (London: Clarendon, 1972).
28. B. K. Carpenter, *Determination of Organic Reaction Mechanism* (New York: Wiley-Interscience, 1984).
29. B. Russell, *The Problems of Philosophy* (Oxford: Oxford University, 1980), p. 97.
30. N. Oreskes, K. Shrader-Frechette, and K. Belitz, Verification, Validation, and Confirmation of Numerical Models in the Earth Sciences. *Science* 263 (1994): 641, endnote 25.

31. V. I. Minkin, L. E. Nivoroshkin, and M. S. Korobov, Stereodynamics and Degenerate Ligand Exchange in Solutions of Tetracoordinate Chelate Complexes of Nontransition Metals. *Russian Chemical Reviews, Uspekhi Khimii* (1994): 303.

32. According to the original definition, pericyclic reactions are those "in which all first order changes in bonding relationships take place in concert on a closed curve" (R. B. Woodward and R. Hoffmann, *The Conservation of Orbital Symmetry* [Weinheim: Verlag Chemie, 1970]).

33. K. N. Houk, Y. Li, and J. D. Evanseck, Transition Structures of Hydrocarbon Pericyclic Reactions. *Angew. Chem. Int. Ed. Engl.* 31 (1992): 682; M. J. S. Dewar and C. Jie, Mechanisms of Pericyclic Reactions: The Role of Quantitative Theory in the Study of Reaction Mechanisms. *Accounts of Chemical Research* 25 (1992): 537. Of special interest are the contradictory conclusions (in some cases) arising from *ab initio* vs. or semiempirical methods of calculation. For an up-to-date and most readable account of this neverending story see K. N. Houk, J. González, and Y. Li, Pericyclic Reaction Transition States: Passions and Punctilios, 1935-1995. *Accounts of Chemical Research* 28 (1995): 81.

34. For a leading reference see H. Schwarz, The Mechanism of Fullerene Formation. *Angew. Chem. Int. Ed. Engl.* 32 (1993): 1412.

35. R. E. Smalley, Self-Assembly of the Fullerenes. *Accounts of Chemical Research* 25 (1992): 98; J. M. Hunter et al., Annealing Carbon Cluster Ions: A Mechanism for Fullerene Synthesis. *Journal of Physical Chemistry* 98 (1994): 1810; G. von Helden et al., Experimental Evidence for the Formation of Fullerenes by Collisional Heating of Carbon Rings in the Gas Phase. *Nature* 363 (1993): 60; J. Hunter et al., Annealing C_{60}^+: Synthesis of Fullerenes and Large Carbon Rings. *Science* 260 (1993): 784; R. L. Murray et al., Role of sp^3 Carbon and 7-Membered Rings in Fullerene Annealing and Fragmentation. *Nature* 366 (1993): 665; C. Xu and G. E. Scuseria, Tight-Binding Molecular Dynamics Simulations of Fullerene Annealing and Fragmentation. *Physical Review Letters* 72 (1994): 669. See also Note 34 and the work of T. Belz et al., On the Mechanism of Fullerene Formation. *Angew. Chem. Int. Ed. Engl.* 33 (1994): 1866.

36. See R. E. Smalley (Note 35).

37. Quantum mechanical tunnelling is a widespread reaction mechanism responsible for occurrence of many proton and electron transfer reactions. Tunnelling also occurs, as we have only recently learned, for heavier atoms; B. K. Carpenter, Heavy Atom Tunneling As the Dominant Pathway in a Solution-Phase Reaction? Bond Shift in Antiaromatic Annulenes. *J. Am. Chem. Soc.* 105 (1983): 1700.

38. J. Platt, Strong Inference. *Science* 146 (1964): 347.

39. Of course it takes six for sex: L. B. Bjostad et al., Identification of New Sex Pheromone Components in *Trichoplusia ni*, Predicted from Biosynthetic Precursors. *Journal of Chemical Ecology* 10 (1984): 1309; W. L. Roelofs and T. Glover in *Chemical Senses*, vol. 3, ed. by C. J. Wysocki and M. R. Kare (London: Dekker, 1991). For more detail, see Chapter 8 in this book.

40. The provisional nature of Ockham's Razor based arguments has been stressed by R. O. Kapp, Ockham's Razor and the Unification of Physical Science. *The British Journal for the Philosophy of Science* 8 (1958): 265.

41. The definition of the term "degrees of freedom" here differs from that commonly used in statistical mechanics.

42. The choice becomes a little less clear-cut when one starts to inquire about models for which the line comes close to, but does not pass exactly through all of the points. In statistical practice, this situation is dealt with by applying the Fisher *F* test (see, for example, J. C. Miller and J. N. Miller, *Statistics for Analytical Chemistry*, 2nd ed. [New York: Ellis Horwood, 1988], p. 60). The sum of the squares of the deviations $(y - ym)^2$ is computed for each model (y is the experimentally measured response of the system under study to a given value of x; ym is the response predicted by the model) and then divided by the appropriate number of degrees of freedom. This calculation provides the variance for each model. The ratio of the variances is compared with tabulated values that allow one to decide, with a specified level of confidence, whether the more complex model has made a statistically significant improvement to the fit. If it has not, one will generally opt for the simpler model because, again, it has the larger number of degrees of freedom.

43. See also the clear arguments here of H. Jeffreys, *Theory of Probability* (Oxford: Oxford University, 1939); W. H. Jefferys and J. O. Berger, *American Scientist* 80 (1992): 64.

44. E. R. Malinowski and D. G. Howery, *Factor Analysis in Chemistry* (New York: Wiley-Interscience, 1990).

45. See also H. J. Gauch Jr., Prediction, Parsimony, and Noise. *American Scientist* 81 (1993): 468.

46. T. Auf der Heyde, Representation, Comparison, and Analysis of Molecular Deformations. *South African Journal of Chemistry* 46 (1993): 45.

47. A. R. Katritzky et al., Aromaticity As a Quantitative Concept. 1. A Statistical Demonstration of the Orthogonality of Classical and Magnetic Aromaticity in Five- and Six-Membered Heterocycles. *J. Am. Chem. Soc.* 111 (1989): 7.

48. See V. I. Minkin, M. N. Glukhovtsev, and B. Y. Simkin, *Aromaticity und Antiaromaticity* (New York: Wiley, 1994), for leading references.

49. Recently Katritzky et al.'s conclusion has been questioned by P. v. R. Schleyer et al., Aromaticity and Antiaromaticity in Five-Membered C_4H_4X Ring Systems: "Classical" and "Magnetic" Concepts May Not Be "Orthogonal." *Angew. Chem. Int. Ed. Engl.* 34 (1995): 337.

50. R. Murray-Rust and W. D. S. Motherwell, Computer Retrieval and Analysis of Molecular Geometry. III. Geometry of the β-1'-Aminofuranoside Fragment. *Acta Crystallographica - Section B* 34 (1978): 2534.

51. T. P. E. Auf der Heyde and H. B. Bürgi, Molecular Geometry of d^8 Five-Coordination. 1. Data Search, Description of Conformation, and Preliminary Statistics. *Inorganic Chemistry* 28 (1989): 3960; T. P. E. Auf der Heyde and H. B. Bürgi, Molecular Geometry of d^8 Five-Coordination. 2. Cluster Analysis, Archetypal Geometries, and Cluster Statistics. *Inorganic Chemistry* 28 (1989): 3970; T. P. E. Auf der Heyde and H. B. Bürgi, Molecular Geometry of d^8 Five-Coordination. 3. Factor Analysis, Static Deformations, and Reaction Coordinates. *Inorganic Chemistry* 28 (1989): 3982.

52. G. Basu et al., A Collective Motion Description of the 3^{10}-/.α.-Helix Transition: Implications for a Natural Reaction Coordinate. *J. Am. Chem. Soc.* 116 (1994): 6307.

53. M. E. Fisher, personal communication.

54. J. M. Harris, D. N. Rhodes, and B. B. Chrystall, Meat Texture: I. Subjective Assessment of the Texture of Cooked Beef. *Journal of Texture Studies* 3 (1972): 101; see also the analysis of sensory characteristics of ham and their relationships with composition, visco-elasticity and strength: G. R. Nute et al., Sensory Characteristics of Ham and Their Relationships with Composition, Visco-Elasticity and Strength. *International Journal of Food Science and Technology* 22 (1987): 461. Our source for these references on sensory properties and preferences of meat products is a perceptive overview by Einar Risvik, Sensory Properties and Preferences. *Meat Science*, 36 (1994): 67. Also see E. Risvik., *New Approaches to Analysis and Understanding of Data in Sensory Science*. Dr. Agric. thesis, Agricultural University of Norway.

55. H. K. Sivertsen and E. Risvik, A Study of Sample and Assessor Variation: A Multivariate Study of Wine Profiles. *Journal of Sensory Studies* 9 (1994): 293 have carried out a detailed multivariate study of French wine profiles. They had eight panelists rate thirty wines for 17 specific and integrated sensory attributes. Two Principal Components accounted for 65% of the variation. The first of these was a fruity aroma axis, "going from berry aromas to vegetative aromas and attributes representing a more ripened aroma, like 'animal' and 'vanilla.' PC2 could be described as a mouth feel and color axis, going from 'suppleness' to 'astringency,' 'color intensity,' 'potential,' and 'color tone.'" Sivertsen and Risvik point to an amusing mapping of the wine varieties separating along these Principal Components onto the geography of France wine-producing regions. See also A. C. Noble, A. A. Williams, and S. P. Langdon, Descriptive Analysis and Quality Ratings of 1976 Wines from Four Bordeaux Communes. *Journal of the Science of Food and Agriculture* 35 (1984): 88.

56. H. Jeffreys, *Theory of Probability* (Oxford: Oxford University Press, 1939). See also articles by S. F. Gull in *Maximum Entropy and Bayesian Methods in Science and Engineering, Vol. 1: Foundations*, ed. by G. J. Erickson and C. R. Smith (Dordrecht: Kluwer, 1988), pp. 53–74; and in J. Skilling (ed.), *Maximum Entropy and Bayesian Methods* (Cambridge: Cambridge University Press, 1988; Dordrecht: Kluwer, 1989, pp. 53–71).

57. D. J. C. MacKay, Bayesian Interpolation. *Neural Computation* 4 (1992): 415.
58. W. H. Jefferys and J. O. Berger, Ockham's Razor and Bayesian Analysis. *American Scientist* 80 (1992): 64; see also H. G. Gauch Jr., Prediction, Parsimony, and Noise. *American Scientist* 81 (1993): 468.
59. H. G. Schuster, *Deterministic Chaos* (Weinheim: Verlag Chemie, 1984).
60. D. A. Hrovat, K. Morokuma, and W. T. Borden, The Cope Rearrangement Revisited Again: Results of Ab Initio Calculations Beyond the CASSCF Level. *J. Am. Chem. Soc.* 116 (1994): 1072. See also P. M. Kozlowski, M. Dupuis, and E. R. Davidson, The Cope Rearrangement Revisited with Multireference Perturbation Theory. *J. Am. Chem. Soc.* 117 (1995): 774; H. Jiao and P. v. R. Schleyer, The Cope Rearrangement Transition Structure Is Not Diradicaloid, but Is It Aromatic? *Angew. Chem. Int. Ed. Engl.* 34 (1995): 334; and the account of Houk, González and Li (Note 23).
61. B. K. Carpenter, Intramolecular Dynamics for the Organic Chemist. *Accounts of Chemical Research* 25 (1992): 520.
62. C. J. Suckling, R. E. Suckling, and C. W. Suckling, *Chemistry Through Models. Concepts and Applications of Modelling in Chemical Science, Technology and Industry* (Cambridge: Cambridge University Press, 1980).
63. For a contrary view, see H. A. Simon, *The Sciences of the Artificial*, 2nd ed. (Cambridge, MA: MIT Press, 1981).
64. M. Jammer, *The Conceptual Development of Quantum Mechanics*; 2nd ed. (College Park, MD: American Institute of Physics, 1989). J. Mehra and H. Rechenberg, *The Historical Development of Quantum Theory* (New York: Springer, 1982), vol. 1, part 1, p. 150; see also T. S. Kuhn, *Blackbody Theory, and the Quantum Discontinuity* (Oxford: Oxford University Press, 1978), part 2, pp. 616–617.
65. H. J. Hoffmann, personal communication.
66. R. Hoffmann, *The Same and Not the Same* (New York: Columbia University Press, 1965).
67. M. E. Fisher, personal communication.
68. It is not all that easy to define simplicity: "The notion of simplicity, like truth, beauty and effective process is an intuitive one calling for a more objective characterization, *i.e.*, formalization, before we can even hope to agree about the relative complexities of different theories," J. J. Casti, *Searching for Certainty* (New York: Morrow, 1990).
69. B. Tierney, personal communication
70. Isaac Newton, *The Mathematical Principles of Natural Philosophy*, trans. A. Motte (Benjamin Motte, 1729).
71. The source of this quotation cannot be traced. Wikipedia, in its entry on Albert Einstein, gives the following approximation: "It can scarcely be denied that the supreme goal of all theory is to make the irreducible basic elements as simple and as few as possible without having to surrender the adequate representation of a single datum of experience." *On the Method of Theoretical Physics*, The Herbert Spencer Lecture, delivered at Oxford (10 June 1933); also published in *Philosophy of Science* 1 (April 1934): 163. The Wikipedia entry is thanks to Dr. Techie @ www.wordorigins.org and JSTOR.

7

Qualitative Thinking in the Age of Modern Computational Chemistry, or What Lionel Salem Knows[1]

ROALD HOFFMANN

1

The achievements of modern computational chemistry are astounding. It is reasonable today to handle billions of configurations, and to achieve chemical accuracy, kilocalories say, in calculating binding energies and geometries, in ground and transition states of reasonably complex molecules. There is no question that the enterprise of computational theoretical chemistry is *successful*.

Now Lionel Salem and I grew up and developed scientifically in the climate of the very same computer which made all this possible. Russ Pitzer taught me to punch cards; I still miss the sound of the key punch. The extended Hückel method, which several of us developed in the Lipscomb group, would have been impossible without modern computers. But I took a different turn, moving from being a calculator in the framework of semiempirical theory, to being an explainer, the builder of simple molecular orbital models. I was and am still doing calculations, but my abiding interest is in the construction of explanations. And also in thinking up moderately unreasonable things for experimentalists to try.

In existing as a scientist, meaning that my work was of continuing interest to other chemists, I was helped in that I moved into whatever part of chemistry I did, just a little ahead of the heavy guns of computational chemistry. So I switched from organic to inorganic molecules just when organic molecules became reasonably calculable. Recently I've been less fortunate, for when I moved to solids and surfaces I came back into heavy fire—physicists had been doing calculations on these materials for a long time. And they were (are) hardly likely to believe that one-electron calculations and a chemical viewpoint are of value.

I want to make some observations on computational quantum chemistry, perforce influenced by my prejudices. Given the advances in the field, any molecule

I can calculate (without geometrical optimization), with the simplest extended Hückel approximation, can be done so much better by most computational chemists. So why don't I feel threatened; why is there a role for people of my ilk? Or for Lionel.

Actually, I do feel threatened and bypassed! But that's just an emotional reaction, and my aging figures in it. But when I think about it quasi-rationally, I don't feel very threatened. Or maybe I put on a brave face, for here ...

It's even more complicated: Part of me feels bypassed, and it feels so good, for the knowledge that a calculation is reliable and can predict the anomalous, weird geometry of, say, Si_2H_2 before the molecule is made, that is really great! Part of me is bypassed, and it doesn't feel good, because something I said may be just wrong, not supported by better calculation.

But in general while I am bypassed, I feel that actually there is a deeper need than before for the kind of work I or Lionel Salem do. The factors I see shaping this need are many, at least four. The first, I would claim, is the difference between predictability and understanding. A second has to do with human-machine interactions; a third with the peculiar dialogue of experiment and theory in chemistry. And a fourth with the special features of theory in science in general.

Let me discuss these in turn, and I will do so in a personal way, as provocatively as I can, asking you to disagree.

2

"Understanding" means different things to different people. To some, understanding has a reductionist meaning; understanding in chemistry is viewed as resolving the underlying physics. Elsewhere, I have written of my dissatisfaction with reductionism[2]—I believe that the notion is philosophically flawed, and within the practice of science mainly invoked as self-justification. I have argued that understanding is both vertical (reductionist), and horizontal—through a set of concepts of roughly equal complexity to the explanatory question being asked and answered. Understanding in chemistry I see as a mix of both vertical and horizontal thinking.

Even if we were to accept the vertical mode as dominant, I think there is a problem that modern computational approaches to chemistry face. Let me explain by putting forward one definition of understanding in physics and chemistry that most people I think will agree on, and which follows the reductionist program: We understand an observable if we know the mix of different physical mechanisms leading to it (there may be more than one) and can make a semiquantitative estimate of the contributions of each mechanism.

In thinking about explanations this way, experimentalists invoke contributions or factors with a venerable history: electrostatics (charges, dipoles), donor-acceptor interactions, orbitals. The computational chemist, on the other hand, is content with getting the Hamiltonian right and then proceeding as expeditiously and accurately as possible to a solution of the wavefunction and a calculation of the desired observable. The calculations, whatever formalism is followed, are elaborate. What happens next is an imaginary dialogue between the experimentalist and the theorist:

The experimentalist asks: "What is the bond angle of water?" You, the theorist, plug it into the best programs available and you get it right to three significant figures. Everyone is happy. Then the experimentalist asks the same question for TeH_2. You say "Wait a minute, I have to calculate it..." And you get that right. And you get right Li_2O and F_2O as well. But if that's all you do, no matter how well you do it, the experimentalist will grow increasingly unhappy. Because you haven't provided him/her with a simple, portable explanation, one based on electronegativity, or relative energies of s and p orbitals, or donor or acceptor character or whatever set of factors he or she feels comfortable with. The experimentalist will think "This theorist is only good at simulating experiment." Or, less charitably, "The computer understands, this theorist doesn't." Which is not too bad. Sometimes that predictability is needed, sometimes it's quicker, or as Fritz Schaefer so aptly calls it in a paper, "odorless." Sometimes it even gets the facts right before experiment—the story of the ground and excited states of Si_2H_2 is an exemplary case. In many interesting areas of chemistry we are approaching predictability, but...I would claim, not understanding.

Let me digress to give a simple pedagogic strategy which I teach for eliciting understanding. This is to alternate the computational and the "understanding" roles. I define understanding (operationally, if faced with a really good computer program and a complex problem), as being able to predict qualitatively (this forces you to think before) the result of a calculation before that calculation is carried out. If the calculational result differs from what your understanding gives you, well—then it is time to think again, do numerical experiments until you rationalize (that's also to "understand") the results. Until the explanation is so clear that you could kick yourself in the butt for not having seen why...But don't stop, iterate the process, go on. Understanding will build if you follow this way of analysis.

Back to the program with perfect predictability. Does it, or you, its programmer, understand chemistry? I would say "No, it doesn't, it simulates it." It is capable of making predictions, but it is not capable of telling you trends. Because it is the human being who defines, probably in a horizontal way of thinking, what makes for a trend—donor, acceptor, base, acid... The program also lacks that most human and chemical quality of intelligence, of seeing connections, relationships, metaphors. I ask you to think about what you consider intelligence in a person, a student or mentor. Is it the ability to calculate or to make connections?

I realize full well that I tread a minefield full of angry controversy—namely that of artificial intelligence, of what thought and consciousness mean. Francis Crick, Roger Penrose, Marvin Minsky, Joseph Weizenbaum will not agree with me or each other. I just give you my view—one reason I survive is that I build explanations (deceptive, oversimplified, as they may be); in doing so I answer the desperate plea from the experimentalist, a spiritual plea, for understanding and not simulation.

3

Now let me talk about another problem of computational molecular science. This is the psychology of man-machine interactions. From such interactions arise non-trivial psychological barriers to successful chemical theory. For instance there is the inherent game-playing aspect of debugging, which enthralls and eventually enslaves. As you try to make that program work, as you curse and cajole it, you may get so involved with it that you lose contact with chemical reality, the problem to be solved.

Another concern is a giving in to fancy graphics. The computer can spew out a near infinity of contour diagrams of orbitals. If you put in 20 of them in a paper, no one is going to read the paper; it's absolute folly to think that anyone will be impressed! Reporting research has an abiding pedagogical component; those 20 overly detailed diagrams clearly serve as a sign that you don't want anyone to understand or worse, maybe you think this is a way to hide lack of understanding, which is usually, like revelation, bright and intense.

In my opinion graphical presentation has actually deteriorated with the advent of computers. For the ornery software never gives you what you want, and eventually we accept compromises we never would accept in a human drawing. Often the published lettering on the axes of a graph, or the scale markings, is weird. This is a give-away of a compromise born of laziness or desperation.

But these are trivia. More important is the inherent complexity of the enterprise of computational chemistry. Suppose you have that perfect program. It does require an incredible effort to write the code. The results have oscillated before with increasingly faithful accounts of correlation. Now you are sure the problem is solved. For any molecule the code will give you the correct geometry, polarizability, spectrum. That's the grail to which we have aspired.

But it's complex. It has to be complex, for nature is (only the wild dreams of theoreticians of the Dirac school make nature simple). Our minds are simple, not nature. And given that complexity there is a natural tendency on the part of the computational theorist to think that there can be no single, simplified explanation (in a way a curious conclusion for a reductionist who believes in Dirac's diction). And to be loathe to give such an explanation to experimentalists. Especially in their own language, based as it is on vague, time-honored contexts which may bear no relation to what you calculate.

I can only give you the advice: Try, please try. Take the existential plunge, the hazard of providing an explanation. Do numerical experiments to probe your wild notion, to be sure. But don't be afraid, forward a simple explanation.

I know that I am not consistent in advocating a complex universe, but simple explanations.

4

A third reason why I'm not too worried about being bypassed has to do with the curious dance of experiment and theory.

Chemistry is an experimental subject (85% of chemists in academia, 99.5% in industry). So is physics, which has a real psychological problem, a kind of schizophrenia, for it is an experimental field but (because of its buying whole-hog reductionist logic) one which has opted to have its heroes theoreticians (the only exceptions are Fermi and Rutherford). Experimentalists use theory or theoretical frameworks in certain ways, they use them a lot; there are no facts without a framework to think about them. The theories experimentalists use are often simple—often so simple they're ashamed to put them into print, so they invent grander reasons than those that initially moved them.

One such theory is the orbital model, learned not without pain by experimentalists, taught to them by great English teachers such as Coulson and Orgel. Orbitals have served chemistry well—I have lived off this model, and pushed it on. One does need to go beyond orbitals, that is clear. A problem that I see with modern CI methods is that people have not sufficiently stressed why certain configurations mix. Actually VB ideas are helpful here. Much more conceptual work is needed in just this area.

There is also a real problem with density functional theory, in that the originators of the method, and most of its first generation implementors, have denied the attribution of any significance to the Kohn-Sham orbitals. This is now being repaired, but there is a difficulty here in that experimentalists are being denied a part of their intuition. And nothing is offered in exchange except "Trust me." And why should they trust theoreticians?

5

Now I want to mention some special features of theory in science and ask how does modern computational chemistry fare with respect to taking advantage of these. Let me enumerate some of the circumstances in which chemical or physical theory just might have an inherent advantage over experiment:

1. Theory allows calculations on unstable molecules, unstable conformations of stable molecules, transition states for reactions not followed, to be performed with as little or as much difficulty as the corresponding computations on

stable molecules. The purpose of such calculations is not merely to quantify the energetic misery of these unstable species, but to discern from the calculations what are the factors responsible for their instability. If we understand these factors, we will be able to devise a strategy for moving the molecules to lower energy. Theory serves uniquely here, for the constraint of the Boltzmann factor makes unlikely the experimental probing of such metastable species.

2. Chemistry is discontinuous, but theory allows and often dictates continuous variation. One example is the Karplus curve for vicinal proton-proton coupling constants, derived theoretically as a continuous function, probed experimentally by discrete conformationally fixed molecules. Another example is the concept of a continuously varying dimensionality in the advances in critical phenomena, contrasted with the reality of three dimensions and the modeled "reality" of one or two dimensions.

3. Observables in chemistry may be the resultant of several simultaneously operative physical mechanisms. A measurement cannot resolve these mechanisms (though a series of observations on related molecules may provide that resolution—witness the elegant dissection of through-space and through-bond interactions in Heilbronner's photoelectron studies). But theory has no problem in resolving mechanisms. One can calculate the contribution of each physical factor, or if all factors are already in the calculation one can throw away certain matrix elements, keep others, thus manipulating the theory to isolate the separate effects.

4. Theory can simplify. The strong dictates of thermodynamics often prevent the observation of the simplest version of a reaction type or of a molecule. Substituents may modify a reaction path very slightly, and yet the parent reaction with no substituents may be masked by an entirely different process. Theory is not hampered by the reactions observed and can, in fact, examine the simplest variant of a reaction. Of course, this is a curse as well as a blessing. Substituents may be what makes an important reaction go, but the theoretician has an innate tendency to throw away those perturbations and to idealize the problem to the soluble stage. On occasion the explicitly soluble may be totally impractical. The stage is set for a classical dialectical contest between the practical experimentalist and the oversimplifying theorist. Both will gain if they persist in their struggle.

6

The most important role of theory is to make connections. I do believe that to see "my nature singing in me is your nature" (to quote Archie Ammons) is our purpose. There is no greater joy and no greater contribution than to make people see the unity of this world (which is what connections form), and yet be at peace with the diversity and richness of the universe.

Notes

1. This essay was originally given as a lecture at a Cambridge meeting honoring two great pioneers of computational quantum chemistry, Frank Boys and Isaiah Shavitt. At that time the paper was entitled "How Nice to Be Bypassed by What Boys and Shavitt Have Wrought." At the urging of the organizer of the conference, my friend Fritz Schaefer, the paper was submitted to *The Journal of Physical Chemistry*, where the other papers presented at the Cambridge meeting were published. But the style of the paper proved too much for the editors and reviewers of *J. Phys. Chem.*, who rejected it because it was not a typical scientific paper.

   Lionel Salem is a great theoretical chemist, popularizer of science, and friend. This essay is dedicated to him.

2. See R. Hoffmann, *The Same and Not the Same* (New York: Columbia University Press, 1995), Chapter 4; also R. Hoffmann, What Might Philosophy of Science Look Like If Chemists Built It? *Synthèse* 155(3) (2007): 321, Chapter 3 in this book.

8

Narrative

ROALD HOFFMANN

What does a political campaign have to do with tetrahedrane, a beautiful yet unstable hydrocarbon with four CH groups at the corners of a tetrahedron? Just look and listen (you'll have a hard time not doing so) to the onslaught of masterfully crafted oversimplifications thrown at you in any campaign, by all the parties. And think about why people—no, not you, of course—succumb to it. It has something to do with the reasons why we lust for the elegantly simple molecule in the shape of a Platonic solid or the beautiful (and preferably, soluble) equation. I want to think about what gives us satisfaction in science when simplicity fails us, as it must, in a real world.

The Loveliest of Prejudices

If one can make any generalization about the human mind, it is that it craves simple answers. The ideology of the simple reigns in science, as it does in politics. So we have the romantic dreams of theoreticians (for example, Dirac) preferring simple and/or beautiful equations. And the moment Richard Smalley, Harold Kroto, Robert Curl and their coworkers intuited that the C_{60} peak in their laser-ablated carbon mass spectrum came from a molecule that should grace the flag of Brazil, I believed it. It could not be otherwise. And they were right.

Simplicity, symmetry, and order ride a straight ray into our souls. I wonder why? Perhaps (this is far out) we have evolved a psychobiological predilection for the qualities of the world that rationalize our existence as locally contraentropic creatures that build molecules and poems. And I am a little unfair to the creative force implicit in the psychological imperative for the simple. The cult of mathematical simplicity as beauty is a reaching for essences that parallels the compact truth-telling of poetry. This is what Dalton, Dirac, and Einstein aspired to. And this perspective has led to "the majesty, subtlety, and grace of science, and her deepest insights and discoveries," as Michael Fisher so aptly put it.

But what if the world is determined by us, by scientific us, to be complex, unsymmetrical, and moderately chaotic? How do we find satisfaction, and I do mean psychological satisfaction, in such a world?

Narrative

I think the answer is simple (I'm smiling). We construct with ease an aesthetic of the complicated, by adumbrating reasons and causes. We do so by structuring a narrative to make up for the lack of simplicity. And then we delight in the telling of the story. Nearly every seminar I go to brings evidence of this joy of storytelling.

I suggest that narrative becomes the substitute for soaring simplicity in the operative aesthetic structures of chemists, and I think it's the same even for the most hard-core reductionist physicist. Continuing the story is the motive force for experimentation and the weaving of theories.

Three Short Stories of the Real World

By way of example, here are three tales of chemical discovery:

INSECTS ARE THE GREATEST CHEMISTS

They use pretty simple chemicals in communication, mating, defense and predation. In 1966 R. S. Berger identified the main sex pheromone of the cabbage looper moth, *Trichoplusia ni* (*Noctuidae*), shown in Figure 8-1 in its caterpillar stage, as "(Z)-7-dodecenyl acetate," a pretty simple molecule related to some fatty acids in all living things. This pheromone is also illustrated in Figure 8-1.

Those were the halcyon days of early pheromone chemistry; everyone was happy with one molecule (as they were with one gene for each trait). Thirteen years later L. B. Bjostad et al. identified a second component, important especially in close-range courtship behavior. Then the same group began to think through the biosynthetic relations between these two components and other molecules observed in the pheromone gland. Obviously, enzymes that do various transformations— shorten molecules, remove hydrogens, add various atoms, all the wondrous machinery of the living—are at work. I show below a complex graph from one of their papers (Figure 8-2), indicating the biochemical relations between the various kinds of fatty acids in the moth.[1] Here's the story, a biochemical story that moved Bjostad et al., which they confide to us. A blend of six components, *suggested by their analysis*, elicited complete courting flights against a stiff breeze in a wind tunnel. Would a human master perfumer be surprised?

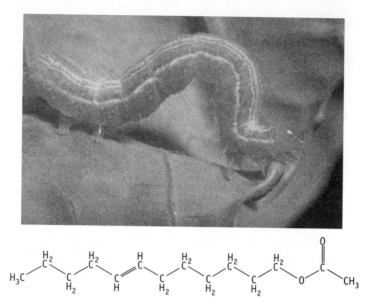

Figure 8-1 The cabbage looper moth in its caterpillar stage, and the first identified component of its sex pheromone, wafted by the female moth. Any resemblance?

The story is told with sufficient verve in the Bjostad, Linn, Du, and Roelofs paper that even I, an outsider to the field, am pulled in by it. More than just an analysis of pheromone glands, the biochemical relations are clever. I am intrigued by their tale and begin to think of its sequel—how do the females evolve that blend? How do the males evolve the receptors to it? Thomas C. Baker and his coworkers at Iowa State University have actually located separate compartments for the six components (and one so-called antagonist, a molecule that acts to negate the physiological reaction to the pheromone) near where the male antenna input is first processed.[2] Extending the story is life-enhancing. And not just in the *Thousand and One Nights*.

BLOOD RED

Hemoglobin is the stumbling block to the simpleminded; there is a story in every turn in this best known of proteins. I will pick one of the oldest, that of the cooperativity of oxygen uptake by this molecule. Hemoglobin has four subunits, two consisting of 146 amino acids, two of 141, each cradling a porphyrin where an oxygen molecule binds to an iron atom. Oxygenation of one subunit makes the subsequent oxygenation of another easier. An important early phenomenological theory accounted for the kinetics. But how does it happen on the molecular level, over a distance of 25–30 Ångströms between iron atoms?

Max Perutz, whose perseverance and talent first gave us the structure of hemoglobin, also built a bold theory of the cooperativity. It begins with an iron atom on one

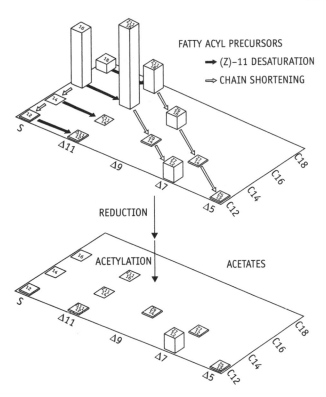

FATTY ACYL PRECURSORS

→ (Z)–11 DESATURATION

⇒ CHAIN SHORTENING

Figure 8-2 Biosynthetic clues to pheromone mixture. One axis is the length of the carbon chain, the other specifies the position of the double bond in the chain. This drawing is reproduced from Bjostad et al., Identification of New Sex Pheromone Components in *Trichoplusia ni*, Predicted from Biosynthetic Processes. *Journal of Chemical Ecology* 10 (1984): 1309.

subunit moving into the heme plane upon oxygenation, pulling a proximal histidine with it. Movement of the histidine shifts a helix, it is suggested; eventually a geometry change at the subunit interface ensues, where a salt bridge between subunits is broken. A net conformational change in the protein occurs, influencing the binding of the next oxygen. Is this a Rube Goldberg (Heath Robinson in England) machine? Rube Goldberg studied some chemistry at Berkeley—maybe he learned something about reaction mechanisms. It is a mechanism, a story well told, remarkably convincing. And permitting elaboration, as work by Martin Karplus and his collaborators shows.[3]

THE ROAD TO CARIPORIDE[4]

Once upon a time (1986), in a pharmaceutical company (Hoechst, now Aventis), the chemists Hans-Jochen Lang and Heinrich Englert became interested in the sodium (Na$^+$)–hydrogen (H$^+$) exchange (therefore called NHE) system, a fine biochemical machine for moving about protons and sodium ions, and thereby regulating cellular acidity. NHE had been first described in 1976 by Swiss physiologist Heini Murer as an ion-transport system in the proximal tubule of the kidney. There were many speculations about the role of this device, present in virtually every type of mammalian cell. For instance, might NHE affect the pathophysiology of brain edema caused by stroke?

Pharmacologists and chemists started looking for NHE inhibitors. As is often the case in drug development, the problem is not so much the chemical compounds to test, for chemists have certainly learned the lesson of Genesis, that we have been put on this earth to create. No, the problem is so often the assay. In the case at hand, a promising one, using renal membrane vesicles, proved insensitive.

At the same time Hoechst pharmacologist Wolfgang Scholz was working on a completely different system, the use of ion transport in red blood cells as an assay for the identification of diuretics. One day he was asked by a cardiovascular research group to test the red blood cells of rabbits on a high-cholesterol diet for possible changes in their ion-transport mechanisms. Remarkably, whereas NHE activity is quite low in red blood cells under normal conditions, there was an approximately tenfold increase caused by the special cholesterol-rich diet.

Whatever the reason for the original experiment (rabbits emulating American junk-food consumers?), the Hoechst scientists saw an opportunity—these red blood cells provided an exquisitely sensitive NHE assay, 1,000 times as sensitive as the kidney membrane vesicles. There was now momentum for synthetic chemists to ply their art. New classes of compounds were tried. One pharmacologist was reading a paper in the Russian literature, on a totally different subject, when he came across the statement that a sodium ion was roughly of the same size as a guanidinium group. Now that turns out to be somewhat farfetched, but no matter, it gave impetus to the synthesis (and testing with the new assay) of a variety of guanidine derivatives. Some of these compounds, the benzoyl guanidines, turned out to be potent and specific enough to test them for reduction of brain edema. The results were quite disappointing. Thus, in late 1988, Lang, Englert and Scholz had in hand a new class of ion-transport inhibitors. The only problem was that there were no known clinical indications for them! It was then decided to test one of the best compounds in a broader range of pharmacological models. One of them was the isolated working rat heart in the lab of the pharmacologist Wolfgang Linz. When a benzoyl guanidine code-named HOE 694 was tested in this model, Linz was amazed to find it was about the most protective compound in cases of cardiac ischemia/reperfusion (blood vessel constriction and blood resupply) that he had ever seen.

At this point molecular biology kicks in. In 1989 the group of Jacques Pouyssegure in Nice cloned the NHE gene. In the following years, several subtypes of NHE were identified. A collaboration between the Hoechst team and Pouyssegure's group soon determined that NHE subtype 1 is not only ubiquitous but also the predominant subtype in the heart and blood cells. It was found that HOE 694 (see Figure 8-3) and most of the related benzoyl guanidine compounds were quite selective inhibitors of NHE subtype 1. The predominant subtype in the proximal tubule of the kidney was NHE-3. Compounds like HOE 694 were about 1,000 times more effective on

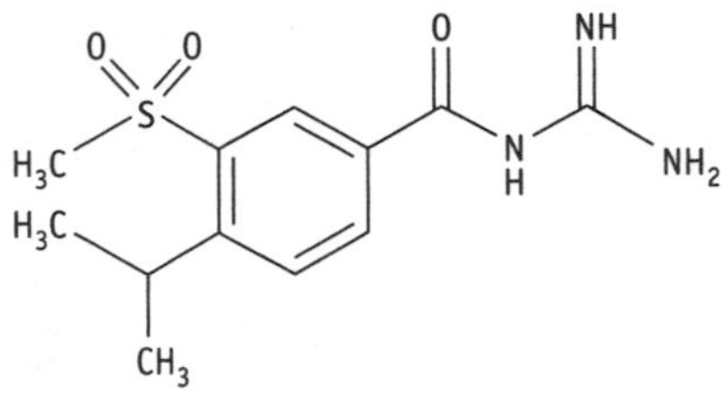

Figure 8-3 The chemical structure of cariporide.

NHE-1 than on NHE-3. So, finally, it was understood why the red-blood-cell assay had worked and the renal membranes had not!

All the laws that characterize the infinity of failures facing human beings apply to pharmaceutical research as well. Within weeks it was revealed that HOE 694 formed a metabolite that concentrated in the rat kidneys and precipitated in the tubular system, where it caused obstruction and inflammation. A strategy to construct compounds metabolized in a different way led to a new compound, HOE 642, synthesized by chemist Andreas Weichert. Now, HOE 642 (generic cariporide) has reached a late stage of clinical development and is the subject of ongoing studies.

Real Tales

I have recounted three tales of discovery. They start simple, yet in each, the ornery complexity of the real is parlayed by the protagonist chemists into a delightful, deeper story.

Everywhere one looks in science, there are stories (see Figure 8-4). I could have recounted the grand ones, of the inflationary universe, of evolution, of continental drift, of Fermat's theorem. I could have told about smaller ones, no less thrilling—the quest for octanitrocubane, the European duplication of Chinese porcelain or the discovery of sulfa drugs. I could have recounted Primo Levi's stories, of a "solitary chemistry, unarmed and on foot, at the measure of man." All of these stories have the hallmarks that literary theorists have seen in narratives, small and grand:

- Temporality: a peaceful beginning, a disequilibrated, tense middle, and a resolution that often sets the world upside down.
- Causation: essential in science, the most deterministic of narrative genres. Everything must have reason, or why would you tell it to your peers?
- Human interest: Reflect—how much more interesting are lectures than articles? Our microsociety's ossified strictures on what should go into a paradigmatic article are relaxed in seminars. One tells a story, and the audience drinks it up, for it sees the why and wherefore. And the speaker naturally tells a heroic tale of blind alleys, serendipity, obstacles overcome, and all-conquering logic . Who needs a samurai epic when I can hear Sam Danishefsky or K. C. Nicolau struggling with the synthesis of calicheamycin?

Figure 8-4 In his 1921 painting, "Tale à la Hoffmann," Paul Klee constructs witty and intense stories out of representational elements and color fields. Reprinted by permission of the Metropolitan Museum of Art, New York. © 2011 Artists Rights Society (ARS), New York.

Not Fear, but a Bond

At times, when I've spoken of narrative as a motive force to scientists, I've encountered a certain queasiness. Could it be that if we admit we tell a story of our research, that we get uncomfortably close to "just so" stories, inventions, fiction? Or, God forbid, that we should render support to relativists, that nefarious social construction of a science gang?

Relax. What we study is real. Yet we live in a mansion furnished with real things and an infinity of mirrors. Modern science is a successful social invention for acquiring not truth, but reliable knowledge (to borrow a phrase from John M. Ziman). An essential part of the structure of science is a built-in alternation of flights of wild theoretical and narrative fancy with experimental probing of some underlying reality. The fancy is not unfettered. In the pursuit of the art, craft, science and business of chemistry, there are numerous checks with reality. To be sure, each is individually deconstructible, but their totality shapes a pretty reliable network of knowledge.

But this in no way precludes tall, fancy and mythical stories that fit into absolutely every category of folktale you have, for it is human beings that seek reliable knowledge. So we clothe the oral and written reports of our curious exploration with the fabric of narrative. Narrative is absolutely indestructible; it looms just under the surface in the driest chemical article. And I am so happy that I am privy to the codes, so that I may see the myth (and the approach to reliable knowledge) underneath.

John Polanyi has recently described the close relationship between science and storytelling:

> *Scientia is knowledge. It is only in the popular mind that it is equated with facts. This is, of course, flattering, since facts are incontrovertible. But it is also demeaning, since facts are meaningless. They contain no narrative. Science, by contrast, is story-telling. That is evident in the way we use our primary scientific instrument, the eye. The eye searches for shapes. It searches for a beginning, a middle, and an end.*[5]

The power of stories may indeed exceed that of facts. As Walter Benjamin has written:

> *The value of information does not survive the moment when it was new. It lives only at that moment; it has to surrender to it completely and explain itself to it without losing any time. A story is different. It does not expend itself. It preserves and concentrates its strength and is capable of releasing it even after a long time.*[6]

In telling the story of scientific discovery, we form a praiseworthy bond with literature and myth, all the other ways that human beings have of telling stories. Yes, there are times when the story has to be told simply, the fire engine sent the shortest route to the fire. But a world without stories is fundamentally inhuman. It is a world where nothing is imagined. Could a chemist be creative in such a world?

Acknowledgments

I am grateful to Wolfgang Scholz and Hans-Jochen Lang for telling me the story of cariporide, and Wendell Roelofs for that of the *T. ni* pheromone.

Notes

1. J. C. Bjostad, C. E.T. Linn, J.-W. Du and W. L. Roelofs, Identification of New Sex Pheromone Components in *Trichoplusia ni*, Predicted from Biosynthetic Processes. *Journal of Chemical Ecology* 10 (1984): 1309.
2. J. L. Todd and T. C. Baker, The Cutting Edge of Insect Olfaction. *American Entomologist* 43 (1997):174.

3. W. A. Eaton, E. R. Henry, J. Hofrichter and A. Mozzarelli, Is Cooperative Oxygen Binding by Hemoglobin Really Understood? *Nature Structural Biology* 6 (1999): 351.
4. M. Karmazyn, The Role of the Myocardial Sodium-Hydrogen Exchanger in Mediating Ischemic and Reperfusion Injury: From Amiloride to Cariporide. *Annals of the New York Academy of Sciences* 874 (1999): 326.
5. J. Polanyi, Science, Scientists and Society. *Queen's Quarterly* 107 (2000): 31.
6. W. Benjamin, The Storyteller: Reflections on the Works of Nikolai Leskov. In *Illuminations*, trans. Harry Zohn (Boston: Harcourt, Brace & World, 1968).

9

Learning from Molecules in Distress

ROALD HOFFMANN AND HENNING HOPF

From the time we first got an inkling of the geometries and metrics of molecules, the literature of organic chemistry has contained characterizations of molecules as unstable, strained, distorted, sterically hindered, bent and battered.[1] Such molecules are hardly seen as dull; on the contrary, they are perceived as worthwhile synthetic goals, and their synthesis, or evidence of their fleeting existence, acclaimed.

What is going on here? Why this obsession with abnormal molecules? Is this molecular science sadistic at its core?

Let's approach these questions, first describing what is normal for molecules, so we can define the deviance chemists perceive. After a digression into the anthropomorphic language chemists generally use, and the psychology of creation in science, we will turn to the underlying, more serious concern: "What is the value of contemplating (or creating) deviance within science?"

The Denumerable, Flexible Chemical Universe

As many as 366,319 different eicosanes $(C_{20}H_{42})$ are conceivable, not counting optical isomers. And an enumeration of the components of a reasonably constrained universe of all compounds with up to 11 C, N, O, F atoms comes to >26 million compounds.[2] An important feature of the chemical universe is that the tree of possible structures is denumerable. At the same time, the playground of chemical structures is subject to systematic elaboration, through the decoration of an underlying skeleton by functional groups of some stability. Very quickly a multitude turns into a universe. Of structure, and of function.

Thinking of these molecules as fixed, rigid structures is natural—don't they look like olive and toothpick assemblages, prettied up by computer rendering? And one can certainly get a long way in organic chemistry in the classical, mechanical mode. But the atoms in a molecule move continually, deviating, oscillating, as if held by springs, around an average position. The honey-comb structure of the benzene ring (a molecular tile, seemingly ever so flat and rigid as the one on your bathroom floor) has become an icon of chemistry just as the angled water molecule. Yet that tile is not rigid, it moves—and one can *see* the deformations/deviations by looking at its vibrational (what a telling name!) spectrum.

Recognizing the Abnormal

Chemistry is more than graph theory; it is graph theory with a metric (bond lengths, bond angles). From the time structural theory was established, and progressing into the 20th century, normal behavior—tetrahedral four-coordinate carbon, the coplanarity of the six atoms in ethylene, the planar benzene hexagon—was established. Bonding theories consonant with that normality—Lewis structures, valence bond pictures—gained currency. And measures of the cost of departing from the norm were obtained, both experimentally (force constants for those vibrations) and later, as calculations became reliable, theoretically. From that knowledge, mainly coming into our hands in the second half of the 20th century, derives our perception of the "normal" molecule. And, by contrast, an intuition for what is unusual.

The isomers of benzene (**1**) are a case in point. Of the $(CH)_6$ graphs, Dewar benzene (**2**),[3] prismane (**3**),[4] benzvalene (**4**),[5] as well as bicyclopropenyl (**5**)[6] were perceived as "makeable," or "not too unstable" (see Figure 9-1). And, in remarkable synthetic achievements, they *were* made.

Other isomers were also recognized, no assistance needed from computations, as just plain impossible, as **6** (Figure 9-2), a "realization" of the Claus formula of benzene.[7] And still other $(CH)_6$ isomers, such as the bis-carbene **7** (or any molecule derived by breaking a single or double bond in the benzene isomers **1–5**) are admitted by chemists as possible metastable species.[8]

The impediments to stability can be quantified energetically. A preliminary note is in order here: in chemistry, especially organic chemistry, real stability is relatively unimportant, and metastability (kinetic persistence) more than suffices. To put it

1	**2**	**3**	**4**	**5**

Figure 9-1 The isolable $(CH)_6$ isomers.

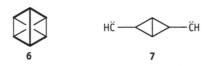

6	**7**

Figure 9-2 An "impossible" and a highly reactive $(CH)_6$ isomer.

one way, every organic molecule in our bodies is thermodynamically unstable in the presence of oxygen, every one can (and will!) be oxidized to water and carbon dioxide. But we burn only figuratively, with passion. The barriers to many exothermic chemical reactions are high—you can read this paper without fear that it will start burning in your hands—and this ensures persistence under ambient terrestrial conditions for the time a slow chemist (or life) requires.

Focusing on thermodynamics, one could compare, for instance, the energetics of hydrogenation of cyclopropane and cyclohexane and find the former is more exothermic by 27 kcal/mol.[9] Or we can estimate departures from the normal by other measures: So one can look at the elongation of the linking single bonds in the anthracene photodimer **8**,[10] or the drift of electrons away from the methyl groups of 4,5-dimethylphenanthrene (**9**),[11] gauged by the NMR chemical shifts, the bending away from each other of the phenyl groups of 1,8-diphenylnaphthalene (**10**),[12] the boatlike deformation of the benzene ring of a [*n*]paracyclophane (**11**).[13] Or the eponymic structural essence of a helicene (**12**, [8]helicene,[14] Figure 9-3) and the [1.1.1]propellane (**13**) with the four C-C bonds of a tetrahedral carbon atom folded into one hemisphere, "looking" prohibitively strained, yet a stable compound.[15]

The reader will have noticed that in our sampling of stressed molecules we have focused primarily on hydrocarbons. We could have chosen problematic heterocycles, or reactive molecules such as vinyl alcohol. Without heteroatom reactivity there would be no life, nor would many (if any) of the exemplary hydrocarbons we cite have been made. Chemical reactivity is set by the differences in functional groups. And the vast majority of these contain N, O, S, and in their interactions and peculiarities set as many obstacles to persistence as the hydrocarbons we cite. But we'd like to start somewhere in a journey through "unhappy" molecules, and the archetypal hydrocarbons are a good place...

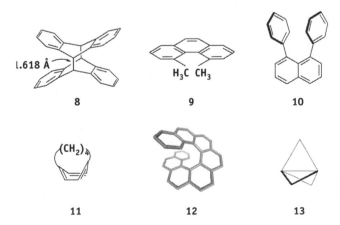

Figure 9-3 A (small) selection of distorted hydrocarbons.

So some molecules emerge as bent and battered. But why should they be of interest, and why turn to such descriptors in their characterization? Is it telling us something about the molecules, or about us? Let's talk first about the language.

Anthropomorphisms (and Metaphors) Are OK

The colloquial and anthropomorphic nature of the descriptors used for the molecules of interest—strained, hindered, battered, unstable—is revealing. And we suspect that such language makes some of our colleagues uncomfortable. It shouldn't. Scientists think words don't matter, that equations, formulas, spectra do. But the facts are mute; without words no sense could be made of this world. And, as Wittgenstein said: *"Die Grenzen meiner Sprache bedeuten die Grenzen meiner Welt."*[16] Words, first of all, are friends; they humanize the inanimate world, form a liaison, a bond with a human being. Words mislead much less than they encourage, for it is just through their anthropomorphism that they provide a rationale for the often tedious labor of chemistry.

Words—rather than physical formulae—are also very well suited to describe the dynamic aspects of chemistry: contrast "backside attack" *vs.* "calculated trajectory." Or consider the phrase "an eclipsed conformation" —what a clear and convincing description of a certain steric situation! The metaphor is astronomical; were we to give the Cartesian coordinates of the respective atoms in that eclipsed conformation, this would be "more precise" but perhaps even tell us less, especially since the exact location of atoms is often not needed in chemical reasoning.

If you haven't thought through the science underneath plain language, words can cause confusion. But if you understand, then colloquial, anthropomorphic, colorful expression makes inanimate matter spring to life.

Reasons for Making Molecules in Distress

Back to our chemical universe, replete with stable and metastable molecules. From the 19th century on, the synthesis of molecules that deviated from the norm took a special place in the imagination of chemists. There are many examples to point to, whether "Bredt rule violations,"[17] the classical double bond rule, or whether a stereotypical organometallic ligand CO could bond to a transition metal through its oxygen. Once theoreticians could contribute more or less reliably, they took to this game with a vengeance—one of the authors' strategy for stabilizing square planar carbon is a good example, having generated a veritable menagerie of (largely) hypothetical molecules.[18]

But why do this? In the first instance the reasons are psychological—the molecules are there, they are perceived as intriguing to weird, so people want to make them. So it makes sense to inquire the way the psyche enters into creation.

Psychological Reasons

Psychology does not find a comfortable place in the ritualized, ossified format of a typical scientific article. But, as one of us has repeatedly argued,[19] the subconscious forces in our psyche are the motive force. And if they aren't entirely savory, they are part of the beauty of human creation. Do we have to list all the ways angelic science and art has been and is made by men and women who are far from angels?

This is not an excuse for being unethical—creation demands that we consider its consequences—it is just an awareness of the reality of making the new in art and science.

One reason for synthesizing some pretty unhappy molecules is simply the desire to do what has not been done before. And to be praised for it. Ideas and actions are our stock in trade—a curious thing in a way, since science is after universals. If $E = mc^2$ is true and benzvalene persists in air, does it matter in the long run who came up with the equation or made the molecule? Oh, it does. Scientists are driven as much by emotions as by reason—otherwise whence the overpowering wish to be the first? To be quoted? To be the most often quoted? Science is done by scientists ... not machines ... and scientists, as human beings, crave recognition[20] for their ideas and the fruit of their handiwork.

All of us of a certain age remember the inner front and back covers of the Cram and Hammond textbook of organic chemistry, with its drawings of molecules made and not made at the time of writing (Figure 9-4).[21]

And Hilbert's theorems have played a similar, long-lasting role in mathematics, as challenges.[22] It is in the nature of human beings to try to do what has not yet been done. One of the authors (R. H.) resisted R. B. Woodward putting in the original orbital symmetry paper, referring to exceptions, the phrase "There are none!"[23] R. H. was wrong; the phrase was a creative provocation.

Another motive force, always at work in scientists, is curiosity. With no thought of reward, no reaching after putative praise. Are there space-filling networks of carbon other than graphite, in which every atom is trigonal (three bonds going off at 120°)?[24] If CO and N_2 are common or known organometallic ligands, why not BF?[25] Can one make C-C bonds really short?[26]

It's just so much fun to explore that chemical playground, asking "What if?" or "Why not?" questions. One can feel guilty, for the time spent in following one's curiosity does not deal with that other aim of science—melioration of the human condition. Through technology we can change and have changed the world. For the better, by and large. But more needs to be done for humanity; why play games?

There are games and there are games. We are *homo ludens*;[27] to outlaw games would probably banish creation altogether. Games pleasure people, and games advance chemistry. Time and time again, we see a use materializing in molecules made for no great purpose; consider also that the world of useful natural products did not evolve for medicinal chemists.

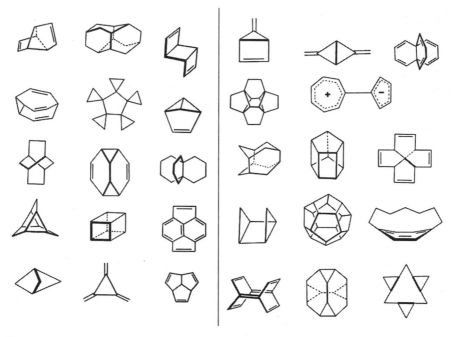

Figure 9-4 Cram's and Hammond's 1970 challenge to hydrocarbon chemists.

The reason for playing these experimental and theoretical games is more than curiosity. We probe the limits, so as to learn. As Jerry Berson says,

> *Something drives us to go deeper and push the limits. That is the expectation— or at least the hope—that something new will emerge when we explore those far regions.*[28]

Elsewhere Berson has described aptly the mix of psychological factors motivating synthesis of unnatural molecules:

> *As chemists find themselves stirred by the mysterious allure of the symmetrical and the beautiful, aesthetic and self-challenging motivations also become apparent in many such instances. Like the impulse driving the heroic geographical expeditions, the urge to explore is often mixed with a sheer will to surmount risk or hazard in order to triumph over adversity. The parallel that comes to mind is George Mallory's famous answer when he was asked why he wanted to climb Mt. Everest: "Because it is there."*[29]

The recognition of psychological factors—of *reaching for praise and priority*, of *curiosity* of how things work and why, of *exploration in the search for understanding*, of *game-playing*, is important. For when we couple these motive forces with the natural anthropomorphism that enters our language and thought as we deal with molecules—they are "strained," they do or do not "want to" react in these ways—we enter into a psychological relationship with an inanimate object. And this is perfectly OK.

In the course of doing normal (good!) science, curiosity, a desire to understand become mixed up with forces in our psyche that power creation. Of course the hagiographic ideology of science tries to deal those forces out of the picture in its creation of an (artificial) Apollonian universe. The strategy for doing so in any activity as imaginative, as so inherently human, as passionate as science, can only be suppression.

A more balanced view of the creative work of science would find a place for the streams Nietzsche called Apollonian and Dionysian.[30]

Molecular Sadism?

Louis-Donatien-Alphonse-François, the Marquis de Sade, is a striking and contentious figure straddling the 18th and 19th centuries. His name labels forever a despicable human trait, of taking satisfaction if not delight in inflicting physical suffering on others.

The characterization of what chemists do to strained molecules as molecular sadism is no more than what the expression aims to be—a cute turn of phrase. As a description it is facile if not puerile, and does not stand up under consideration.

Yet attempts at being funny often do betray underlying tensions. The relationship of human beings and the objects of their creation or contemplation (here compounds/molecules) is never simple. Into it enter our irrepressible tendency to anthropomorphize, and the satisfactions and travails of the creation. While we don't think sadism comes into the equation, it is nevertheless interesting to look at Sade, his writings, and their interpretation, and see if there is something other than mental illness in them.

The eventful story of the Marquis de Sade has been told several times;[31,32] yet the interpretation of his life and writing remains controversial.[33-36] Born in an aristocratic family, Sade grew up to be a libertine. In an age where the privileged could escape the consequences of their actions if their victims were of a lower class, his were excessive. And he had a persistent and vengeful mother-in law, Mme. de Montreuil. A series of imprisonments ensued. In the last of these, in the Bastille and at Charenton, he began to write. Sade's accommodations to the revolution ran afoul first of Jacobin, then of Napoleonic, mores. All in all, the Marquis spent some 30 of his 74 years confined.

The characteristic features of Sade's philosophy, obsessively shaped by his personality, include an opposition to all authority, atheism, sexual freedom (including an unusual recognition, for its time, of female sexuality), and extreme libertarianism.

And, in his actions, and in his imaginative fictions, a penchant for a mix of cruelty and sex. This celebration (perhaps that's not the right word, for much of the repeated fornication in Sade's writings is joyless) of inflicted pain has rightly earned the Marquis the ill fame of lowercase sadism.[37]

Strangely enough, one can even find in Sade's writings passages that seem close to science. But then we read on, and this Heraclitean bent, so close in spirit to the eternal change that underlies chemistry, is put in the service of…rationalizing murder:

> So this is what murder is! A little organized matter disorganized; a few compositional changes, the combination of some molecules disturbed and broken, those molecules tossed into the crucible of Nature, who, re-employing the selfsame materials, will cast them into something else so that in but a day or so they shall reappear in the world again, only guised a little differently; this is what they call murder—truly now, in all seriousness, I ask myself, where is the wrong in murder?[38]

This is Juliette speaking, yes. But it is also Sade. Moreover, this passage is not an exception, but a piece of a repetitive litany of perversity. There is no way that any rational or ethical human being can follow Sade here.

Let Sade, sick as he was, rest. The synthesis of real molecules, even ones that are not "normal," has precious little to do with sadism. The ethical considerations that should accompany bringing a new compound into the world preclude that.

Still, there are things in Sade's philosophy that have intrigued 20th century writers from Roland Barthes to Simone de Beauvoir. For the Marquis, that destroyer of normalcy, was the border violator *par excellence*. He was an intellectual libertine, as much as he was a sexual one—one who believed that everything was possible for the human spirit. And that the human condition was in fact one of ringing all the changes on creation.

There is another connection of science and Sade. Pierre Laszlo has remarked that

> …the monstrous is a major theme of Sade. And the Promethean push by scientists for the limits of their science, whatever field they are engaged in, and notwithstanding Mary Shelley, draws from the Early Modern mentality. Western science is underlined by its sense of wonder. Monsters are and remain an integral part of it. Wondering at monsters marks early science and carries into modern science. Such seeking for the abnormal is blatant in chemistry, in the exploration of unusual natural products and in the drive to make stressed molecules.[39]

In no way descending with Sade into perversity, we think that this makes it less than absurd to contemplate the ideas of the poor Marquis (not sadism) in the context of chemistry.

A Better Metaphor

The purpose of studying unhappy molecules is *not* delight in their squirming under stress. We learn, or try to learn, from the abnormal. Actually, if you want a metaphor, bringing succor might serve as well as torture—the moment one of us looked, really looked, at that poor square planar carbon atom, he and his coauthors were thinking of a strategy to stabilize it. To give it a chance, just a chance, for existence.

So, Will Any Molecule That Can Be Imagined Be Made?

The understanding we have of molecules is everywhere partial and incomplete, as many example from the history of chemistry show: future generations of chemists will look at our DFT-calculated electron distributions in molecules in the same way we look at Kekulé's "sausage structures." And often a theory describes something fortuitously; probing it at the extremes helps one see its limitations.

Repeatedly, the making of molecules that are untypical or abnormal tests our understanding of that fundamental yet fuzzy entity—the chemical bond.[40] And it stretches the limits of the efficiency of our laboratory techniques. An unprotected silicon-silicon double bond, an ethylene substituted by donors and acceptors in a push-pull pattern, these are all wonderful probes of our understanding, of the factors that make normal chemistry such a productive enterprise.

With time there arose a simple way to account for bonding in molecules, providing reasonable guidelines for stability—draw me a Lewis structure and it can be made. There are striking, yet understandable exceptions—small molecules for which you can draw a Lewis structure, but which have so far avoided synthesis. We can think of molecules such as cyclic ozone (**14**), dicarbondioxide (**15**, OCCO), hexaazabenzene (**16**), and hexaprismane (**17**, Figure 9-5).

From each of these potentially transgressive molecules we learn something:

- Cyclic ozone **14** is unstable relative to normal ozone by ~30 kcal/mole (all those lone pairs crammed into a small space). But it has a substantial calculated barrier to falling apart into its open isomer, because that process is a forbidden reaction.[41]
- OCCO **15** is a dimer of carbon monoxide, and that would explain, it might seem, its nonexistence. But hold on, it should be a triplet ground state, for the same reasons that O_2 is.[42] And that may (or may not) make a difference.

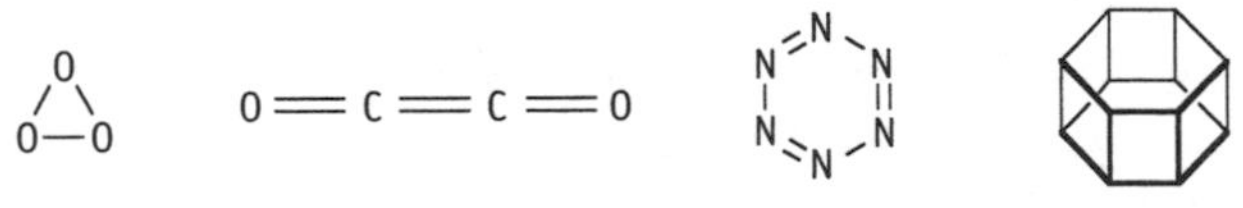

14 **15** **16** **17**

Figure 9-5 Can you make a compound when you can draw its Lewis structure?

- The decomposition of hexaazabenzene, **16**, cyclic N_6, to three N_2 molecules is highly exothermic. And it is an allowed reaction. The computed barrier to fragmentation is tiny.[43]

We note that the first three molecules in our list are perceived as problems mainly from the perspective of our impoverished representations. Lewis structures, otherwise a remarkably effective heuristic paper tool in chemistry, do not describe well the impediments to persistence that face (or protect) these molecules. For that, a quantum mechanical perspective is needed.

Then there is (or isn't) hexaprismane, **17**. The molecule should possess a strong tendency to dissociate into its two halves, two benzenes rings—the tiles from above! —however, we would also expect a substantial kinetic barrier here since the cleavage of a cyclobutane ring in such a cage molecule is a forbidden process. It is our limited synthetic methodology that so far has prevented the preparation of this molecule, not its (presumably) excessive strain.[44]

Let us look at how another field uses extremes, and then return to chemistry.

Why Probe Limits? Philosophy

An interest in extremes characterizes that jewel of contemplation, philosophy. In logic, for instance, paradoxes play a special role—the Cretan who always lies, or Zeno's paradox, for example. Philosophical texts across the discipline consistently take up arcane conundrums at the periphery of their fields, testing in excruciating detail the strength of definitions. And not just in philosophy—the famous Einstein, Podolsky, and Rosen paper probed the limits of the paradigmatic Copenhagen formulation of quantum mechanics.[45]

Philosophers relish Gedanken-experiments that set up constrained or extreme conditions, and use these to clarify a concept. We quote here first two amusing (and influential) examples of this genre from the philosophy of mind; in both chemistry figures.

In an important 1973 paper, Hilary Putnam tells the following story:[46]

> ...we shall suppose that somewhere there is a planet we shall call Twin Earth. Twin Earth is very much like Earth: in fact, people on Twin Earth even speak English...
>
> One of the peculiarities of Twin Earth is that the liquid called "water" is not H_2O, but a different liquid whose chemical formula is very long and complicated. I shall abbreviate this chemical formula simply as XYZ. I shall suppose that XYZ is indistinguishable from water at normal temperatures and pressures.

Putnam's story goes on in some detail. Assuming that Twin Earthlings call XYZ "water" in their English, he then asks whether when an earthling, say Oscar, and his twin on Twin Earth say "water" do they mean the same thing? For Putnam's response, and how he used it to shape a position called "externalism," you will have to read his paper.

Another round in the debate that followed Putnam's paper used more chemistry in a Gedanken-experiment mode, *vide* Donald Davidson's *Swampman*. Davidson goes hiking in a swamp:[47]

Suppose lightning strikes a dead tree in a swamp; I am standing nearby. My body is reduced to its elements, while entirely by coincidence (and out of different molecules) the tree is turned into my physical replica. My replica, The Swampman, moves exactly as I did; according to its nature it departs the swamp, encounters and seems to recognize my friends. It moves into my house and seems to write articles on radical interpretation. No one can tell the difference.

Chemical practice might have something to say about the plausibility of these arguments, but we are not going to go there. The scenarios of *Twin Earth* and *Swampman* definitely help philosophers sharpen their ideas. As distorted molecules help the chemist.

Ethics is replete with constructed dilemmas that illuminate. Here is Philippa Foot's Trolley Problem:[48]

[Supposing that a person] is the driver of a runaway tram which he can only steer from one narrow track on to another; five men are working on one track and one man on the other; anyone on the track he enters will be killed.

Which track should he steer to? The Wikipedia article on *The Trolley Problem* describes a number of ingenious variations on this problem.

A general (and recommended) strategy for discussing the ethics of any action is to set up a range of cases, so to clarify for oneself what the criteria for action might be.

At times, philosophy's concern with strange exceptions makes one want to scream: "You are obsessed with the periphery; think about the center!" But considering extremes is often the most direct way to challenge accepted, yet perhaps not well-thought-through notions. The fringes are a frame; they define the center.

Vexing Nature

In a seminal text of the ideology of science, "Of the Advancement of Learning," Francis Bacon writes in 1605:

For like a man's disposition is never well known till he be crossed, nor Proteus ever changed shapes till he was straitened and held fast; so the passages and

*variations of nature cannot appear so fully in the liberty of nature, as in the
trials and vexations of art.*[49]

By "art" Bacon here means experiment. And he repeats the argument for inter-
rogating nature through experiment in his 1620 *Novum Organum*.

The potential connection to sadism is clear. In a discussion of the Protean metaphor and the tension of "invasive" (destructive?) and "noninvasive" (nondestructive?)
techniques in chemical experimentation, Pierre Laszlo and one of the authors write
elsewhere:

> *Bacon was accused of being the first of a long series of villains to "put nature on
> the rack," a rationalizer of torture in the service of science. Goethe's revulsion at
> Newton's incarcerating passage of light through a slit (and its subsequent analysis
> into the component colors by a prism) is emblematic. As are Donne, Wordsworth
> and Ruskin's impassioned denouncements of science and of the attendant indus-
> trial revolution. The line continues, to some (hardly all) of the environmentalist
> and animal rights critiques of the interventionist nature of science.*[50]

The tickling of a molecule to get it to send us signals from within may be very,
very light. No bonds are made or broken. But the quantum strings must be plucked.
Chemical experiment so often reaches beyond analysis, and thrives on perturbation
and intervention. And if we left things where they were, there would be no chemical
industry. For real change is effected by perturbing equilibrium, by transforming the
natural. In art as well as science.

Extreme Conditions

Another way in which the normal is perturbed in science is by exposing matter to
extreme conditions—high or low pressure or temperature, high magnetic or electric
fields, high levels of radiation, extreme salinity or concentrations of one or another
chemical.

Heating, of course, is at the heart of chemistry. One could not imagine our science without the motive force of first fire, and then its surrogates. Yet extreme heating does away with chemistry; molecules don't have a life on the surface of the sun.
On the other hand, low temperature turns off entropy, so to speak. The temperature
range of existence of that marvel of systems chemistry, terrestrial life, is wider than
we think—we continue to be surprised by extremophiles that flourish in temperatures higher than that of boiling water at 1 atm, and lower than that of freezing water.
The strategies life adopts under extreme conditions are just fascinating, from anti-
freeze proteins, to the lipids of thermophiles, to the repair mechanisms of *Deinococcus
radiodurans*.

Each set of circumstances has interesting consequences for physics, chemistry and life; let's discuss in detail one of these, high pressure, a field which one of the authors has recently entered.[51] High pressures are the norm in the interior of planets. Such pressures can also be attained in controlled fashion in the laboratory, between diamond anvils.

The only imperative under high pressure is "get denser." The average dimensions of a crystal may shrink by a linear factor of 1.7, and the PV term in the enthalpy of any reaction approaches 10eV at pressures of the order of 350GPa (greater than the strength of any bond).

Here are some of the incredible things that happen at high pressure: Everything eventually turns metallic. In the range of pressures accessible in the laboratory people have made Xe metallic. Iodine also becomes metallic under pressure, and the diatomic bonds "dissolve" into square sheets of I atoms. Not yet NaCl, but CsI and BaTe, pretty ionic solids, can be metallized.

If not metallization, there is coordination alchemy under high pressure. Two of our best thermodynamic sinks, CO_2 and N_2, are transformed at high pressure into structures resembling forms of quartz and elemental P or As, respectively—the advantage of multiple bonding gives in to the necessity of compactification. If someone insists on cramming you in, tighter than sardines in a can, then you had better form as many bonds as possible with your neighbors.[51]

Even our cherished notion of close-packing, as obvious to the sellers of oranges in ancient Egypt as to us, has to be modified. *Every* alkali metal and alkaline earth element under pressure has been shown in the last decade to go out of the familiar close-packed hcp or fcc structures into incommensurate and commensurate not-close-packed (but denser) structures.

What's normal depends on the niche, and your perspective—the center of the earth (not to speak of that of Jupiter) is not the world of 1 atm. And squeezing the hell out of molecules is hardly sadistic—it leads to new chemistry, new ideas. Obviously, our understanding of molecules under high pressure probes the state of our understanding of molecules under ambient conditions.

Mastery

It could be that the central role of synthesis in chemistry is related to the desire to make unusual molecules. To make such molecules is, of course, far from the only reason people make compounds, whether it is those occurring in nature, or unnatural ones. Molecules were synthesized in the past to confirm analysis, i.e., to show that the scant clues as to their structure were interpreted correctly. Today they are more often made because they may display a function of interest or because they may provide a starting material for the usual medicinal chemical exploration of increasing activity, decreasing toxicity. They may be made simply because they are Everests of evolutionary complexity in Nature, so why not do it in the lab? Berson has written cogently of all the non-Popperian reasons for making molecules.[52]

All the time, in organic chemistry classes, our apprentices are asked to design syntheses. As we pointed out in a previous section and above, there are many reasons, psychological, intellectual, and aesthetic, for this fundamental chemical activity. But making molecules, natural or unnatural, is the way understanding of nature is demonstrated.

The easy ones are made first, then increasingly more difficult ones. It makes sense that the goalposts be always placed further away—more complex, more asymmetric centers if it is a natural product, less stable if one has in mind an unnatural product. In an experimental science, especially in the science where making it is paramount, the very human desire to understand is quite naturally parlayed into a search for the extreme.

Not that this outlook is without a certain measure of human arrogance. Alain Sevin has put it well:

> The incredible richness and fantasy of Nature is an act of defiance to Man, as if he had to do better in any domain. Flying faster than birds, diving deeper than whales... We are promethean characters in an endless play which is now in its molecular act.[53]

Liberté

We have laid out some of the intellectual motives for preparing "borderline," or even "beyond the pale" molecules. But the urge to reach the extreme goes deeper, to where we came from. It paid to be an extremist—the faster runner, the healthier mother, the better hunter and their off-spring had the better chance to survive, especially when the niche changed. Deviance, the father of existence under extreme conditions, is thus absolutely essential for evolution.

Society needs border-crossing too. For knowledge evolves, just as organisms do. And the economies of the world, and our ability to deal with the crises of our own doing, depend on innovation. Transgressive (yet ethical) research should not only be tolerated, but actively embraced. Support is needed for way-out research projects that apparently do not lead to application. The periphery is the zone where innovation occurs. And it is the rim from which we understand the center better: normality in all its importance, but also in its limitations.

It is said that the Marquis de Sade shouting from his cell played a small role in inciting the assault on the Bastille. The slogans of the French Revolution—Liberté, Egalité, Fraternité—actually provide a better signpost than Sade's vision. They carry in them the tension of human creation: the normal, and common good, contending with individual freedom. The desire to make the molecule that violates the norm is part of that human struggle.

Acknowledgments

The authors thank for comments, references, and discussion Jerry Berson, Jennifer Cleland, Sylvie Coyaud, Pierre Laszlo, Errol Lewars, Joel Liebman, Georgios Markopoulos, Petra Mischnick, and Michael Weisberg.

Notes

1. The last phrase comes from D. J. Cram and J. M. Cram, Cyclophane Chemistry: Bent and Battered Benzene Rings. *Acc. Chem. Res.* 4 (1971): 204. A recent paper describes molecules suffering from "molecular frustration": H. Dong, S. E. Paramonov, L. Aulisa, E. L. Bakota, and J. D. Hartgerink, Self-Assembly of Multidomain Peptides: Balancing Molecular Frustration Controls Conformation and Nanostructure. *J. Am. Chem. Soc.* 129 (2007): 12468.
2. (a) T. Fink and J.-L. Reymond, Virtual Exploration of the Chemical Universe up to 11 Atoms of C, N, O, F: Assembly of 26.4 Million Structures (110.9 Million Stereoisomers) and Analysis for New Ring Systems, Stereochemistry, Physicochemical Properties, Compound Classes, and Drug Discovery. *J. Chem. Inf. Model.* 47 (2007): 342; (b) T. Fink, H. Bruggesser, and J.-L. Reymond, Virtual Exploration of the Small-Molecule Chemical Universe Below 160 Daltons. *Angew. Chem.* 117 (2005): 1528; *Angew. Chem. Int. Ed. Engl.* 44 (2005): 1504.
3. E. E. van Tamelen and S. P. Pappas, Bicyclo [2.2.0]hexa-2,5-diene. *J. Am. Chem. Soc.* 85 (1963): 3297.
4. T. J. Katz and N. Acton, Synthesis of Prismane. *J. Am. Chem. Soc.* 95 (1973): 2738.
5. T. J. Katz, E. J. Wang, and N. Acton, Benzvalene Synthesis. *J. Am. Chem. Soc.* 93 (1971): 3782.
6. W. E. Billups and M. M. Haley, Bicycloprop-2-enyl (C_6H_6). *Angew. Chem.* 101 (1989): 1735; *Angew. Chem. Int. Ed. Engl.* 28 (1989): 1711.
7. A. Claus, *Theoretische Betrachtungen und deren Anwendungen zur Systematik der organischen Chemie* (Freiburg: Freiburg, 1867), p. 207; *cf.* A. Kekulé, Ueber Einige Condensationsproducte des Aldehyds. *Liebigs Ann. Chem.* 162 (1872): 77.
8. The molecules we discuss have at least normal Lewis or Kekulé structures (more on this below); once we leave these, we encounter a fascinating variety of still other structures, not even within the imagination of the 19th century chemist: J. A. Berson, A New Class of Non-Kekulé Molecules with Tunable Singlet–Triplet Energy Spacings. *Acc. Chem. Res.* 30 (1997): 238; H. Hopf, *Classics in Hydrocarbon Chemistry* (Weinheim: Wiley-VCH, 2000), chapter 16.2, p. 492.
9. The heats of hydrogenation were calculated from the different experimentally determined heats of formation of the respective hydrocarbons.
10. Anthracene photodimer 8: K. A. Abboud, S. H. Simonsen, and R. M. Roberts, Redetermination of the Structure of bi(9,10-dihydro-9,10-anthracenediyl) at 198 K. *Acta Cryst.* C46 (1990): 2494. The marked C-C-bond in 8 is not the longest known carbon-carbon single bond in a hydrocarbon; for the presumably longest (1.781 Å) so far observed single bond see G. Fritz, S. Wartanessian, E. Matern, W. Hönle, and H. G. v. Schnering, Bildung Siliciumorganischer Verbindungen. 85 [1]. Bildung, Reaktionen und Struktur des 1,1,3,3-Tetramethyl-2,4-bis(trimethylsilyl)-1,3-disilabicyclo[1, 1, 0]butans. *Z. anorg. allg. Chem.* 475 (1981): 87.
11. J. B. Stothers, C. T. Tan, and N. K. Wilson, ^{13}C N.M.R. Studies of Some Phenanthrene and Fluorene Derivatives. *Org. Magn. Res.* 9 (1977): 408.
12. X-ray structural analysis of 1,8-diphenylnaphthalene (10): R. Tsuji, K. Komatsu, K. Takeuchi, M. Shiro, S. Cohen, and M. Rabinovitz, Structural Study on 1-phenyl- and 1-(2-naphthyl)-8-tropylionaphthalene Hexafluoroantimonates. *J. Phys. Org. Chem.* 6 (1993): 435 and references cited therein.
13. Most recent review on [n]cyclophanes: H. Hopf in *Beyond van 't Hoff and Le Bel*, ed. H. Dodziuk (Weinheim: Wiley-VCH, 2008).

14. Reviews: (a) R. H. Martin, *The Helicenes. Angew. Chem.* 86 (1974): 727; *Angew. Chem. Int. Ed. Engl.* 13 (1974): 649; (b) H. Hopf, *Classics in Hydrocarbon Chemistry* (Weinheim: Wiley-VCH, 2000), chapter 12.1, p. 323.

15. K. B. Wiberg and F. H. Walker, [1.1.1]Propellane. *J. Am. Chem. Soc.* 104 (1982): 5239; *cf.* K. Semmler, G. Szeimies, and J. Belzner, Tetracyclo[5.1.0.0^{1,6}.0^{2,7}]octane, a [1.1.1]Propellane Derivative, and a New Route to the Parent Hydrocarbon. *J. Am. Chem. Soc.* 107 (1985): 6410.

16. L. Wittgenstein, *Tractatus Logico-Philosophicus* (Berlin: Edition Suhrkamp, 1963), Satz 5.6, p. 89. See also in this context, the Sapir-Whorf hypothesis, that a worldview is determined by the structure of one's language. The idea goes back to Humboldt: W. v. Humboldt, *Über die Verschiedenheit des menschlichen Sprachbaues und ihren Einfluss auf die geistige Entwickelung des Menschengeschlechts* (Königl. Academie der Wissenschaften,1836). Benjamin Lee Whorf, who contributed importantly here, was a chemist by education and profession: "Benjamin Lee Whorf: once a Chemist..." R. Hoffmann and Pierre Laszlo, *Interdisciplinary Science Reviews* 26 (2001): 15; in French in *Alliage* 47 (2001): 59.

17. (a) J. Bredt, J. Houben, and P. Levy, Ueber Isomere Dehydrocamphersäuren, Lauronolsäuren und Bihydrolauro-Lactone. *Ber. Dtsch. Chem. Ges.* 35 (1902): 1286; (b) J. Bredt, *Liebigs Ann. Chem.* 437 (1924): 1.

18. R. Hoffmann, R. W. Alder, and C. F. Wilcox, Jr., Planar Tetracoordinate Carbon. *J. Amer. Chem. Soc.* 92 (1970): 4992; K. Sorger and P. v. R. Schleyer, Planar and Inherently Non-Tetrahedral Tetracoordinate Carbon: A Status Report. *J. Mol. Str. Theochem.* 338 (1995): 317. A recent review is by R. Keese, Carbon Flatland: Planar Tetracoordinate Carbon and Fenestranes. *Chem. Rev.* 106 (2006): 4787.

19. R. Hoffmann, *The Same and Not the Same* (New York: Columbia University Press, 1995).

20. This is almost a quote from Carl Djerassi and Roald Hoffmann, *Oxygen* (Weinheim: Wiley-VCH, 2001), p. 109.

21. J. B. Hendrickson, D. J. Cram, and G. S. Hammond, *Organic Chemistry* (New York: McGraw Hill Book Company, 1970). Scheme 4 is a copy of the front pages of the 3rd edition. Left of the vertical line molecules are shown that had been prepared up to the time of publication of the text book. The molecules on the right were unknown at that time; some of the shown compounds have been synthesized since then.

22. D. Hilbert, *Mathematische Probleme* (Vortrag, gehalten auf dem internationalen Mathematiker-Kongreß zu Paris, 1900).

23. R. B. Woodward and R. Hoffmann, The Conservation of Orbital Symmetry. *Angew. Chem.* 81 (1969): 797; *Angew. Chem. Int. Ed. Engl.* 8 (1969): 781.

24. R. Hoffmann, T. Hughbanks, M. Kertesz, and P. H. Bird, A Hypothetical Metallic Allotrope of Carbon. *J. Amer. Chem. Soc.* 105 (1983): 4831.

25. F. M. Bickelhaupt, U. Radius, A. W. Ehlers, R. Hoffmann, and E. J. Baerends, Might BF and BNR$_2$ Be Alternatives to CO? A Theoretical Quest for New Ligands in Organometallic Chemistry. *New J. Chem.* (1998): 1; U. Radius, F. M. Bickelhaupt, A. W. Ehlers, N. Goldberg, and R. Hoffmann, Is CO a Special Ligand in Organometallic Chemistry? Theoretical Investigation of AB, Fe(CO)$_4$AB, and Fe(AB)$_5$ (AB = N$_2$, CO, BF, SiO). *Inorg. Chem.* 37 (1989): 1080.

26. D. Huntley, G. Markopoulos, P. M. Donovan, L. T. Scott, and R. Hoffmann, Squeezing CC Bonds. *Angew. Chem.* 117 (2005): 7721; *Angew. Chem. Int. Ed.* 44 (2005): 7549.

27. J. Huizinga, *Homo Ludens* (Amsterdam: Pantheon, 1939).

28. J. A. Berson, personal communication, Nov. 2007.

29. J. A. Berson, *Chemical Discovery and the Logicians' Program: A Problematic Pairing* (Weinheim: Wiley-VCH, 2003), p. 128. One of the greatest achievements in modern hydrocarbon synthesis, the preparation of the Platonic hydrocarbon dodecahedrane (by Leo Paquette and coworkers), has been termed "the Mt. Everest of hydrocarbon chemistry": *Nachr. Chemie, Labor, Technik* 25 (1977): 59.

30. F. Nietzsche, *Die Geburt der Tragödie aus dem Geiste der Musik* (Leipzig, 1872); F. Nietzsche, *The Birth of Tragedy from the Spirit of Music* (New York: Penguin, 1993).

31. F. du Plessix Gray, *At Home with the Marquis de Sade* (New York: Simon & Schuster, 1999).

32. G. Lely, *The Marquis de Sade*, tr. Alec Brown (New York: Grove Press, 1961).

33. G. Gorer, *The Life and Ideas of the Marquis de Sade*, 2nd ed. (London: Peter Owen Ltd., 1953).
34. S. de Beauvoir, *The Marquis de Sade* (New York: Grove Press, 1953).
35. R. Barthes, *Sade, Fournier, Loyola*, tr. Richard Miller (Baltimore, MD: Johns Hopkins University Press, 1976).
36. Ph. Sollers, *Sade Contre L'Être Suprême* (Paris: Gallimard, 1996).
37. The term sadism, as commonly used today, was introduced by Richard von Krafft-Ebing, who, according to Gorer, "…with a mixture of impropriety and ignorance took de Sade's name for one of the perversions he described and defined Sadism as 'sexual emotion associated with the wish to inflict pain and use violence;' with even greater impertinence he took the name of a living second-rate novelist, Sader-Masoch, to give the name Masochism to 'the desire to be treated harshly, humiliated, and ill-used.'" G. Gorer, loc. cit., p. 191.
38. M. de Sade, *Juliette*, tr. A. Wainhouse (New York: Grove Press, 1968), p. 415.
39. P. Laszlo, personal communication, Nov. 2007; L. Daston and K. Park, *Wonders and the Order of Nature, 1150-1750* (Cambridge, MA: Zone Books, 1998).
40. (a) H. Dodziuk, *Unusual Saturated Hydrocarbons: Interaction Between Theoretical and Synthetic Chemistry, Topics in Stereochemistry*, Vol. 21, eds. E. Eliel and S. H. Wilen (New York: Wiley, 1994), p. 351; (b) A. Greenberg and J. F. Liebman, *Strained Organic Molecules* (New York: Academic Press, 1978); (c) H. Dodziuk, *Modern Conformational Analysis* (Weinheim: VCH, 1995).
41. B. Flemmig, P. T. Wolczanski and R. Hoffmann, Transition Metal Complexes of Cyclic and Open Ozone and Thiozone. *J. Am. Chem. Soc.* 127 (2005): 1278 gives references to the many studies of cyclic ozone.
42. (a) D. M. Birney, J. A. Berson, W. P. Dailey, III, and J. F. Liebman, in *Molecular Structure and Energetics*, eds. J. F. Liebman and A. Greenberg (Weinheim: VCH Publishers, 1988); (b) D. Schröder, C. Heinemann, H. Schwarz, J. N. Harvey, S. Dua, S. J. Blanksby, and J. H. Bowie, Ethylenedione: An Intrinsically Short-Lived Molecule. *Chemistry - A European Journal* 4 (1998): 2550, and references therein.
43. P. Saxe and H. F. Schaefer, III, Cyclic D^{6h} Hexaazabenzene—A Relative Minimum on the Hexaazabenzene Potential Energy Hypersurface? *J. Am. Chem. Soc.* 105 (1983): 1760; (b) M. N. Glukhovtsev and P. v. R. Schleyer, Structures, Bonding and Energies of N$_6$ Isomers. *Chem. Phys. Lett.* 198 (1992): 547; J. Fabian and E. Lewars, Azabenzenes (azines), The Nitrogen Derivatives of Benzene with One to Six N Atoms: Stability, Homodesmotic Stabilization Energy, Electron Distribution, and Magnetic Ring Current; A Computational Study. *Can. J. Chem.* 82 (2004): 50, and references therein.
44. G. Mehta and S. Padma, Synthetic Studies Towards Prismanes: Seco-[6]-prismane. *Tetrahedron* 47 (1991): 7783.
45. A. Einstein, B. Podolsky and N. Rosen, Can Quantum-Mechanical Description of Physical Reality Be Considered Complete? *Physical Review (ser. 2)* 47 (1935): 777.
46. H. Putnam, Meaning and Reference. *Journal of Philosophy* 70 (1973): 699.
47. D. Davidson, Knowing One's Own Mind. *Proceedings and Addresses of the American Philosophical Association* 60 (1986): 441.
48. P. Foot, The Problem of Abortion and the Doctrine of the Double Effect, in *Virtues and Vices* (Oxford: Basil Blackwell, 1978).
49. F. Bacon, *Of the Advancement of Learning*, ed. G. W. Kitchin (London: J. M. Dent, 1915), p. 73.
50. R. Hoffmann and P. Laszlo, Protean. *Angew. Chem.* 113 (2001): 1065; *Angew. Chem. Int. Ed. Engl.* 40 (2001): 1033. Chapter 11 in this book.
51. W. Grochala, R. Hoffmann, J. Feng, and N. W. Ashcroft, The Chemical Imagination at Work in Very Tight Places. *Angew. Chem.* 119 (2007); *Angew. Chem. Int. Ed. Engl.* 46 (2007): 3620.
52. J. A. Berson, *Chemical Discovery and the Logicians' Program: A Problematic Pairing* (Weinheim: Wiley-VCH, 2003), pp. 128–130.
53. A. Sevin, personal communication, cited in R. Hoffmann, How Should Chemists Think? *Scientific American* February (1993): 66. Chapter 12 in this book.

10

Why Think Up New Molecules?

ROALD HOFFMANN

Some theoreticians in chemistry, myself included, like to think about molecules that do not (yet) exist. I use the simple word "think" advisedly, for the design need not use fancy-schmancy computer-intensive "first-principles" calculations. We conjure up the chemical future in so many ways—simple model building, qualitative thinking, and from ever-more-reliable quantum chemical calculations.[1] Even in dreams, as Henning Hopf reminded me, Kekulé's ouroboric benzene in mind.

But why do we try to imagine new molecules? Aren't there enough molecules already on earth, be they natural or synthetic? A potpourri of reasons follows.

A Stake in Creation

Synthesis, the making of molecules, is at the heart of chemistry—the art, craft, business, and science of substances (molecules at the microscopic level) and their transformations. Of course you need to know what substances are, so analysis is a parallel, lively enterprise. As is figuring out why molecules have the colors or other properties they do, why they react in certain ways and not others.

Chemists make the objects of their own contemplation. And, of course, study the beautiful evolved world around and within them. By being as much (if not more) in the work of creation as discovery, chemistry is close to art. And lest we get too puffed up on that, creation brings chemistry also close to engineering (which certainly can have artistic elements in it!).

I love explaining. But as a theoretician, I also want to take part in the work of creation. I can do so by thinking up interesting molecules not yet made. Maybe, just maybe, an experimentalist will try to make the molecule. Actually, given human nature, a hypothetical molecule will be made more expeditiously if it is thought up by the synthesizer, rather than by me.

So What's Interesting?

Since chemistry is a semi-infinite macrocosm of structure, there are many interesting molecules waiting to be made. And still many more that might as well wait a while longer. Few of the 355 dodecanes ($C_{12}H_{26}$) are extant. For good reasons—new principles, new properties are most unlikely to be found among them.

So it's not just predicting any molecule that does not exist, it's predicting one that's in some way "interesting." That loose word has both cognitive and emotional sides to it, and is definitely subjective. Nevertheless, I find "interesting" works very well, in evoking the psychological mix that makes the intelligent graduate student's mind hop to. Some examples follow.

M. M. Balakrishnarajan in my group thought up a kind of 3-dimensional analogue to one of the very best oxidation/reduction couples in organic chemistry—quinone/hydroquinone.[2] Polyhedral boron cages, such as the octahedron shown below (Figure 10-1), can do it twice over, accepting two and four electrons (the intermediate stage of oxidation/reduction is shown) with correlated changes in geometry. The molecule will breathe as it sops up electrons.

To move away from my work, wonderful predictions were made of two variants of Si_2H_2, a simple molecule that is not likely to fill any glass bottles, but is nonetheless detectable. Wonderful, because completely unexpected—the molecule was calculated by H. Lischka and H. Köhler NOT to have the expected acetylenic HSiSiH connectivity, but instead to feature two bridging hydrogens and a folded geometry, as at left in Figure 10-2:[3]

And it does![4] Then B. T. Colegrove and H. F. Schaefer predicted a second "isomer" to be metastable (Figure 10-2 right).[5] And this, too, was found.

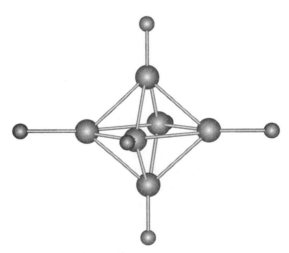

Figure 10-1 A borane-cage analogue of a quinone.

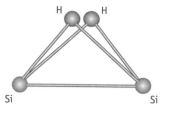 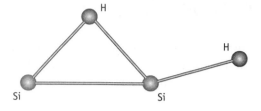

Figure 10-2 Two isomers of Si_2H_2.

Design and Saltation

Compounds/molecules are often useful, ergo the vast transformative chemical industries (and the reasonably populated chemistry and chemical engineering departments of the world). Properties make for function. Be they materials for electronics, polymers with specific properties, pharmaceutical activity, adhesives working under extreme conditions—molecules perform tasks. Never as adequately and cheaply as we desire, of course. So there is a need for design of molecules with specific properties.

Design is often a matter of tuning. Say one has a pharmaceutical lead, a compound with antitumor activity. The chances are that even as it works, the compound is toxic at some level, and that its biological efficacy can be improved. One wants to change a methyl group here for a fluorine, add a hydroxyl group there. This quantized perturbation of an underlying molecular skeleton is our métier, the craftsmanship explicit in dye as drug design. But, tuning is not random change. One needs a way to think about a property—the thinking need not be computational, it may be qualitative—before one sets to the work of synthesis. Here is a great place for theory.

While modifying molecular bits here and there works, one lesson of chemistry has been that really new properties or functions come from big jumps, in structure and electronic properties. I'm thinking as much of liquid crystals and nylon, as I am of fullerenes and metal-metal bonds. Here there is a still more significant role for theory. For small extrapolations are easy, you can calibrate your calculation; predicting the properties of a really different, unusual molecule is risky.

There are pitfalls, psychological ones, in the service of design. The designer's role is at times exaggerated in the process of seeking patronage, a forgivable sin. It is also what journalists think their readers/viewers want to hear—fables of superminds whose predictions are always right. And people—scientists for sure—fall too easily into an excessive valuation of their own mastery of design, to burnish an impression of their rationality. An age-old problem for science, or rather for scientists...

Everything has antecedents and lush interconnections. Still, there is a real opportunity for theoretical chemistry in the creation of new classes of molecules. Instead of being servants to reductionism, we can signal emergence.

Testing Understanding

Theories are fecund webs of understanding. They are accepted by scientific communities for many reasons, but certainly making predictions, preferably risky ones, is important in why people buy into theories.[6]

Could one imagine a better probe of our comprehension of protein catalytic function than the design of an enzyme that, say, flips on its head the selectivity of a catalyzed reaction, or that turns over molecules faster than the natural one? If you have a theory of superconductivity, a super way to demonstrate your understanding would be not only to explain the observed isotope effects and the symmetry properties of the phenomenon, but also to predict what chemical compound should be a high T_c superconductor. We're not there.

The consistent theoretical prediction of viable molecules with unusual structures or properties is the best test of the degree of understanding science has achieved.

Help, Where It Is Needed

The universe is teeming with extreme conditions, from the interior of a big planet, to the cold and near vacuum surrounding a grain in an interstellar cloud. In the course of a chemical reaction, molecular "intermediates" may have only the most fleeting of existences. Their properties, their very existence, may simply not be measurable—out there, or in the laboratory.

Here theoretical chemistry may really be of help. Quantum mechanics has already proven sufficiently reliable to calculate correctly the spectroscopic signatures of molecules in interstellar space. This is how a molecule relatively abundant out there, but hardly persistent in the laboratory, cyclopropenylidene (structure in Figure 10-3), was detected.

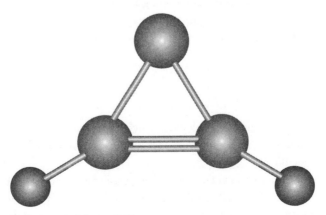

Figure 10-3 Cyclopropenylidene, C_3H_2.

So it is for the metastable intermediates of chemical reactions—there is every reason to think that calculations of their properties are as easy (or is it "as difficult"?) as the properties of stable molecules.

In a more general way, quantum chemistry has proven itself to be "a gentle reductionist companion to experiment," as Fritz Schaefer has aptly written.[7]

Friendly Sparring

Theoreticians are a minority in chemistry, which remains an experimental science. Our experimental friends and that molecular cornucopia, the evolved biosphere, are so productive. We have been given and have made millions of molecules, an incredible diversity in their properties. Some are just ringing changes on a theme, some pose real puzzles. So there we are, theoreticians, in a reactive mode, continually asked to explain.

I don't mind explaining. It is one good test of theory, for sure. But it is nice to turn the tables once in a while, and predict.

As it is, experimentalists and theorists have a certain love/hate relationship within any field, not just chemistry. The stereotypes are clear: to the experimentalist, theorists build castles in the air, don't deign to explain what bothers experimentalists, simplify the world so that it is rendered unreal (the "spherical cow" model). To the theorists, experimentalists complexify matters, vary too many factors simultaneously, never measure the observable the theorists calculate.

It's a game; the "love" part is that both theory and experiment desperately need each other. For the facts are mute. And science, our way of knowing, depends on the coupling of imaginative flights of theoretical fancy with people probing down-to-earth physical and chemical reality.

So... it's fun to make moderately unreasonable predictions of viable or fleeting molecules.[8] The operative part of the phrase is "moderately unreasonable." For if the prospect of synthesis is perceived as entailing a task taking twenty man- or woman-years, the synthesis will not be attempted. The quantum of commitment might be a single Ph.D. student's productive lifetime in graduate school.

Homo Ludens

Science is a marvelous quest for reliable knowledge. Knowing pleasures in and of itself. So does creation. As does sharing that knowledge, and yes, being thought of well for what one does.

The predictor leaves the safety of known molecules and properties for the unknown. He or she takes a risk. And, in a way, is flirting—in a game of interest and synthesis—with the experimentalist. Predicting new molecules is simply great fun.

Notes

1. Not that it's easy to get speculations about molecules culled from simple thinking published, unless it is in the context of a synthesis (or an attempted synthesis, or a review article). Too bad; there should be a journal of speculations in chemistry.
2. M. Balakrishnarajan and R. Hoffmann, Polyhedral Boranes with Exo Multiple Bonds: Three-Dimensional Inorganic Analogues of Quinones. *Angew. Chem. Int. Ed.* 42(32) (2003): 3777.
3. H. Lischka and H.-J. Köhler, Ab Initio Investigation on the Lowest Singlet and Triplet State of Disilyne (Si_2H_2). *J. Amer. Chem. Soc.* 105 (1983): 6646.
4. M. Bogey, H. Bolvin, C. Demuynck, and J. L. Destombes, Nonclassical Double-Bridged Structure in Silicon-Containing Molecules: Experimental Evidence in Si_2H_2 from Its Submillimeter-Wave Spectrum. *Phys. Rev. Lett.* 66 (1991): 413.
5. B. T. Colegrove and H. F. Schaefer, Disilyne (Si_2H_2) Revisited. *J. Phys. Chem.* 94 (1990): 5593.
6. R. Hoffmann, Why Buy That Theory? *Amer. Scientist* 91 (2003): 9. Also Chapter 2 in this book.
7. H. F. Schaefer, III, Odorless Chemistry: A Gentle Reductionist Companion to Experiment. *J. Chinese Chem. Soc.* 43 (1996): 109.
8. For cautionary words on predicting stability, see R. Hoffmann, P. von Ragué Schleyer, and H. F. Schaefer III, Predicting Molecules--More Realism, Please! *Angew. Chem. Int. Ed. Engl.* 47 (2008): 7164.

11

Protean

ROALD HOFFMANN AND PIERRE LASZLO

Science often advances upon willful transgression of a seeming interdiction. Examples which leap to a chemist's mind are noble gas compounds, strained hydrocarbons such as tetrahedranes, activation (by organometallics) of even methane, and, to mention just one brilliant, more recent achievement, inclusion of an allene within the confines of a six-membered ring while preventing its conversion into a benzenoid.[1] Such feats put all the cunning of a scientist into coaxing and, yes, coercing the system at hand to obey instructions from one's daring imagination.[2]

As always, it is hard. Not for nothing is our playroom called a laboratory. And when the task is done and the time arrives to convey to others (who might not be privy to the anguish of the work) all that struggle and the majesty of the achievement, the scientist quite naturally lapses into metaphor. One such, founded in male 19th century language as much as in history, is some more or less prurient variant of "Unveiling the Secrets of Nature."[3] Another, evoking the thorny, twisted path to understanding and the long hours of toil in the laboratory, is "Wrestling with Nature."

The latter metaphor has been central to experimental science at least since the Elizabethan Age, and is the subject of this small essay. While the roots of the metaphor lie in Greek myth, it makes a striking debut in a seminal brief for experiment in science. This arresting phrase also marks a bifurcation in the way science is viewed by nonscientists, even—and especially so—in our day.

The proof text here is that of Francis Bacon (1561–1626), in his 1605 *Of the Advancement of Learning*. Bacon writes:

> *For like a man's disposition is never well known till he be crossed, nor Proteus ever changed shapes till he was straitened and held fast; so the passages and variations of nature cannot appear so fully in the liberty of nature, as in the trials and vexations of art.*[4]

He repeats the imagery in his remarkable 1620 *Novum Organum*. Bacon's 1620 book was a clarion call to replace what passed as Aristotelian reasoning about the world[5] with experiment. Francis Bacon likens matter to Proteus, the Old Man of the Sea, and describes the scientific endeavor, as it strives to discipline matter, as follows:

> *And the manipulations of art are like the bonds and shackles of Proteus,*
> *which reveal the ultimate strivings and struggles of matter.*[6]

By "art" Bacon clearly means experiment, as this passage makes clear:

> *...every interpretation of nature which has a chance to be true is achieved by*
> *instances, and suitable and relevant experiments, in which sense only gives a*
> *judgment on the experiment, while the experiment gives a judgment on nature*
> *and the thing itself.*[7]

Three hundred and sixty-five years later, Primo Levi, in his wonderful chapter on hydrogen in *The Periodic Table*, writes:

> *We would be chemists, Enrico and I. We would dredge the bowels of mystery with*
> *our strength, our talent: we would grab Proteus by the throat, cut short his incon-*
> *clusive metamorphoses from Plato to Augustine, from Augustine to Thomas,*
> *from Thomas to Hegel, from Hegel to Croce. We would force him to speak.*[8]

What is going on here, and who is Proteus? Here is a pithy definition out of a dictionary:

> *Proteus is a prophetic sea divinity, son of either Poseidon or Oceanus. He usu-*
> *ally stays on the Island of Pharos, near Egypt, where he herds the seals of*
> *Poseidon. He will foretell the future to those who can seize him, but when*
> *caught he assumes all possible varying forms to avoid prophesying. When*
> *held fast despite his struggles, he will assume his usual form of an old man*
> *and tell the future.*[9]

The episode of Menelaos wrestling with Proteus in the *Odyssey* is one of the *topoi* of literature. The old man from the sea, when the hero and his companions try to catch him in their arms, first becomes a bearded lion, followed by—the sequence of creatures the magician turns himself is not unlike the list of the years in the Chinese calendar!—a snake, a leopard, a huge boar, flowing water, and a high and leafy tree. But Proteus has to submit to brute force from Menelaos and his companions; and, firmly held, he finally consents to act as a seer—as someone with the uncanny ability of the poet, *viz.* to describe in vivid and memorable words the images which flock into the mind.[10] Indeed, Proteus is an inspired speaker: when Socrates teases Euthyphro, in their dialog, he tells him that "like Proteus, you must be held until you speak!"[11]

Figure 11-1 shows an emblematic artist's take on the struggle with of Proteus.[12]

Figure 11-1 The struggle between Proteus and Menelaos as drawn by Patten Wilson.

But how did we get from a wrestling match with Proteus (definitely male) to one with Nature (mostly feminine in European culture)? Might there be an element of covering up anything remotely homoerotic as one moves on the trajectory out of the Renaissance and toward Victorian times? Is it just an erosion of classical culture as science leads the democratization of the professions? Bacon resorts also to the other metaphor:

> ... *nature reveals herself more through the harassment of art than in her own proper freedom.*[13]

Whatever the reasons, by our own time, excepting stylists such as Primo Levi, Proteus is left in peace in his cave, and nature alone is cross-examined.[14]

Francis Bacon, as an Elizabethan mind, was a master at crafting metaphors—think of his idols of the tribe, idols of the marketplace.[15] He also compared experimental scientists, good at collecting data, with ants; while theorists, he said, are more akin to spiders, "who make cobwebs of their own substance."

Not everyone has been happy with The Lord Chancellor. So one of our icons, Justus von Liebig, had this to say:

> *In all investigations Bacon attaches a great deal of value to experiments in research. But he understands their meaning not at all...An experiment not preceded by theory, i.e. by an idea, bears the same relation to scientific research as the sound of a child's rattle does to music.[16]*

Back to chemistry: we are indeed familiar with the real world being unruly, unyielding to our efforts to make it behave, and apparently uncontrollable. Quite a few chemical reactions ("metamorphoses" as they were termed in former times) seem to us protean like Proteus in their changes, whether those are predictable or not. Cascade reactions, which one of us was perhaps instrumental in naming,[17] come to mind. We know also quite a few tricks to stop in his tracks the Old Man of the Sea, when he comes to haunt chemical dynamics. The trapping of reaction intermediates is such a tactic. So, to return to that cyclohexa[1,2,4]triene—evidence for the fleeting existence of the parent hydrocarbon came from such trapping.[18] Use of radical traps to expose certain chemical species (or to divert them from their appointed rounds) is a common trick.

Turning from reactivity to structure, there also we find an uncanny ability of matter to masquerade in an unpredictable diversity of shapes. Jack Dunitz and Joel Bernstein have written a wonderful account of vanishing polymorphs, crystal forms which were once synthesized, but then could not be recreated. Or so it seemed.[19] Seed crystals have mysterious ways of propagating. The phenomenon is of great importance in the pharmaceutical industry, with respect to the differing efficacies of polymorphs, production control, and their patentability. Polymorphs belong to the realm of the same and not the same.[20]

The "wrestling with nature" metaphor brought about a debate that roils the waters of Proteus' father to this day. Bacon was accused of being the first of a long series of villains to "put nature on the rack," a rationalizer of torture in the service of science.[21] Goethe's revulsion at Newton's incarcerating passage of light through a slit (and its subsequent analysis into the component colors by a prism) is emblematic. As are Donne, Wordsworth, and Ruskin's impassioned denouncements of science and of the attendant industrial revolution. The line continues, to some (hardly all) of the environmentalist and animal rights critiques of the interventionist nature of science.

This is not the place to enter this debate. Our opinion is that modern commerce and technology, with the collusion of science, has some to be ashamed of, such as use of animals in the devising of new cosmetics. And much more to be proud of, our truly democratizing science.

As a community, we have become aware that voyeurism of nature is neither innocent nor totally unobtrusive. We now know that it brings about its own retribution, in like manner perhaps as Acteon is direly punished for his having watched Artemis in the nude. Werner Heisenberg has taught us that any measurement interferes, however slightly, with the system being studied. A limitation to our knowledge? Yes, and no. To be sure, a more courteous approach to reality, forcing us to question macroworld intuition thoughtlessly applied to the microworld, forces us to be as delicate (as we should have been from the beginning) in asking our questions of nature. Take modern chemical dynamics, that remarkable crossroads of classical and quantal. The indeterminacy principle (for that is closer to what Heisenberg called it) has nontrivial consequences in the new kinetic art. It makes us think: To build and shape a femtosecond pulse, what range of energies does it take? And if we want to inject a precise dollop of energy into a bond, what does that entail about the time scale of the process?[22]

Is intervention an absolute correlate of experimental science? The question merits a lengthier discussion. Surprisingly (given Bacon's startling metaphors and the resistance to them), it is only recently that philosophers have begun to explore the subject. An important examination of the question is found in Ian Hacking's idiosyncratic *Representing and Intervening*.[23] In an important essay on "Experimentation and Scientific Realism," Hacking writes: "Interference and interaction are the stuff of reality."[24] J. E. Tiles, in an essay, "Experiment as Intervention," says of Hacking's analysis that it was "received in some quarters with a mixture of incomprehension and hostility."[25] This is a minor surprise to us. We agree with Tiles that philosophers of science ought to consider that experiment does not only follow from observation (as both Aristotelians and positivist classical empiricists mistakenly and narrowly assume; pace Liebig too). At times, experiment is driven by an overtly interventionist stance.

"Nature's narrative of herself" is a fiction. Any acquisition of knowledge involves intrusion. Only the size of the perturbation changes. Accordingly, we have learned to care. We do not want our curiosity to devastate its object, butterfly-like in its fragility. Instead of roughhouse "wrestling with nature," regardless of the consequences, we have made ourselves adepts at gentler sports. Examples are various "non-invasive" spectroscopic techniques, spin tickling for instance in nuclear magnetic resonance, or resorting to isotopes in order to effect as light a perturbation as we know how.[26] And, of course, using all that fine information isotope effects give us...

We try to make our intrusions as delicate as a caress or a kiss. Though, God knows, some kisses can upset a world.

And, yes, we are part of nature too. When trying to understand a complex system or phenomenon—electrocyclic reactions, say, before they were understood—one's mind, one's art seem too limited. We cope with a complex skein of imprecise facts. We wrestle with how to explain phenomena.

Lack of sleep from contending with the truth of the matter is reminiscent of another wrestling match on the world's spiritual stage. It is the story of Jacob wrestling with someone throughout the night. Here is how *Genesis* 32 tells it

> *Jacob was left alone. And a man wrestled with him until the break of dawn. When he saw he had not prevailed against him, he wrenched Jacob's hip at its socket, so that the socket of the hip was strained as he wrestled with him. Then he said, "Let me go, for dawn is breaking." But he answered, "I will not let you go, unless you bless me." Said the other, "What is your name?" He replied, "Jacob." Said he, "Your name shall no longer be Jacob, but Israel, for you have striven with beings divine and human, and have prevailed." Jacob asked, "Pray tell me your name." But he said, "You must not ask my name!"[27]*

As we begin to see things for what they are in this world of chemistry, we are overwhelmed with the feeling of having striven with concepts both divine and human.

Proteus in time became an adjective. To be protean is to be variable in form, to take on or exist in various shapes. Was there ever a science that was more about change than chemistry? And is not the richness of our enterprise, the multitude of molecules we can make, the necessary complexity of the living state due to isomerism, so many chemical incarnations of Proteus? The adjective that serves as the title of this essay is an apt descriptor for chemistry itself.

Notes

1. M. A. Hofmann, U. Bergsträsser, G. J. Reiss, L. Nyulászi, and M. Regitz, Synthesis of an Isolable Diphosphaisobenzene and a Stable Cyclic Allene with Six Ring Atoms. *Angew. Chem. Int. Ed. Engl.* 39 (2000): 1261.
2. See in this context R. Hoffmann and H. Hopf, Learning from Molecules in Distress. *Angew. Chem. Int. Ed. Engl.* 47 (2008): 4474. This is Chapter 9 in this book.
3. For a lovely example, see the poem by one of our heroes, Humphry Davy, cited by D. Knight, *Humphry Davy: Science and Power* (London: Blackwell, 1993). The metaphor presumes the enterprise of science is discovery. But chemists especially know that our business is in substantive part creation; see Chapter 19 in R. Hoffmann, *The Same and Not the Same* (New York: Columbia University Press, 1995).
4. Francis Bacon, *Of the Advancement of Learning* (1605), G. W. Kitchin and J. M. Dent, eds. (London: J. M. Dent, 1915), p. 73.
5. We put it this way out of respect for Aristotle. Bacon and his era reacted against the Scholastics, who "merely polished the idol of Aristotle's investigations—thousands of years after his time." Aristotle was (occasionally) a close observer of nature, as his *Historia Animalium* shows; his was "the boldest scientific thinking of his day." The Baconesque quotations in this endnote are from Michael Weisberg, private communication.

6. Francis Bacon, *The New Organon, Book I*, tr. and ed. by L. Jardine and M. Silverthorne (Cambridge: Cambridge University Press, 2000), p. 227.

7. Francis Bacon, *The New Organon, Book I*, tr. and ed. by L. Jardine and M. Silverthorne (Cambridge: Cambridge University Press, 2000), p. 45.

8. Primo Levi, *The Periodic Table*, tr. by R. Rosenthal (New York: Schocken Books, 1984), p. 23.

9. www.pantheon.org/mythica/areas/greek; Copyright © 1999 Encyclopedia Mythica. All rights reserved. Protected by the copyright laws of the United States and International treaties.

10. Homer, *Odyssey*, trans. by H. B. Cotterill (London: George G. Harrup, 1911): 4.398–460.

11. Plato, *Euthyphro*, 15d, in *Plato*, trans. by H. N. Fowler (New York: McMillan, 1913), p. 59.

12. "The struggle with Proteus." An illustration by Patten Wilson for Homer, *Odyssey*, tr. by H. B. Cotterill (London: George G. Harrap, 1911). There is another image of Proteus drawn by Giulio Buonasone, for Achille Bocchi, *Symbolicae quaestiones* (Bologna, 1574). This is Symbolum LXI.

13. Francis Bacon, *The New Organon*, tr. and ed. by L. Jardine and M. Silverthorne (Cambridge: Cambridge University Press, 2000), p. 21.

14. There is a related, but we think different, metaphor that has been loose in the philosophical literature. This is of "carving nature at its joints," brought to our attention by Michael Weisberg. It goes back to Plato's *Phaedrus* (265e), and we think it refers to scientists (or philosophers) getting classifications or categories "right."

15. For a reminder of the lasting intelligence of Bacon's metaphor, see S. J. Gould, Deconstructing the "Science Wars" by Reconstructing an Old Mold. *Science* 287 (2000): 253.

16. Liebig has much more to say, mostly uncomplimentary, about Bacon. He devotes one lecture and two long polemical essays (published in the Augsburger Allgemeine Zeitung in 1863 and 4) to the other Baron: Justus von Liebig, *Über Francis Bacon von Verulam und die Methode der Naturforschung*, 1863, p. 49. This quotation came to us through I. Hacking, *Representing and Intervening* (Cambridge: Cambridge University Press, 1983), and *Reden und Abhandlungen* (Leipzig: C. F. Winter, 1874), pp. 220–295. See also discussion of Liebig's essay in G. Bachelard, *La formation de l'esprit scientifique* (Paris: J. Vrin, 1969), pp. 58–59.

17. A. Cornélis and P. Laszlo, Addition Modes of Cyanoolefins. III. Reactional Cascade. Addition of Tetracyanoethylene to Dicyclopropylfulvene. *J. Am. Chem. Soc.* 97 (1975): 244.

18. M. Christl, M. Braun, and G. Müller, 1,2,4-Cyclohexatriene, an Isobenzene, and Bicyclo[4.4.0] deca-1,3,5,7,8-pentaene, an Isonaphthalene: Generation and Trapping Reactions. *Angew. Chem.* 104 (1992): 471; *Angew. Chem. Int. Ed. Engl.* 31 (1992): 473. Isobenzene is also formed during the cycloaromatization of 1,3-hexadiene-5-yne to benzene as a reactive intermediate, cf. W. R. Roth, H. Hopf, and C. Horn, 1,3,5-Cyclohexatrien-1,4-diyl und 2,4-Cyclohexadien-1,4-diyl. *Chem. Ber.* 127 (1994): 1765, a process first described in 1969: H. Hopf and H. Musso, Preparation of Benzene by Pyrolysis of cis- and trans-1,3-Hexadien-5-yne. *Angew. Chem.* 81 (1969): 704; *Angew. Chem. Int. Ed. Engl.* 8 (1969): 680. Henning Hopf informs us he proposed the intermediate for the 1,5-hexadiene-5-yne ring closure in his 1972 Habilitationschrift, but didn't dare to mention it in print at the time.

19. J. D. Dunitz and J. Bernstein, Disappearing Polymorphs. *Accts. Chem. Res.* 28 (1995): 193. One of us was moved to write a poem by this article; R. Hoffmann, "On First Sight," unpublished, available on request from the author.

20. R. Hoffmann, "On First Sight," unpublished, available on request from the author.

21. P. Pesic, *Isis* 90 (1999): 81. Pesic's article, entitled "Wrestling with Proteus; Francis Bacon and the 'Torture' of Nature," on which we draw heavily in this essay, lists a number of such interpretations of Bacon's metaphor. The Latin word Bacon used—*vexatio, vexare*—is best translated, as Pesic suggests, by "vexation." In the usage of the time, and Bacon's, it was quite distinct from torture. The earlier (1605) use of "trial or vexation" *in English* by Bacon (our first quote) in a parallel context, is clear evidence for Pesic's point.

22. R. Hoffmann, Pulse, Pump, and Probe. *American Scientist* 87 (1999): 308; R. Hoffmann, Exquisite Control. *American Scientist* 88 (2000): 14.

23. I. Hacking, *Representing and Intervening* (Cambridge: Cambridge University Press, 1983).

24. I. Hacking, in *Scientific Realism*, ed. Jarrett Leplin (Berkeley: Univ. of California Press, 1984), pp. 154–172.

25. J. E. Tiles, Experimentation As Intervention. *Brit. J. Phil. Sci.* 44 (1993): 463. We are grateful to Michael Weisberg for bringing this article to our attention.

26. Hans Christian von Baeyer has written an insightful, meditative essay on how one might and does study nature incisively, but non-destructively. H. C. von Baeyer, A Commanding View. *The Sciences* (Sept./Oct.1989): 6; reprinted in H. C. von Baeyer, *The Fermi Solution* (New York: Random House, 1990).

27. *Genesis* 32: 25–30 (Philadelphia: The Jewish Publication Society of America, 1962).

12

How Should Chemists Think?

ROALD HOFFMANN

The Vatican holds a fresco by Raphael entitled *The School of Athens.* Plato and Aristotle stride toward us. Plato's hand points to the heavens, Aristotle's outward, along the plane of the earth (Figure 12-1). The message is consistent with their philosophies—whereas Plato had a geometric prototheory of the chemistry of matter, Aristotle described in reliable detail how Tyrian purple (now known to be mainly indigo and dibromoindigo) was extracted from rock murex snails. Plato searched for the ideal; Aristotle looked to nature.

Modern chemistry faces the quandary that Raphael's fresco epitomizes. Should it follow the hand sign of Aristotle or that of Plato? Is nature as fertile a source for new materials as some assert it to be? Can we, for example, hope to make better composites by mimicking the microstructure of a feather or of a strand of spider's silk? Are chemists better advised to seek their inspiration in ideal mathematical forms, in icosahedra and in soccer balls? Or should we hazard chance?

To some, the division between natural and unnatural is arbitrary; they would argue that man and woman are patently natural, and so are all their transformations. Such a view is understandable and has a venerable history, but it does away with a distinction that troubles ordinary and thoughtful people. So I will distinguish between the actions, mostly intended, of human beings and those of animals, plants and the inanimate world around us. A sunset is natural; a sulfuric acid factory is not. The 1.3 billion head of cattle in this world pose an interesting problem for any definition. Most of them are both natural and unnatural—the product of breeding controlled by humans.

The molecules that exist naturally on the earth emerged over billions of years as rocks cooled, oceans formed, gases escaped, and life evolved. The number of natural molecules is immense; perhaps a few hundred thousand have been separated, purified, and identified. The vast majority of the compounds that fit into the unnatural category were created during the past three centuries. Chemists have added some 70 million well-characterized molecules to nature's bounty.

Figure 12-1 A detail from Raphael's *School of Athens* in the Stanza della Segnatura in the Vatican. Reproduced by permission of the Vatican.

To everything of this world, be it living or not, there is structure. Deep down are molecules, persistent groupings of atoms associated with other atoms. There is water in the distilled form in the laboratory, in slightly dirty and acid snow, in the waters associated with our protein molecules. All are H_2O. When chemistry was groping for understanding, there was a reasonable reluctance to merge the animate and inanimate worlds. Friedrich Wöhler convinced many people that the worlds were not separate by synthesizing, in 1828, organic urea from inorganic silver cyanate and ammonium chloride.

How are molecules made in nature—penicillin in a mold or a precursor of indigo in a rock murex snail? How are they made in glass-glittery laboratories—those acres of food wrap, those billion pills of aspirin? By a common process—synthesis.

Chemistry is the science of molecules and their transformations. Be it natural or human-steered, the outcome of transformation, A → B, is a new substance. Chemical synthesis, the making of the new, is patently a creative act. It is as much an affirmation of humanity as a new poem by A. R. Ammons or the construction of democracy in Russia. Yet creation is always risky. A new sedative may be effective, but it also may induce fetal malformation. A Heberto Padilla poem may be "counterrevolutionary" to a Cuban apparatchik. Some people in Russia still don't like democracy.

Wöhler mixed together two substances, heated them, and obtained an unexpected result. Much has happened since 1828. To convey what the making of molecules is like today and to relate how the natural intermingles with the unnatural in this creative activity, let me tell you about the synthesis of two substances: Primaxin and the ferric wheel.

Primaxin is one of the most effective antibiotics on the market, a prime moneymaker for Merck & Co. The pharmaceutical is not a single molecule but a designed mixture of two compounds, imipenem and cilastatin (see Figure 12-2). These are their "trivial" names. The "systematic" names are a bit longer; for instance, imipenem is [5R-6S]-3-[[2-(formidoylamino)ethyl]thio]-7-oxo-l-azabicyclo[3.2.0] hept-2-ene-2-car-boxylic acid.

Primaxin was created by a bit of unnatural tinkering, emulating the natural working of evolution. Imipenem by itself is a fine antibiotic. But it is degraded rapidly in the kidney by an enzyme. This would give the drug limited use for urinary tract infections. The Merck chemists found in their sample collection a promising compound, synthesized in the 1940s, that inhibited that ornery enzyme. Modified for greater activity, this became cilastatin. It was obvious to try the combination of the antibiotic and the enzyme inhibitor, and the mix worked.

Imipenem derives from a natural product; cilastatin does not. Both are made synthetically in the commercial process. I will return to this after tracing further the history of one of the components.

Imipenem was developed in the 1970s by a team of Merck chemists led by Burton G. Christensen. It is a slightly modified form of another antibiotic, thienamycin. That, in turn, was discovered while screening soil samples from New Jersey. It is produced by a mold, *Streptomyces cattleya,* so named because its lavender color resembles that of the cattleya orchid. The mold is a veritable drug factory, producing thienamycin and several other varieties of antibiotics.

Unfortunately, thienamycin was not chemically stable at high concentrations. And, to quote one of the Merck crew, "The lovely orchid-colored organism was too stingy." The usual fermentation processes, perfected by the pharmaceutical industry over the past century, did not produce enough of the molecule. So the workers decided to produce greater quantities of thienamycin in the laboratory.[1]

The production of thienamycin required 21 major steps, each involving several physical operations: dissolution, heating, filtration, crystallization. Between the starting material—a common amino acid, L-aspartic acid—and the desired product—thienamycin—20 other molecules were isolated and purified. Of these, only eight are shown in the condensed "reaction scheme" in Figure 12-2.

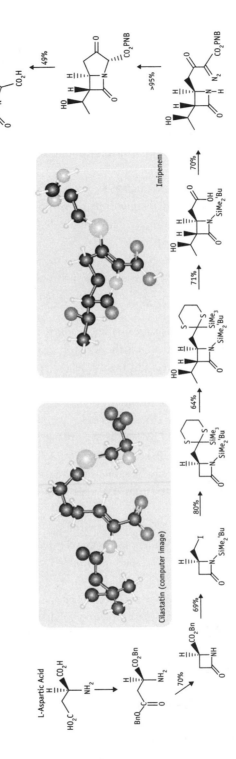

Figure 12-2 Cilastatin (left) and imipenem (right), the components of primaxin. The synthesis of thienamycin, a direct precursor of imipenem is shown around the drawings. Reprinted by permission of Boris Starosta.

The first impression that one gets is of complexity. That intricacy is essential, a laboratory counterpoint to the biochemical complexity of bacteria and us. We would like there to be "magic bullets" of abiding simplicity. The real world is complicated and beautiful. We had better come to terms with that richness.

To get a feeling for the sweat, if not the blood and tears, of the process, we need to turn to the experimental section of the paper reporting the synthesis. Here is an excerpt of that experimental protocol, describing a critical, inventive step in the synthesis—the transformation from compound **8** to **9**:

> A suspension of diazo keto ester **8** (3.98g, 10.58 mmol) and rhodium(II) acetate dimer (0.04 g, 0.09 mmol) in anhydrous toluene (250 mL) was thoroughly purged with nitrogen, and then heated with stirring in an oil bath maintained at 80°C. After heating for two hours, the reaction mixture was removed from the bath and filtered while warm through a pad of anhydrous magnesium sulfate. The filtrate was evaporated under vacuum to afford the bicyclic keto ester **9** (3.27 g, 89%) as an off-white solid...[1]

You can be sure that this jargon-laden account of an experimental procedure is a sanitized, too linear narrative; it is the way things were at the end: neat, optimized. Not the way it first happened. Putting that aside, you feel work, a sequence of operations that take time and effort. Sometimes, just as in our romantic notions of words springing from the brow of inspired poets, we forget the sheer labor of creation. Even the Creator rested on the seventh day.

You might be interested to see the way these experimental procedures change when the very same process is scaled up. You can't make hundreds of millions of dollars' worth of thienamycin the same way you make a few grams in the laboratory. Here is the description of the industrial synthesis, for the very same step:

> The solids containing 200 kg of **8** are dropped into 476 gallons of $MeCl_2$ in tank TA-1432. Meanwhile, the reactor ST-1510 is cleaned out by a 200-gallon $MeCl_2$ boilout. The slurry is transferred to ST-1510, followed by a 50-gallon $MeCl_2$ line flush. An additional 400 gallons of dry $MeCl_2$ are added to ST-1510, and hot water (65°C) is applied to the jackets to concentrate the batch to 545 gallons where the slurry KF (Karl Fischer) is approximately 0.5 g/1 H_2O. Distillates are condensed and collected in another tank.[2]

Making veal stroganoff for a thousand people is not the same as cooking at home for four.

The synthesis of thienamycin is a building process, proceeding from simple pieces to the complex goal. It shares many features with architecture. For instance, a necessary intermediate structure may be more complicated than either the beginning or end; think of scaffolding. Chemical synthesis is a local defeat of entropy, just as our buildings and cities are. The analogy to architecture is so strong that one for-gets how different, how marvelous, this kind of construction is. In a flask there may be 10^{23} molecules, moving rapidly, colliding often. Hands off, following only the strong dictates of thermodynamics, they proceed to shuffle their electrons, break and make bonds, do our bidding. If we're lucky, 99 percent of them do.

Chemists can easily calculate, given a certain number of grams of starting mate-rial, how much product one should get. That is the theoretical yield. The actual amount obtained is the experimental yield. There is no way to get something out of nothing but many ways to get less than you theoretically could. One way to achieve a 50-percent yield is to spill half the solution on the floor. This will impress no one. But even if you perform each transfer as neatly as possible, nature may not give you what you desire but instead transform 70 percent into black gunk. This is also not impressive, for it does not demonstrate control of mind over matter. Experimental yields are criteria not only of efficiency, essential to the industrial enterprise, but also of elegance and control.

There is more, much more, to say about the planned organic synthesis.[3] But let me go on to my second case study: the ferric wheel.

Stephen J. Lippard and Kingsley L. Taft of the Massachusetts Institute of Technology synthesized the ferric wheel, also known as $[Fe(OCH_3)_2(O_2CCH_2Cl)]_{10}$ (see Figure 12-3). They discovered this exquisite molecule while studying model molecules for inorganic reactions that occur in biological systems. For instance, a cluster of iron and oxygen atoms is at the core of several important proteins, such as hemerythrin, ribonucleotide reductase, methane monooxygenase and ferritin (not household words these, but essential to life). In the course of their broad attack on such compounds, Lippard and Taft performed a deceptively simple reaction. Just how simple it seems may be seen from their experimental section, reproduced in its entirety:

> Compound 1 was prepared by allowing the monochloroacetate analogue of basic iron acetate, $[Fe_3O(O_2CCH_2Cl)_6(H_2O)_3](NO_3)$ (0.315 g, 0.366 mmol), to react with 3 equiv of $Fe(NO_3)_3 \cdot 9H_2O$ (0.444 g, 1.10 mmol) in 65 mL of methanol. Diffusion of ether into the green-brown solution gave a yellow solution, from which both gold-brown crystals of 1 and a yellow precipitate deposited after several days.[4]

Using x-ray diffraction on the gold-brown crystals, Lippard and Taft determined the arrangement of atoms in the molecule. The structure consists of 10 ferric ions (iron in oxidation state three) in a near circular array. Each iron atom is joined to its neighbors by methoxide and carboxylate bridges, "forming a molecular ferric wheel," to quote its makers.

No one will deny the visual beauty of this molecule. It does not have the annual sales of Primaxin, estimated to be $500 million (at the time of writing). On the

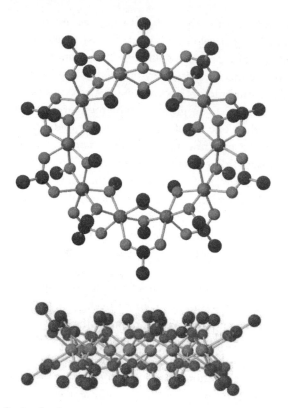

Figure 12-3 The ferric wheel, in two views. Reprinted by permission of Boris Starosta.

contrary, it probably cost the U.S. taxpayer several thousand dollars to make it. But I do not know a single curmudgeonly chemist who would not respond positively to this lovely creation. Perhaps someday the ferric wheel will find a use; perhaps it will form a link in explaining the function of iron-containing proteins. I do not really care— for me, this molecule provides a spiritual high akin to hearing a Haydn piano trio I like.

Why is this molecule beautiful? Because its symmetry reaches directly into the soul. It plays a note on a Platonic ideal. Perhaps I should have compared it to Judy Collins singing "Amazing Grace" rather than the Haydn trio. The melodic lines of the trio indeed sing, but the piece works its effect through counterpoint, the tools of complexity. The ferric wheel is pure melody.

Were we to write out the synthesis of the ferric wheel, there would be but a single arrow, from the iron chloroacetate and ferric nitrate to the product. This is a very different type of synthesis—the product essentially self-assembles to its final glory. When I see such a process, much more typical of inorganic systems than organic ones, I immediately wonder what I'm missing. The Swedish chemist Sture Forsén has aptly expressed the frustration in not being able to observe the intermediate stages of a reaction:

The problem facing the scientist has been compared with that of a spectator of a drastically shortened version of a classical drama—"Hamlet," say—where he or she is only shown the opening scenes of the first act and the last scene of the finale. The main characters are introduced, then the curtain falls for change of scenery, and as it rises again we see on the scene floor a considerable number of "dead" bodies and a few survivors. Not an easy task for the inexperienced to unravel what actually took place in between.[5]

Wheels, ferric or ferris, don't really self-assemble in one fell swoop. It remains for us to learn in the future how those bridges and irons come together.

Some chemists, especially those who practice the mentally demanding, intellectually exhilarating many-step, planned synthesis of the thienamycin type look askance at one-step self-assembly. Such one-fell-swoop syntheses are especially common in solid-state chemistry, in the formation of materials extended infinitely in one, two or three dimensions. The high-temperature superconductors are a good example of molecules made just this way. Their synthesis does not appear to show control of mind over matter. It looks like magic.

I exaggerate, but this is one strand of thought in the community. If I could corner my straw-man scoffer at self-assembly, typically an organic chemist, and engage him or her in a Socratic dialogue, I would begin with the question "When have you made any diamond for me lately?" Diamond is a beautifully simple three-dimensional structure (natural!). It contains in it six-membered rings, the bread-and-butter of organic chemistry. Such rings of carbon atoms are easy to make in a discrete molecule. But diamond can be made only by techniques organic chemists find unsporting, by discharges forming a plasma in methane or by pressing graphite.

Organic chemists are masterful at exercising control in zero dimensions. To one piece of carbon, perhaps asymmetric, they add another piece. Slowly, painstakingly, a complex edifice emerges. (Thienamycin is pretty simple compared to what you can do today.) One subculture of organic chemists has learned to exercise control in one dimension. These are polymer chemists, the chain builders. Although they may not have as much honor in organic chemistry as they should, they do earn a good bit of money.

But in two or three dimensions, it's a synthetic wasteland. The methodology for exercising control so that one can make unstable but persistent extended structures on demand is nearly absent. Or to put it in a positive way—this is a certain growth point of the chemistry of the future.

Syntheses, like human beings, do not lend themselves to typology. Each one is different; each has virtues and shortcomings. From each we learn. I will stop, however reluctantly, with primaxin and the ferric wheel and turn to some general questions they pose, especially about the natural and the unnatural.

Two paradoxes are hidden in the art of synthesis. The first is that the act of synthesis is explicitly human and therefore unnatural, even if one is trying to make a product of nature. The second is that in the synthesis of ideal molecules, where doing what comes unnaturally might seem just the thing, one sometimes has to give in to nature. Let me explain in the context of the two syntheses I have just discussed.

Imipenem, one component of the successful Merck antibiotic, is made from thienamycin. The thienamycin is natural, to be sure, but an economic and chemical decision dictated that in its commercial production thienamycin be made synthetically.

There is no doubt in this case that the natural molecule served as an inspiration for the synthetic chemists. But, of course, they did not make thienamycin in the laboratory the way it is made by the mold. The organism has its own intricate chemical factories, enzymes shaped by evolution. Only recently have we learned to use genetic engineering to harness those factories, even whole organisms, for our own purposes.

We have grown proficient at simpler, laboratory chemistries than those evolved by biological organisms. There is no way that Christensen and his team would set out to mimic a mold enzyme in detail. They did have confidence that they could carry out a very limited piece of what the lowly mold does, to make thienamycin, by doing it differently in the laboratory. Their goal was natural, but their process was not.

To make thienamycin, Christensen and his co-workers used a multitude of natural and synthetic reagents. For instance, one of their transformations in Figure 12-2 uses a magnesium compound, $(CH_3)_3CMgCl$, known as a Grignard reagent. Magnesium compounds are abundant in nature (witness Epsom salt and chlorophyll). But the reagent in question, a ubiquitous tool of the synthetic chemist, was concocted by Victor Grignard some time around the turn of the century. The creation of the compound in question also requires treatment with hydrochloric acid and ammonium chloride, both natural products. (Your stomach has a marginally lower concentration of hydrochloric acid than that used in this reaction, and ammonium chloride is the alchemist's sal ammoniac.) But even though these molecules occur in nature, they are far easier to make in a chemical plant.

Because everything in the end does come from the earth, air or water, every unnatural reagent used in the synthesis ultimately derives from natural organic or inorganic precursors. The very starting material in the synthesis of imipenem is an amino acid, aspartic acid.

Now consider the most unnatural and beautiful ferric wheel. It was made simply by reacting two synthetic molecules, the iron monochloroacetate and ferric nitrate, in methanol, a natural solvent. The methanol was probably made synthetically; the two iron-containing reagents derive from reactions of iron metal, which in turn is extracted from iron ores. And the final wrinkle is the method of assembly: the pieces of the molecule seem to just fall into place (self-assembly). What could be more natural than letting things happen spontaneously, giving in to the strong dictates of entropy?

It is clear that in the unnatural making of a natural molecule (thienamycin) or of an unnatural one (the ferric wheel), natural and synthetic reagents and solvents are used in a complex, intertwined theater of letting things be and of helping them along. About the only constant is change, transformation.

We may still wonder about the psychology of chemical creation. Which molecules should we expend our energies in making? Isn't there something inherently better in trying to make the absolutely new?

Four beautiful polyhedra of carbon have piqued the interest of synthetic organic chemists during the past 40 years: tetrahedrane (C_4H_4), cubane (C_8H_8), dodecahedrane ($C_{20}H_{20}$), and buckminsterfullerene (C_{60}) (Figure 12-4). Cubane is quite unstable because of the strain imposed at each carbon. (In cubane the angle between any three carbon atoms is 90 degrees, but each carbon would "prefer" to form angles of 109.5 degrees with its neighbors.) C_{60} is also somewhat strained because of both its nonplanarity and its five-membered rings. Tetrahedrane is particularly unstable. One has to create special conditions of temperature and solvent to see it; even then, the parent molecule has not yet been made, only a "substituted derivative," in which hydrogen is replaced by a bulky organic group.

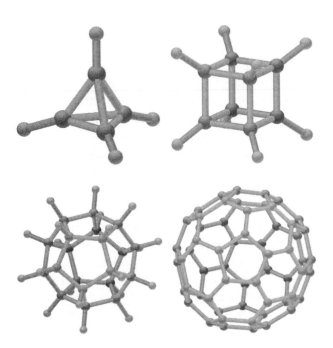

Figure 12-4 Tetrahedrane, cubane, dodecahedrane, and buckminsterfullerene. Reprinted by permission of Boris Starosta.

As far as we know, tetrahedrane, cubane and dodecahedrane do not exist naturally on the earth. C_{60} has been found in old soot and a carbon-rich ancient rock, shungite. It may turn up elsewhere. Be that as it may, all four molecules were recognized as synthetic targets at least 20 years, in some cases 50 years, before they were made. Some of the best chemists in the world tried to make them and failed. The syntheses of cubane and especially dodecahedrane were monumental achievements in unnatural product chemistry.

C_{60} was different. The pleasing polyhedral shape was first noted by some theoreticians. Their calculations indicated some stability; such indications as the theoreticians had at their command were sometimes unreliable. These theoreticians' dreams were ignored by the experimentalists and by other theoreticians. It is sometimes difficult to see the shoulders of the giants we stand on when we are looking so intently ahead. I myself have suggested a still unsynthesized metallic modification of carbon, different from diamond or graphite, and even though I have substantially more visibility among chemists than the proposers of buckminsterfullerene, no one has paid much attention to my pipe dream either, probably for good reason. We see what we want to see.

One organic chemist I know, a very good one, Orville L. Chapman of the University of California at Los Angeles, independently thought up the structure and devoted much time to the planned, systematic making of C_{60}. After all, this was a "simple" molecule, not an extended material like the repeating lattice of carbon atoms that make up a diamond. So it should be possible to make it. Despite persistent efforts over a 10-year period, Chapman and his students failed in their effort.

The first evidence, indirect but definitive, for C_{60} was obtained from a very different branch of our science, physical chemistry. The credit for the discovery belongs properly to Richard E. Smalley and Robert F. Curl of Rice University and Harold W. Kroto of the University of Sussex and their coworkers.[6] They obtained hard evidence for tiny amounts of C_{60} in the gas phase, assigned the molecule its name and, more important, deduced its structure. Did they make it? Absolutely. It did not matter to me or to other believers in their evidence that they had made "just" 10^{10} molecules instead of the 10^{20} we need to see in a tiny crystal. But there were doubters, many I suspect, in the organic community. One wanted to see the stuff.

Grams of buckminsterfullerene were provided by a synthesis by Donald R. Huffman of the University of Arizona and Wolfgang Krätschmer and Konstantinos Fostiropoulos of the Max Planck Institute for Nuclear Physics in Heidelberg. Striking a carbon arc in a helium atmosphere (which is what they did) is about as unsporting as firing a laser at graphite (the Smalley-Kroto-Curl synthesis). But it certainly makes plenty of C_{60}, enough of the molecule to determine its structure by typical organic methods, enough to convince any chemist that it has the soccer-ball structure.

I think many chemists wished C_{60} had been made in a planned, unnatural way. I am happy that—just to make the world slightly less rational than we would like it to be—it was made in a serendipitous way.

Serendipity—a word invented by Horace Walpole—has come to mean "a discovery by chance." Yet whether it is chemical synthesis or a Japanese master potter piling organic matter around the ceramic objects in his Bizen kiln, chance favors the prepared mind. You need to have the knowledge (some call it intuition) to vary the conditions of striking the arc or the arrangement of the leaves in the kiln just so. You need to have the instruments and intuition to deduce structure from a few fuzzy lines in a spectrum and to reject false leads. And you need to have the courage to shatter a vase that didn't come out right and to learn from one firing what to do in the next.

Many chemical syntheses, even if part of a grand design, proceed by steps that are serendipitous. One wants to link up a bond here, but it doesn't work. So one follows a hunch, anything but the codified scientific method. One knows that if a reaction works, one can construct a rationalization for it—an argument spiffy clean enough to make an impression on one's colleagues. Eventually one can make the damned reaction work if it is a necessary step in the design.

Because chance also operates to foil every design, it is almost certain that in the course of any planned synthesis there will be a step that will not work by any known process. So a new one will be invented, adding to the store of the chemists, aiding others around the world facing the same problem. Some synthetic chemists—for instance, E. J. Corey of Harvard University, a grand master of the art—have a special talent for not only making interesting molecules but also using the opportunity of the synthesis to introduce a brilliant, unprecedented methodology, applicable to other syntheses.

When the synthesis is planned, be its aim a natural or unnatural molecule, we suppress the aleatory nature of the enterprise. We want to project an image of mind over matter, of total control. When the molecule made is unanticipated, as the ferric wheel was, we find it very difficult to hide the workings of chance. But hazard—to use the meaning that is dominant in the French root of our word, and secondary in ours—plays an unrecognized and enlivening role in all synthesis.

Let us return to nature and our struggles to emulate it. Or surpass it. Can we make substances that have properties superior to those found in nature? I say "yes" while recognizing that the phrase "superior to nature" is patently value laden and anthropocentric and should immediately evoke ecological concerns.

There is nylon instead of cotton in fishing nets, nylon instead of silk in women's stockings. No one, least of all Third World fishermen, will go back to the old nets. Some people may go back to silk stockings, but they will only be the rich, out to impress. There are new chemical materials and new combinations of old materials for dental restorations. They make a world of difference to older people in this world, and their benefit cannot be dismissed.

Yet the thought that we can do better than nature is provocatively arrogant. As we have attempted to improve on nature (while failing to control the most natural thing about us, our drive to procreate), we have introduced so many transformations and in such measure that we have fouled our nest and intruded into the great cycles of this planet. We must face the reality that natural evolution proceeds far too slowly to cope with our changes. This is a concern that, just as much as utility, should guide the industrial-scale syntheses of the future.

I want to touch on another kind of human arrogance implicit in the intellectual drama of synthesis. A French chemist, Alain Sevin, has put it well:

> *The incredible richness and fantasy of Nature is an act of defiance to Man, as if he had to do better in any domain. Flying faster than birds, diving deeper than whales... We are Promethean characters in an endless play which now is in its molecular act.*[7]

We are driven to transform. We have learned to do it very well. But this play is not a comedy.

Were chemical synthesis in search of a single icon, the outstretched hand of Prometheus bringing fire to humanity would serve well. Prometheus, a name meaning "forethought," represents the element of design, the process of fruitfully taking advantage of chance creation. Fire is appropriate because it drives transformation. The hand of Prometheus is the symbol of creation—the hand of God reaching to Adam in Michelangelo's fresco, the hands in contentious debate in Dürer's *Christ among the Doctors*, the infinite variety of hands that Rodin sculpted. Hands bless, caress and hide, but most of all, they shape.

The sculptor's art itself mimics the complexity of motion of a chemist across the interface between natural and unnatural. Rodin, in his human act of creation, sketches, then shapes by hand (with tools) an out-of-scale yet "realistic" artifact, a sculpture of a hand, out of materials that are synthetic (bronze) but that have natural origins (copper and tin ores). He uses a building process (maquettes, a cast) that is complex in its intermediate stages. The sculptor creates something very real, whose virtue may reside in calling to our minds the ideal.

Margaret Drabble has written that Prometheus is "firmly rooted in the real world of effort, danger and pain." Without chemical synthesis, there would be no aspirin, no cortisone, no birth-control pills, no anesthetics, no dynamite. The achievements of chemical synthesis are firmly bound to our attempt to break the shackles of disease and poverty. In search of an ideal, making real things, the mind and hands engage.

Notes

1. T. N. Salzmann, R. W. Ratcliffe, F. A. Bouffard and B. G. Christensen, A Stereocontrolled, Enantiomerically Specific Total Synthesis of Thienamycin. *Philosophical Transactions of the Royal Society of London*, Series B, 289 (1980): 191.
2. B. G. Christensen, private communication.

3. For further reading on synthesis, much more authoritative than what I have to say, see E. J. Corey and X.-M. Cheng, *The Logic of Chemical Synthesis* (New York: John Wiley and Sons, 1989); R. B. Woodward in *Perspectives in Organic Chemistry*, ed. A. R. Todd (New York: Interscience, 1956).

4. K. L. Taft and S. J. Lippard, Synthesis and Structure of $[Fe(OMe)_2(O_2CCH_2Cl)]_{10}$, A Molecular Ferric Wheel. *J. Am. Chem. Soc.* 112 (1990): 9629.

5. S. Forsén, in *Les Prix Nobel 1986* (Lund: Almqvist & Wiksell, 1986).

6. R. F. Curl and R. E. Smalley, Fullerenes. *Scientific American* (October 1991): 54.

7. Alain Sevin, personal communication.

Part 2

WRITING AND COMMUNICATING IN CHEMISTRY

13

Under the Surface of the Chemical Article

ROALD HOFFMANN

You open an issue of a modem chemical periodical, say the important German *Angewandte Chemie*[1] or the *Journal of the American Chemical Society,* and what do you see? Riches upon riches: reports of new discoveries, marvelous molecules, unmakeable, unthinkable yesterday—made today, reproducibly, with ease. The chemist reads of the incredible properties of novel high-temperature superconductors, organic ferromagnets, and supercritical solvents. New techniques of measurement, quickly equipped with acronyms—EXAFS, INEPT, COCONOESY—allow you to puzzle out more expeditiously the structure of what you make. Information just *flows.* No matter if it's in German, if it's in English. It's chemistry—communicated, exciting, alive.

Let's, however, take another perspective. To the pages of the same journal turns a humanist, a perceptive, intelligent observer who has grappled with *Shakespeare, Pushkin, Joyce,* and *Paul Celan.* I have in mind a person who is interested in what is being written, and also in how and why it is written. My observer notes in the journal short articles, a page to ten pages in length. She notes an abundance of references, trappings familiar to literary scholars, but perhaps in greater density (number of references per line of text) than in scholarly texts in the humanities. She sees a large proportion of the printed page devoted to drawings. Often these seem to be pictures of molecules, yet they are curiously iconic, lacking complete atom designations. The chemist's representations are not isometric projections, nor real perspective drawings, yet they are partially three-dimensional.

My curious observer reads the text, perhaps defocusing from the jargon, perhaps penetrating it with the help of a chemist friend. She notes a ritual form. The first sentences often begin: "The structure, bonding and spectroscopy of molecules of type X have been subjects of intense interest.[a-z]" There is general use of the third person and a passive voice. She finds few overtly expressed personal motivations, and few accounts of historical development. Here and there in the neutered language she glimpses stated claims of achievement or priority—"a novel metabolite," "the first synthesis," "a general strategy," "parameter-free calculations." On studying many papers she finds a mind-deadening similarity. Nevertheless, easy to spot in some of the articles, she also sees style—a distinctive, connected scientific/written/graphic way of looking at the chemical universe.

I want to take a look, now not hiding behind this observer, at the language of my science, as it is expressed in the essential written record, the chemical journal article. I will argue that much more goes on in that article than one imagines at first sight; that what goes on is a kind of dialectical struggle between what a chemist imagines should be said (the paradigm, the normative) and what he or she must say to convince others of his or her argument or achievement. That struggle endows the most innocent-looking article with a lot of suppressed tension. To reveal that tension, I will claim, is not at all a sign of weakness or irrationality, but a recognition of the deep humanity of the creative act in science.

The Scientific Article: A Brief History

There was chemistry before the chemical journal. It was described in books, in pamphlets or broadsides, in letters to secretaries of scientific societies. These societies, for instance the Royal Society in London, chartered in 1662, played a critical role in the dissemination of scientific knowledge. Periodicals published by these societies helped to develop the particular combination of careful measurement and mathematization that shaped the successful new science of the time.[2]

The scientific articles of the period are a curious mixture of personal observation and discussion, with motivation, method, and history often given firsthand. Polemics abound. Cogent arguments for the beginning of a codification of the style of the scientific article in France and England in the 17th century have been given by *Shapin,*[3a] *Dear,*[3b] and *Holmes.*[3c] I think the form of the chemical article rigidified finally in the 1830s and 1840s, and that Germany was the scene of the hardening. The formative struggle was between the founders of modern German chemistry—people such as *Justus von Liebig*—and the *Naturphilosophen.* In that particular period the latter group might be represented by *Goethe's* followers, but their like was present elsewhere in Europe even earlier, in the eighteenth century. The Natural Philosophers had well-formed notions, all-embracing theories, of how Nature should behave, but did not deign to get their hands dirty to find out what Nature actually did. Or they tried to fit Nature to their peculiar philosophical or poetic framework, not caring about what our senses and their extension, our instruments, said. I think the early 19th century scientific article evolved to counter the "pernicious" influence of the Natural Philosophers. The ideal report of scientific investigation should deal with the facts (often labeled explicitly or implicitly as truth; more on this later). The facts had to be believable independent of the identity of the person presenting them. It followed that they should be presented unemotionally (so in the third person) and with no pre-judgment of structure or causality (therefore the agentless or passive voice).

The fruits of this model reportage were immense. An emphasis on experimental facts stressed the reproducible. The conciseness of the German language seemed ideally suited to the developing paradigm. Cadres of chemists were trained. The development of the dyestuff industry that followed in England and Germany is a particularly well studied manifestation of the industrial application of the new, organized chemistry.

The scientific article acquired in this period a canonical or ritual form. In Figure 13-1, I reproduce part of a typical article of that period.[4,5] Note most of the features of a modern article—references, experimental part, discussion, diagrams. All that's lacking is the acknowledgment thanking the Deutsche Forschungsgemeinschaft or the National Science Foundation.

In Figure 13-2, a contemporary (at the time of writing) article, we approach the present. This particular contribution, an important one by *O. J. Scherer* and *T. Bruck*[6] reports a remarkable ferrocene-type system in which one cyclopentadienyl ring has been replaced by P_5. The work is novel and significant, but I want to focus on the mode of presentation rather than the content. How does this article differ from one published a hundred years ago? The dominant language has changed, for interesting geopolitical reasons, to English.[7] Yet it seems to me that there is not much change in the construction or tone of the chemical article. Oh, marvelous, totally new

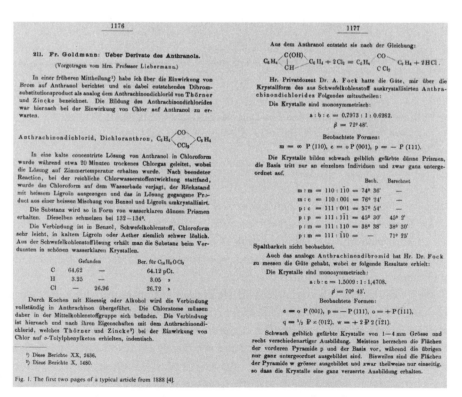

Figure 13-1 The first two pages of a typical article from 1888 (Ref. 4).

[(η^5-P$_5$)Fe(η^5-C$_5$Me$_5$)],
a Pentaphosphaferrocene Derivative**

By *Otto J. Scherer** and *Thomas Brück*

After having demonstrated that *cyclo*-P$_5$ could be stabilized as bridging ligand in the mixed-valence triple-decker complex 1,[1] we then attempted to realize the classical

[(η^5-C$_5$Me$_5$)Cr(μ,η^5·P$_5$)Cr(η^5-C$_5$Me$_5$)]
1

sandwich coordination of this ligand (cyclo-P$_5^0$ as 6π electron donor). Success was achieved upon cothermolysis of 2 with white phosphorus.

Pentamethylpentaphosphaferrocene 3 forms sublimable, green crystals which can be handled in the presence of air and which begin to melt (with partial sublimation and slight decomposition) at 270°C when heated in a sealed tube. 3 is very soluble in dichloromethane, readily soluble in benzene and toluene, and moderately soluble in pentane.

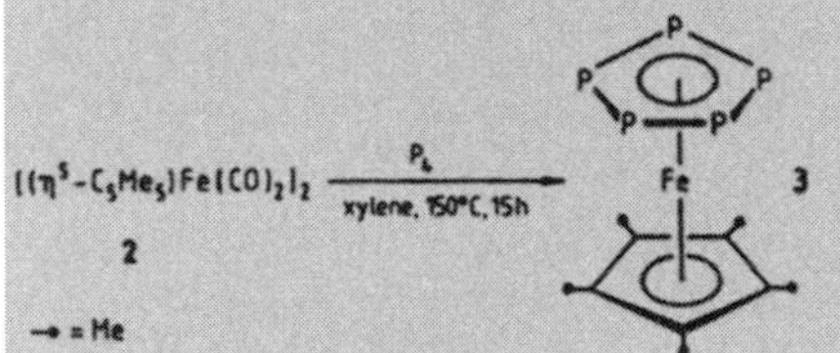

[(η^5-C$_5$Me$_5$)Fe(CO)$_2$]$_2$ $\xrightarrow[\text{xylene, 150°C, 15h}]{P_4}$ 3

2

→ = Me

In the ^{1}H-NMR spectrum (200 MHz, 293 K, C$_6$D$_6$, TMS int.) of 3 a sharp singlet is observed at δ = 1.08 which is shifted 0.6 ppm upfield compared to that of decamethyl-

ferrocene [(η^5-C$_5$Me$_5$)$_2$Fe] 4.[3] In comparison to the ^{13}C{^{1}H}-NMR signals of 4,[2,3] those of 3[2] are shifted slightly downfield. The ^{31}P{^{1}H}-NMR signal shows a continuous downfield shift within the series triple-decker 1 (δ = −290.5[1]), monophosphaferrocene [(η^5-C$_5$H$_5$)Fe(η^5-PC$_4$H$_4$)] 5 (δ = −67.5[4]), 3 (81.01 MHz, C$_6$D$_6$, 85% H$_3$PO$_4$ ext., δ = 153.0 (s)), and Li(P$_5$) (δ = 470[5]). In the mass spectrum,[2] the most intense peak is the molecular peak of 3, followed by the peak for M$^⊕$ − P$_2$. So far, all attempts to prepare a single crystal suitable enough for an X-ray structure analysis have failed, both by sublimation as well as recrystallization. cyclo-P$_5^0$ is probably formed from P$_4^0$ and P$_2$.[6]

Experimental

A mixture of 2 [7] (980 mg, 2.3 mmol) and P$_4$ (1500 mg, 12.1 mmol) in xylene (80 mL) was stirred under reflux for 15 h and the insoluble material removed by filtration on a D3 frit and extracted three times with 80 mL each of CH$_2$Cl$_2$ (393 mg of residue after drying under high vacuum). The solvent was removed from the combined extracts under oil-pump vacuum. After three extractions with 50 mL each of pentane there remained 726 mg of a brown solid whose composition could not be unequivocally established. After removal of the solvent from the green extracts (oil-pump vacuum), the residue remaining behind was sublimed. At 60°C/0.01 torr, excess phosphorus was removed; between 90° and 110°C, green needles sublimed on the wall of the glass vessel. Recrystallization from pentane furnished 175 mg of 3 (yield 11%). Correct elemental analysis.

Received: October 20, 1986;
supplemented: November 7, 1986 [Z 1957 1E]
German version: *Angew. Chem.* 99 (1987) 59

[1] O. J. Scherer, J. Schwalb, G. Wolmershäuser, W. Kaim, R. Gross, *Angew. Chem.* 98 (1986) 349; *Angew. Chem. Int. Ed. Engl.* 25 (1986) 363.
[2] ^{13}C{^{1}H}-NMR (50.28 MHz, C$_6$D$_6$, TMS intern) 3: δ = 90.6 (s; C$_5$Me$_5$), 10.6 (s; CH$_3$); 4: δ = 78.5 (s; C$_5$Me$_5$), 9.8 (s; CH$_3$). EI-MS (70 eV) of 3: m/z 346 (M$^⊕$, I$_{rel}$ = 100%), 284 (M$^⊕$ − P$_2$, 91%), P$_4$ (19.8%), P$_2$ (7.9%), P$_3$ (53%), P (7.8%) and further, weak intensity lines.
[3] Cf. : J. L. Robbins, N. Edelstein, B. Spencer, J. C. Smart, *J. Am. Chem. Soc.* 104 (1982) 1882.
[4] F. Mathey, *Struct. Bonding (Berlin)* 55 (1983) 153.
[5] M. Baudler, *Phosphorus Sulfur*, in press; M. Baudler, D. Düster, D. Ouzounis, *Z. Anorg. Allg. Chem.*, in press.
[6] Cf. also the theoretical studies on N$_5^+$ and its complex stabilization. M. T. Nguyen, M. Sana, G. Leroy, J. Elguéro, *Can. J. Chem.* 61 (1983) 1435; M. T. Nguyen, M. A. McGinn, A. F. Hegarty, J. Elguéro, *Polyhedron* 4 (1985) 1721.
[7] D. Catheline, D. Astruc, *Organometallics* 3 (1984) 1094, and references cited therein.

[*] Prof. Dr. O. J. Scherer, Dipl.-Chem. T. Brück
Fachbereich Chemie der Universität
Erwin-Schrödinger-Strasse, D-6750 Kaiserslautern (FRG)
[**] This work was supported by the Fonds der Chemischen Industrie.

Figure 13-2 A contemporary (1987) article, O. J. Scherer and T. Brück, [(η-^{5}-P$_5$)Fe (η^5-C$_5$Me$_5$)], a Pentaphosphaferrocene Derivative. *Angew. Chem.* 99 (1987): 59; *Angew. Chem. Int. Ed. Engl.* 26 (1987): 59. ©Wiley-VCH Verlag GmbH & Co. KGaA. Reproduced with permission.

things are reported. Measurements that took a lifetime are made in a millisecond. Molecules unthinkable a century ago are easily made, in a flash revealing their identity to knowledgeable us. All communicated, with better graphics and computer typesetting, in a flashier journal, printed probably on poorer paper. But essentially in the same form. Is that good, is that bad?

Well, I think both. The periodical-article system of transmitting knowledge has worked remarkably well for two centuries or more. But I think there are real dangers (to which I will return) implicit in its current canonical form. My primary concern is what is really going on in the writing and reading of a scientific article, which is much more than just communication of facts. To set the stage for a discussion of the journal article, I need to say something about what I think science, chemistry in particular, is.

A Personal View of Chemistry

No one else's than my own, and others will see it in a different way:[8]

1. SCIENCE IS THE ACQUISITION OF KNOWLEDGE ABOUT THE WORLD

That sense is clear in the etymology of the English word, or in the German *Wissenschaft* or *Naturwissenschaft*. Note that "truth" doesn't enter explicitly into that etymology. To be sure, reproducibility, verifiability, reliability, which are essential to any scientific enterprise, depend on *honest* measurement. Scientists would like to think that they acquire truth (morally and ethically valued) and not just knowledge (quite neutral; recall the tree of knowledge of good and evil in Genesis). But I would warn my colleagues that to purvey to the world at large the image of scientists as seekers of truth, rather than reliable knowledge,[9] is dangerous. It makes us out as priests, with the attendant hazards. I suspect much of the exaggerated interest of the public in the rare cases of fraud in science bears some similarity to the prurient interest of the world at large in the moral failings of priests and ministers. I think we gain knowledge, and do that as truthfully as possible.

2. SCIENCE IS PART DISCOVERY, PART CREATION

I take discovery in the sense of revealing some perhaps obscured laws of nature, creation in the sense of making new things. In describing their work most scientists will stress the discovery metaphor, while most artists will emphasize creation. Well, I think much of what we do in science is creation. Especially so in chemistry. The synthesis of molecules not present on earth before is clear evidence of this. Synthesis is a marvelous congeries of discovery and creation that brings chemistry close to the arts—and to engineering.[10]

3. SCIENCE IS DONE BY HUMAN BEINGS AND THEIR TOOLS

Which means that it is done by fallible human beings. The driving forces for acquiring knowledge are, to be sure, curiosity and altruism, rational motives. But creation is just as surely rooted in the irrational, in the dark, murky waters of the psyche, where fears, power, sex, childhood traumas swim in all their hidden, mysterious movements. And spur us on. To follow *Pushkin's Mozart and Salieri* or its modern reincarnation in *Peter Shaffer's* play *Amadeus* and the movie based on it: Angelic music (read chemistry) is brought into this world by such crass vessels. Not only do character and deep-down motivations not matter—their "unsavory" side may well be the driving force of the creative act.

4. SCIENCE PROCEEDS IN PART BY THE RULES

The modern model of the scientific method, associated in this century perhaps with the name of *Karl Popper;* begins with measurable, reproducible observations. One then forms alternative hypotheses or models, explaining these observations. Finally, with the aid of further experiments or reliable theory, the hypotheses are culled, eliminated one by one, until the one most likely to be right remains. Sometimes it works this way,[11] though rather interestingly this happens mostly with everyday and not ground-breaking science. But does the model apply to a modern synthesis of an unnatural molecule, for instance dodecahedrane, a beautiful $C_{20}H_{20}$ molecule shaped like its name? Or the development of the technique of nuclear magnetic resonance, equally useful to chemists seeking the arrangement of atoms in space and to a physician looking for a brain tumor?

5. SCIENCE DEPENDS ON ARGUMENT

"Argument" has several meanings—it could be taken as a simple process of reasoning, a statement of fact; the word also may mean disagreement, the confrontation of opposites (see Figure 13-3). I would argue that both senses are essential to science: dispassionate logical reasoning *and* impassioned conviction that one (model, theory, measurement) is right and another is wrong. I feel that scientific creativity is rooted in the inner tension, within one and the same person, of knowing that he or she is right and knowing that that conviction has to be proven to others' satisfaction. By a journal article.

Figure 13-3 Drawing by Constance Heller from the original publication of this article, *Angew. Chem. Int. Ed. Engl.* 27 (1988): 1596. © Wiley-VCH Verlag GmbH & Co. KGaA. Reproduced with permission.

6. AS A SYSTEM, SCIENCE WORKS

Individual scientists struggle to acquire knowledge, and in their struggle they are driven by many complex motivations. Because researchers are human, they are subject not only to inaccuracy, but sometimes prejudice. Remarkably, the error and prejudice of individual chemists does not matter to the progress of chemistry. Chemistry as a science, the collective activity of the half-million people in the world who are chemists, advances despite mistakes by individual chemists.[12] The science has self-correcting features in abundance: the most important one is that the more interesting the observation or theory, the more likely it is to be checked by someone else. Often for entirely the "wrong" reasons—driven by plain disbelief arising from the conviction that the initial observation must be wrong. It doesn't matter why an individual chemist repeats a critical synthesis, or tries an alternative theory. Chemistry progresses.

And yet, that a chemist *try* to prove someone else's mechanism wrong, or to make a molecule first, does matter. For without the human impulse nothing would get accomplished. It's a curious creation, this science of ours—an incredibly sturdy, exciting, and useful system of knowledge built by imperfect people, and depending for progress on their imperfection.

7. CHEMISTRY IS THE SCIENCE OF MOLECULES

Need I say more? There are limitations to the definition, but molecules, from diatomics to carbonic anhydrase and $YBa_2Cu_3O_{7-x}$ are our business. They're also big business—there are 10^{11} kilograms of sulfuric acid made per year around the world, and more pounds of ammonia than there are human beings.

8. CHEMISTRY IS NOT REDUCIBLE TO PHYSICS

Reductionism is something we've been saddled with for two centuries. By the term I mean the setting up of a hierarchical ordering of the sciences: social sciences, biology, chemistry, physics, and a corollary definition of understanding in one science as a reduction to the next, deeper level.[13] My feeling is that scientists buy this as an ideology, but it does not represent the reality of productive work in any science, and the idea is dangerous. I'll come back to the dangers later. What I think happens in reality is that from every activity of human beings there evolves a set of objects, questions and concepts. We might call these categories. Understanding may be defined vertically, in a reductionist way, but also horizontally, in terms of the concepts and questions of that science. Most of the useful concepts of chemistry (for the chemist: aromaticity, the concept of a functional group, steric effects, and strain) are imprecise. When reduced to physics they tend to disappear.[14] But with them marvelous chemistry has been—is—wrought.

What Really Goes on in a Chemical Paper

One could proceed and list other points of view on science and chemistry, but for me these suffice to begin a discussion of the chemical article and what transpires in it.

On the face of it the article purports to be a communication of facts, perhaps a discussion, always dispassionate and rational, of alternative mechanisms or theories, and a more or less convincing choice between them. Or the demonstration of a new measurement technique, a new theory. And, remarkably, the article works. An experimental procedure detailed in *Angewandte Chemie* in German or English can be reproduced (actually how easily it can be reproduced is another story[15]) by someone with a rudimentary knowledge of either language, working in Okazaki or Krasnoyarsk. This underlying feature of potential or real reproducibility is to me the ultimate proof that science is reliable knowledge.[16]

But in so many ways there is more than meets the eye in the scientific article. I see in it the following themes, many of which are also described and analyzed in a much deeper way in a remarkable book by *David Locke, Science As Writing*.[17]

ART

The chemical article is an artistic creation. Let me expand on what might be viewed as a radical exaggeration. What is art? Many things to many people. One aspect of art is aesthetic, another that it engenders an emotional response. In still another attempt to frame an elusive definition of that life-enhancing human activity, I will say that art is the seeking of the essence of some aspect of nature or of some emotion, by a human being. Art is constructed, human and patently unnatural.

What is written in a scientific periodical is not a true and faithful representation (if such a thing were possible) of what transpired. It is not a laboratory notebook, and one knows that that notebook in turn is only a partially reliable guide to what took place. It is a more or less (one wishes more) carefully constructed, man- or woman-made *text*. Most of the obstacles that were in the way of the synthesis or the building of the spectrometer have been excised from the text. Those that remain serve the rhetorical purpose (no weaker just because it's suppressed) of making us think better of the author. The obstacles that are overcome highlight the success story.

The chemical article is a man-made, constructed abstraction of a chemical activity. If one is lucky, it creates an emotional or aesthetic response in its readers.

Is there something to be ashamed of in acknowledging that our communications are not perfect mirrors, but in substantial part literary texts? I don't think so. In fact, I think that there is something exquisitely beautiful about our texts. These "messages that abandon," to paraphrase *Jacques Derrida*,[18] indeed leave us, are flown to careful readers in every country in the world. There they are read, in their original language, and understood; there they give pleasure *and,* at the same time, they can be turned into chemical reactions, real new things. It would be incredible, were it not happening thousands of times each day.

HISTORY

One of the oft cited distinguishing features of science, relative to the arts, is the more overt sense of chronology in science. It is made explicit in the copious use of references. But is it real history, or a prettified version?

A leading chemical style guide of my time admonished: "…one approach which is to be avoided is narration of the whole chronology of work on a problem. The full story of a research may include an initial wrong guess, a false clue, a misinterpretation of directions, a fortuitous circumstance; such details possibly may have entertainment value in a talk on the research, but they are probably out of place in a formal paper. A paper should present, as directly as possible, the objective of the work, the results, and the conclusions; the chance happenings along the way are of little consequence in the permanent record."[19]

I am in favor of conciseness, an economy of statement. But the advice of this style guide, if followed, leads to real crimes against the humanity of the scientist. In order to present a sanitized, paradigmatic account of a chemical study, one suppresses many of the truly creative acts. Among these are the "fortuitous circumstance"—all of the elements of serendipity, of creative intuition at work.[20]

Taken in another way, the above prescription for good scientific style demonstrates very clearly that the chemical article is *not* a true representation of what transpired or was learned, but a constructed text.

LANGUAGE

Scientists think that what they say is not influenced by the language they use, meaning both the national language (German, French, Chinese) and the words within that language. They think that the words employed, providing they're well-defined, are just representations of an underlying material reality which they, the scientists, have discovered or mathematicized. Because the words are faithful representations of that reality they should be perfectly translatable, into any language.

That position *is* defensible—as soon as the synthesis of the new high-temperature superconductor $YBa_2Cu_3O_{7-x}$, was described, it *was* reproduced, in a hundred laboratories around the globe.

But the real situation is more complex. In another sense words are all we have. And the words we have, in any language, are ill-defined, ambiguous. A dictionary is a deeply circular device—just try and see how quickly a chain of definitions closes upon itself. Reasoning and argument, so essential to communication in science, proceed in words. The more contentious the argument, the simpler and more charged the words.[21]

How does a chemist get out of this? Perhaps by realizing what some of our colleagues in linguistics and literary criticism learned over the last century.[22] The word is a sign, a piece of code. It signifies something, to be sure, but what it signifies must be decoded or interpreted by the reader. If two readers have different decoding mechanisms, then they will get different readings, different meanings. The reason that chemistry works around the world, so that BASF can build a plant in Germany or Brazil and expect it to work, is that chemists have in their education been taught the same set of signs.

I think this accounts in part for what *Carl Friedrich van Weizsäcker* noted in a perceptive article on "The Language of Physics":[23] If one examines a physics (read chemistry) research lecture in detail one finds it to be full of imprecise statements, incomplete sentences, halts, etc. The seminar is usually given extemporaneously, without notes, whereas humanists most often read a text verbatim. The language of physics or chemistry lectures is often imprecise. Yet chemists understand those presentations (well, at least some do). The reason is that the science lecturer invokes a code, a shared set of common knowledge. He or she doesn't have to complete a sentence—most everyone knows what is meant halfway through that sentence.

GRAPHICS

The semiotics of chemistry is most apparent in the structures of molecules that grace most every page of a chemical journal, that identify at a glance a paper as chemical.[24,25] The given, just over a century old, is that the structure of a molecule matters. It's not only what the atomic constituents are. It's also how the component atoms are connected up, how they are arranged in three-dimensional space, and how easily they move from their preferred equilibrium positions. The structure of a molecule, by which I mean the disposition in space of nuclei and electrons, both the static equilibrium structure and its dynamics, determines every physical, chemical, and ultimately biological property of the molecule.

It is crucial for chemists to communicate three-dimensional structural information to each other. The media for that communication are two-dimensional—a sheet of paper, a screen. So one immediately encounters the problem of representation.

Actually that problem is already there. What is a ball-and-stick model of a molecule? Is that reality? Of course not. The model is just one representation of the equilibrium positions of the nuclei, with some further assumptions about bonding. An instructive videotape from the laboratory of *F. P. Brooks. Jr.* at the University of North Carolina is entitled "What Does a Protein Look Like?" It shows 40 different representations (recalling to my mind the woodcut suite of *Hokusai* entitled "36 Views of Mt. Fuji") of one and the same enzyme, superoxide dismutase—a "ball-and-stick" model, a "space-filling" model, the electrostatic field experienced by a probe charge near the molecule, and so on.[26]

Let me return to the communication problem. Chemists are impelled to communicate three-dimensional information in two dimensions. But they're not *talented* at that. Young people are not selected or select themselves to be chemists on the basis of their artistic ability.

So people are required to do what they're not talented at. That's one definition of life! Chemists cope, by the expedient of inventing a code for communicating three-dimensional structure. *And* they train people in that code. The various elements of that code are known to chemists as Fischer or Newman projections, or as the wedge-dash representation. A molecule of ethane, C_2H_6, is represented in two possible geometries, 1 and 2, in Figure 13-4.

Note another piece of code: carbon is that vertex of the graph which has four lines to it, but it's not labeled C. There are hydrogens at the ends of the lines. They're also not labeled. The notation is simple: a solid line is in the plane of the paper, a wedge in front, a dashed line in back. Saying that may be enough to make these structures rise from the page for some people, but the neural networks that control representation are effectively etched in, for life, when one handles (in human hands, not in a computer) a ball-and-stick model of the molecule while looking at this picture.

It's fascinating to see the chemical structures floating on the page of every journal and to realize that from such minimal information people can actually *see* molecules in their mind's eyes. The clues to three-dimensionality are minimal. The molecules do float, and you're usually discouraged from putting in a reference set of planes to help you see them (3 vs. 4).

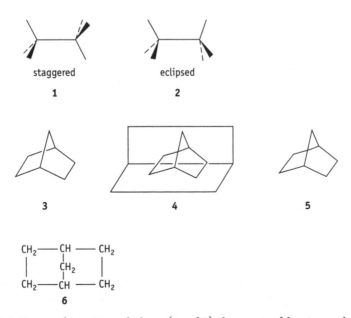

Figure 13-4 Two conformations of ethane (1 and 2), three ways of drawing norbornane (3, 4, 5); the way, 6, norborane was represented in journals until 1950.

Some chemists rely so much on the code that they don't draw **3** but **5**. What's the difference? One line "crossed" instead of "broken." What a trivial clue to three-dimensional reconstruction, that one part of a molecule is behind another one, is given in **3**, absent in **5**! The person who draws **5** is making many assumptions about his audience. I bet he or she hardly thinks about that.

The policies of journals, their economic limitations, and the available technology put constraints not only on what is printed but also on how we *think* about molecules. Take norbornane, C_7H_{12}, molecule **3**. Until about 1950 no journal in the world was prepared to typeset this structure as **3**. Instead you saw it in the journal as **6**. Now everyone knew, since 1874, that carbon was tetrahedral, meaning that the four bonds to it were formed along the four directions radiating out from the center of a tetrahedron to its vertices. You can see this geometry in the two carbon atoms of structures **1** and **2**. Molecular models were available or could be relatively easily built. Yet I suspect that the image or icon of norbornane that a typical chemist had in his mind around 1925 was **6** and not **3**. He was conditioned by what he saw in a journal or textbook—an image. He was moved to act, in synthesizing a derivative of this molecule, for instance, by that unrealistic image.

Maybe it's not that different from the way we approach romance in our lives, equipped with a piecewise reliable set of images from novels and movies.

STYLE

Every manual of scientific writing I've seen exhorts you to use an impersonal, agentless, superrational style. Please give us the facts, gentlemen, and just the facts! Still style rears its head.

My papers are recognizable in a journal just by the proportion of the space given to graphical material, or the way I "line" my orbitals. *R. B. Woodward's* papers were recognizable in the (constructed) elegance of their scientific storyline, and in the matching elegance, the cadence of his words. I can read a paper by *Jack Dunitz,* by *Rolf Huisgen, Rudolf Hoppe,* or *Bill Doering* and hear their voices in these papers just as surely as I hear the voice of *A. R. Ammons,* a great American poet, or of *Bertolt Brecht,* a German one, when I see their work on a printed page.

It is in the nature of creative human beings to have a style. Why should I write my theory the way *Bill Goddard,* a theorist I admire, does, anymore than you expect *Karlheinz Stockhausen* and *Pierre Boulez* to write piano pieces that sound alike?

DIALECTICAL STRUGGLES

A nice, even-toned, scientific article may hide strong emotional undercurrents, rhetorical maneuvering, and claims of power. One has already been mentioned—the desire to convince, to scream "I'm right, all of you are wrong," clashing with the established rules of civility supposedly governing scholarly behavior. Where this balance is struck depends on the individual.

Another dialogue that is unvoiced is between experiment and theory. There is nothing special about the love-hate relationship between experimentalists and theorists in chemistry. You can substitute "writer" and "critic" and talk about literature, or find the analogous characteristics in economics. The lines of the relationship are easily caricatured—experimentalists think theorists are unrealistic, build castles in the sky. Yet they need the frameworks of understanding that theorists provide. Theorists may distrust experiments, wish that people would do that missing experiment, but where would the theorists be without any contact with reality?

An amusing manifestation of the feelings about this issue is to be found in the occasionally extended quasi-theoretical discussion sections of experimental papers. These sections in part contain a true search for understanding, but in part what goes on in them is an attempt to use the accepted reductionist ideal (with its exaggerated hailing of the more mathematical)—so as to impress one's colleagues. On the other side, I often put more references to experimental work in my theoretical papers than I should, because I'm trying to "buy credibility time" from my experimental audience. If I show experimental chemists that I know of their work, perhaps they'll give me a little time, listen to my wild speculations.

Another struggle, related, is between pure and applied chemistry (see Figure 13-5). It's interesting to reflect that this separation also may have had its roots in Germany in the mid-nineteenth century; it seems to this observer that in the other chemical power of that time, Britain, the distinction was less congealed. Quite typical in a pure chemical paper is a reaching out after some justification in terms of industrial use. But at the same time there is a falling back, an unwillingness to deal with the often unruly, tremendously complicated world of, say,

Figure 13-5 Drawing by Constance Heller from the original publication of this article, *Angew. Chem. Int. Ed. Engl.* 27 (1988): 1600. © Wiley-VCH Verlag GmbH & Co. KGaA. Reproduced with permission.

industrial catalysis. And in industrial settings there is a reaching after academic credentials (quite typical, for instance, of the leaders of the chemical industry in Germany).

THE ID WILL OUT

I use the subject word in the psychoanalytical sense, referring to the complex of instinctive desires and terrors that inhabit the collective unconscious. On one hand, these irrational impulses, sex and aggression figuring most prominently, are our dark side. On the other hand, they provide the motive force for creative activity.[27]

The irrational seems to be effectively suppressed in the written scientific word. But of course scientists are human, no matter how much they might pretend in their articles that they are not. Their inner illogical forces push out. Where? If you don't allow them in the light of day, on the printed page, then they will creep out or explode in the night, where things are hidden, and no one can see how nasty you are. I refer, of course, to the anonymous refereeing process, and the incredible irrational responses unleashed in it by perfectly good and otherwise rational scientists. You have to let go sometime...

I actually think that what saves the chemical article from dullness is that its language comes under stress. We are trying to communicate things that perhaps cannot be expressed in words but require other signs—structures, equations, graphs. And we are trying hard to eliminate emotion from what we say. Which is impossible. So the words we use occasionally become supercharged with the tension of everything that's *not* being said.

So, many things go on in a scientific article. I will spare you the charts that I've seen hanging in many laboratories, that translate shop-worn phrases into what they really mean. Like "A regime of slow cooling over a period of four weeks produced a 90% yield of black crystals of..." meaning "I went on vacation and forgot to wash out the flask. When I came back..." And so on.

I want to return to the complexity of deciding what is truth in the scientific process. In a 1985 article, *Harold Weinrich*[28] describes the classic paper of *Watson* and *Crick* on the structure of DNA in *Nature* in 1953. It is succinct, beautifully reasoned, elegant. *David Locke,* in his book, analyzes the rhetorical structure and the use of irony in the same paper.[17] I think most readers were immediately convinced that the Watson and Crick model was the truth, that this was the way it had to be. And so it is (with some minor variation on hydrogen-bonding schemes in unusual forms of DNA). The Watson and Crick model was and is right. One needs little more.

But then in 1968 *Watson* writes a book, *The Double Helix,* in which he tells the story of how it really happened. Of course, it's a self-centered story, and not kind or fair to *Rosalind Franklin* and others. *Watson's* story is much like one of the four views of an event in *Akira Kurosawa's* masterful film *Rashomon.* It needs other views, which some historians have provided. But there is no question that

Watson's account is vibrant, full of life. It tells us how he saw the truth, and I think it's a great book.

So (and here I follow *Weinrich*) what *is* the truth: Is it the 1953 paper of *Watson* and *Crick,* or is it the 1968 book by *Watson!* Does the latter diminish the former? It bears reflection.

What Is to Be Done?

I have tried to deconstruct, to borrow some language from critical theory, the scientific article, ossified in its present form for over a hundred years. I think I could have done this better with a specific text before me, but then I would have run the danger of libel suits or loss of friendship. And it is in the nature of human beings to be unable to criticize, truly, deeply, their own work. So I can't do it to myself. But I'm sure every chemist knows some superb example of what I allude to, especially in the work of one of his less favorite scientists.

The deconstruction is done by me not with malice, but with care. I love this molecular science. I love its richness, and the underlying simplicity, but most of all the rich, life-giving variety and connection of all of chemistry. Let me give you an example in Figure 13-6. I see C_2 in a carbon arc and in the tail of Halley's Comet.[29a] I see it in acetylene, ethylene, and ethane. I see it in the lovely molecular complexes of *P. Wolczanski,*[29b] *M. I. Bruce,*[29c] and *G. Longoni*[29d] and their co-workers, shown in Figure 3. I see it in CaC_2, in *W. Jeitschko's*[29e] and *A. Simon's*[29f] and their co-workers' rare-earth carbides. It's staggering!

I know that that richness was created by human beings. So I'm unhappy to see their humanity suppressed in the way they express themselves in print. I also think that there are some real dangers in the present intellectual stance of scientists and in their ritual mode of communication.

Here are the dangers as I perceive them. Acceptance of a facile reductionist philosophy, a vertical mode of understanding appropriated as the *only* mode of understanding, creates a gap between us and our friends in the arts and humanities. They know very well that there isn't only one way of "understanding" or dealing with the death of a parent, or a political election, or a woodcut by *Ernst Ludwig Kirchner.* The world out there is refractory to reduction, and if we insist that it must be reducible, all that we do is to put ourselves into a nice box. The box is the small class of problems that are susceptible to a reductionist understanding.

A second danger, more specific to the scientific article, is that by dehumanizing our mode of communication, by removing emotion, motivation, the occasionally irrational, we may have in fact done much more than chase away the Natural Philosophers. That we've accomplished, to be sure. But 150 years down the line what we have created is a mechanical, ritualized product that 3×10^5 times per year propagates the notion that scientists are dry and insensitive, that

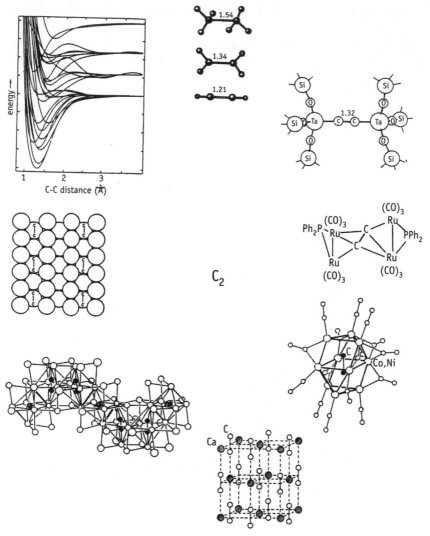

Figure 13-6 Some of the ways C₂ finds its way into the world.

they respond only to wriggles in a spectrum. The public at large types us by the nature of our product. How can it do otherwise when we do not make a sufficient effort to explain to the public what it is that we really do in our jargon-barricaded world?

What is to be done? I would argue for a general humanization of the publication process. Let's relax those strictures, editorial or self-imposed, on portraying in words, in a primary scientific paper, motivation, whether personal and scientific, emotion, historicity, even some of the irrational. So what if it takes a little more space? We *can* keep up with the chemical literature, and tell the mass of hack work from what is truly innovative, without much trouble as it is. The humanizing words will not mislead; they may actually encourage us to read more carefully the substance of what is said. I would plead for a valuation and teaching of style, in the written and spoken language of one's own country, as well as in English. I think chemistry has much to gain from reviving the personal, the emotional, the stylistic core of the struggle to discover and create the molecular world.

Notes

1. *Angewandte Chemie* has been very important to me personally, and it is a pleasure for me to dedicate this lecture to the past and present editorial staffs of one of the world's greatest journals.
2. E. Garfield, *Essays of an Information Scientist* (Wilmington, DE: ISI Press, 1981), pp. 394–400 and references therein.
3. (a) S. Shapin, Pump and Circumstance: Robert Boyle's Literary Technology. *Social Stud. Sci.* 14 (1984): 487; (b) P. Dear, Totius in Verba: Rhetoric and Authority in the Early Royal Society. *Isis* 76 (1985): 145; (c) F. L. Holmes in *The Literary Structure of Scientific Argument: Historical Studies*, ed. P. Dear (Philadelphia: University of Pennsylvania Press, 1991), pp. 164–181.
4. F. Goldmann, Ueber Derivate des Anthranols. *Ber. Deutsche Chem. Ges.* 21 (1888): 1176.
5. For a discussion of the evolution of science writing see B. Coleman, Science Writing: Too Good to Be True. *New York Times Book Review* (September 27, 1987), p. 1; also R. Wallsgrove, Selling Science in the 17th Century. *New Sci.* 116 (1987): 55.
6. O. J. Scherer and T. Brück, $[(\eta^{-5}\text{-}P_5)Fe(\eta^5\text{-}C_5Me_5)]$, a Pentaphosphaferrocene Derivative. *Angew. Chem.* 99 (1987): 59; *Angew. Chem. Int. Ed. Engl.* 26 (1987): 59.
7. The version shown here is in English. *Angewandte Chemie* is unique in printing (since 1963) the same article in English and German versions.
8. For a demonstration of the range of attitudes about what science is, see the discussion in the pages of *Nature* 330 (1987): 308, 689; 331 (1988): 129, 558; following an article by T. Theocharis and M. Psimopoulos, Where Science Has Gone Wrong. *Nature* 329 (1987): 595.
9. The phrase is used after J. Ziman, *Reliable Knowledge* (Cambridge: Cambridge University Press, 1978). I disagree with some points in this book, but there is no better, nor more humanistic, description of what science is and should be than this small volume.
10. M. Berthelot, *Chimie Organique Fondée sur la Synthèse, Tome 2* (Paris: Mallet-Bachelier, 1860). See also J.-P. Malrieu, Du Devoilement au Design: Une Evolution des Sciences ou la Chimie a Cent Ans d'Avance. *L'Actualité Chimique* (1987), Mars: IX; A. F. Bochkov and V. A. Smit, *Organicheskii Sintez* (Organic Synthesis) (Moscow: Nauka, 1987); and R. Hoffmann, In Praise of Synthesis. *American Scientist* 79 (1991): 11.
11. For different views of the way science works, see (a) P. Feyerabend, *Against Method* (NLB, 1975); *Wider den Methodenzwang* (Frankfurt: Suhrkamp, 1976); (b) B. Latour and S. Woolgar, *Laboratory Life* (Princeton: Princeton University Press, 1986); (c) K. Knorr-Cetina, *Die Fabrikation von Erkenntnis* (Frankfurt: Suhrkamp, 1984).
12. I was reminded of the importance of the individual-system distinction through a conversation with *Barry Carpenter.* I'm grateful to him for a discussion of the points raised in this essay, as well as to another colleague, *Bruce Ganem,* for his comments.
13. Several types of reductionism must be distinguished. See the interesting debate between S. Weinberg, Newtonianism, Reductionism and the Art of Congressional Testimony. *Nature* 330 (1987): 443; S. Weinberg, The Limits of Reductionism. *Nature* 331 (1988): 475; and E. Mayr, The Limits of Reductionism. *Nature* 331 (1988): 475 and references therein.
14. See, *inter alia*, K. Mislow and P. Bickart, An Epistemological Note on Chirality. *Israel J. Chem.* 15 (1976/77): I; D. W. Theobald, Some Considerations on the Philosophy of Chemistry. *Chem. Soc. Rev.* 5 (1976): 203.

15. *R. G. Bergman,* in an unpublished lecture on "Values in Science," cites some fascinating raw data on this question, coming from the experience of the journals *Organic Synthesis* and *Inorganic Synthesis.*
16. Latour and Woolgar ([Ref. 11b], p. 183) are rather scornful of the kind of marveling at verification and reproducibility of scientific facts that I engage in here. I think they've gotten caught in the consistently questioning and skeptical ideology of their otherwise incisive anthropological investigation of how scientific facts are constructed. They should take a look at the systematic, worldwide, industrial production of pharmaceuticals, just to take one example of a reproducible experimental activity.
17. D. Locke, *Science As Writing* (New Haven: Yale University Press, 1992).
18. J. Derrida in his essay "Signature Event Context" in *Marges de la Philosophie* (Paris: Editions Minuit, 1972): p. 365; translated by A. Bass as *Margins of Philosophy* (Chicago: University of Chicago Press, 1982), p. 307.
19. L. F. Fieser and M. Fieser, *Style Guide for Chemists* (New York: Reinhold, 1960), p. 51.
20. P. B. Medawar, *Saturday Review* (August 1, 1964), p. 42, also argues that the standard format of the scientific article misrepresents the thought processes that go into discovery.
21. See R. Hoffmann, Unstable. *Am. Sci.* 75 (1987): 619; R. Hoffmann, Nearly Circular Reasoning. *Am. Sci.* 76 (1988): 182. Chapters 4 and 5 in this book.
22. For an introduction to modern literary theories see T. Eagleton, *Literary Theory* (Minneapolis: University of Minnesota Press, 1983).
23. C. F. von Weizsäcker, *Die Einheit der Natur* (dtv, 1974), p. 61.
24. For a description of the geometrical and topological information processing that goes on in organic chemistry, see also N. J. Turro, Geometric and Topological Thinking in Organic Chemistry. *Angew. Chem.* 98 (1986): 872; *Angew. Chem. Int. Ed. Engl.* 25 (1986): 882.
25. Pierre Laszlo has written an illuminating article on technical illustration that is relevant to my discussion: P. Laszlo, Pictures of Science. *Interdisciplinary Science Reviews* 20 (1995): 51.
26. M. Pique, J. S. Richardson, and F. P. Brooks, Jr., *Invited Videotape. 1982 SIGGRAPH Conference.* I am grateful to J. S. Lipscomb for showing this videotape to me.
27. P. B. Medawar says, along the same lines, that "scientists should not be ashamed to admit...that hypotheses appear in their minds along uncharted by-ways of thought..." (see endnote 20).
28. H. Weinrich, Sprache und Wissenschaft. *Merkur* 39 (1985): 469. I'm grateful to P. Gölitz for bringing this paper to my attention.
29. (a) The C_2 potential energy curves are drawn after P. P. Fougere and R. K. Nesbet, Electronic Structure of C_2. *J. Chem. Phys.* 44 (1966): 285; (b) $[({}^tBu_3SiO)_3Ta]_2C_2$: R. E. LaPointe, P. T. Wolczanski, and J. F. Mitchell, Carbon Monoxide Cleavage by (silox)^{3}Ta (silox=tert-Bu^{3}SiO-). *J. Am. Chem. Soc.* 108 (1986): 6382; (c) $[Ru_4C_2 (PPh_2)_2(CO)_{12}]$: M. I. Bruce, M. R. Snow, E. R. T. Tiekink, and M. L. Williams, The First Example of a μ_4-η^1, η^2-acetylide Dianion: Preparation and X-ray Structure of $Ru_4(\mu_4$-η^1,η^2-$C_2)(\mu$-$PPh_2)_2(Co)_{12}$. *J. Chem. Soc. Chem. Commun.* (1986): 701; (d) $[Co_3Ni_7C_2 (CO)_{15}]^{3-}$: G. Longoni, A. Ceriotti, R. Della Pergola, M. Manassero, M. Perego, G. Piro, and M. Sansoni, Iron, Cobalt and Nickel Carbide—Carbonyl Clusters by CO Scission. *Phil. Tran.. R. Soc. London. Ser. A* 308 (1982): 47; (e) DyCoC$_2$: W. Jeitschko and M. H. Gerss, Ternary Carbides of the Rare Earth and Iron Group Metals with CeCoC$_2$- and CeNiC$_2$-type Structure. *J. Less. Common Met.* 116 (1986): 147; (f) Gd$_{12}$C$_6$I$_7$: A. Simon and E. Warkentin, Gd$_{12}$C$_6$I$_{17}$—eine Verbindung mit Kondensierten, C$_2$-gefüllten Gd$_6$I$_{12}$-Clustern. *Z. Anorg. Allg. Chem.* 497 (1983): 79.

14

Representation in Chemistry

ROALD HOFFMANN AND PIERRE LASZLO

Modes of Representation

If you look in old chemistry books
you see
all those line cuts
of laboratory experiments
in cross-section.
The sign for water
is a containing line, the meniscus
(which rarely curls up the walls of the beaker),
and below it
a sea
of straight horizontal dashes
carefully unaligned vertically.
Every cork or rubber stopper
is cutaway.
You can see inside
every vessel
without reflections, without getting wet,
and explore every kink
in a copper condenser.
Flames are outlined cypresses
or a tulip at dawn,
and some Klee arrows
help to move gases and liquids the right way.
Sometimes a disembodied hand
holds up a flask.
Sometimes there is an unblinking observer's eye.
Around 1920
photoengraving
became economically feasible

and took over.
Seven-story distillation columns
(polished up for the occasion),
like giant clarinets,
rose in every text, along
with heaps of chemicals, eventually in color.
Suddenly
water and glass, all reflection
became difficult.
One had to worry about light,
about the sex
and the length of dress or cut of suit
of the person sitting at the controls of this impressive
instrument.
Car models and hairstyles
dated the books more
than the chemistry in them.
Around that time
teachers noted a deterioration
in the students' ability to follow
a simple experimental procedure.

ROALD HOFFMANN

Introduction

Chemical structures are among the trademarks of our profession, as surely chemi-cal as flasks, beakers, and distillation columns. When someone sees one of us busily scribbling formulas or structures, he has no trouble identifying a chemist. Yet these familiar objects, which accompany our work from start to end, from the initial doo-dlings (Figure 14-1) to the final polished artwork in a publication (Figure 14-2), are deceptively simple. They raise interesting and difficult questions about repre-sentation. It is the intent of this article to reflect upon molecular graphics.

We are hooked on these little diagrams, aren't we? Yet what are they? Are they representations of reality, just a simplified two-dimensional version of the models that can be built from interpretation of X-ray diffraction patterns of a molecular crystal—in a word, are they *realistic*? A look at a few papers by others (not ours or yours, of course) will show how far short of realism these structures fall. They mix convention and realism in the most innocent manner. Take the case of the bicyclo[2.2.1]heptane skeleton (**1** in Figure 14-3):

This commonly seen, supposedly three-dimensional representation actu-ally sports an inverted perspective, the vanishing point being not in back, where it should be, but on the viewer's side. Most chemical drawings eschew obvious

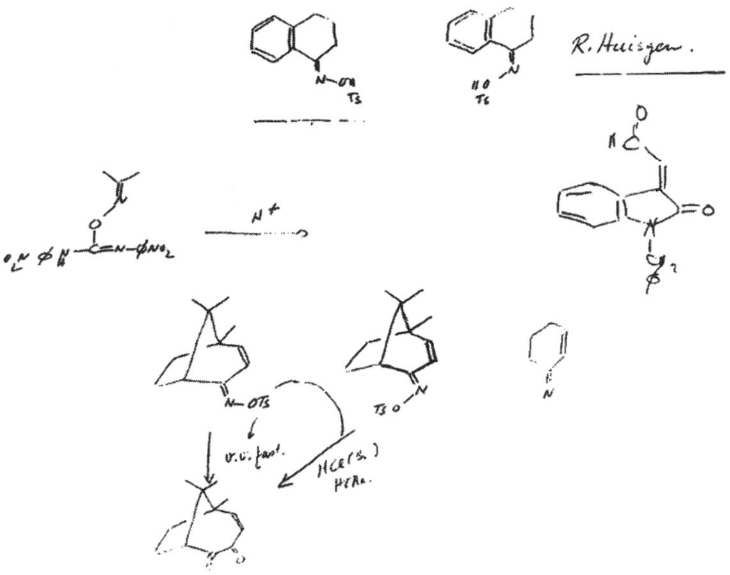

Figure 14-1 A drawing by R. B. Woodward, circa 1966, in the course of a discussion.

primitive artistic stratagems for communicating three-dimensionality; they float in a world of their own.

Perhaps they are "art," then, an abstraction from reality of the essence of norbor-nanes. If so, it's interesting to reflect on what kind of art they are. This is part of what we will do.

Perhaps chemical drawings do not need to be realistic representations because they are symbols, signs that in a chemist's mind are reconstructed into the three-dimen-sional structure, or at least the ball-and-stick model. Chemical structures are then part of a chemical language. What is interesting about language (it doesn't matter whether it is German or English or ...) is that (1) despite its impreciseness, people communi-cate with it and (2) it, language, inevitably brings us complications, ambiguities, and richnesses that we did not expect. Or perhaps that we subconsciously intended.

We both studied chemistry at a time when elongated benzene rings (Figure 14-3, **2a**) went out of fashion. Two arguments can be made about this iconic upheaval. True, the old way did not prevent incredible progress in synthetic organic chem-istry, nor the perceptive theorizing of *E. Hückel.* On the other hand, the new and assuredly more accurate representation by a regular hexagon (**2b**) did not serve only a cosmetic and PR purpose for the novel, powerful physical organic and theoreti-cal chemistries associated with names such as *Saul Winstein* and *Charles Coulson.* It beaconed a change of focus, however subtle. This was the same time, remember, when conformational analysis came into general use. After decades of estrangement, structural chemistry and reactivity studies were coming back under the same roof. The writing of a structure is not innocent. It is ideology-laden. It carries, besides its face value, another message; in this case, the modern reunification of the theoretical and the experimental.

1602

J. CHEM. SOC., CHEM. COMMUN., 1988

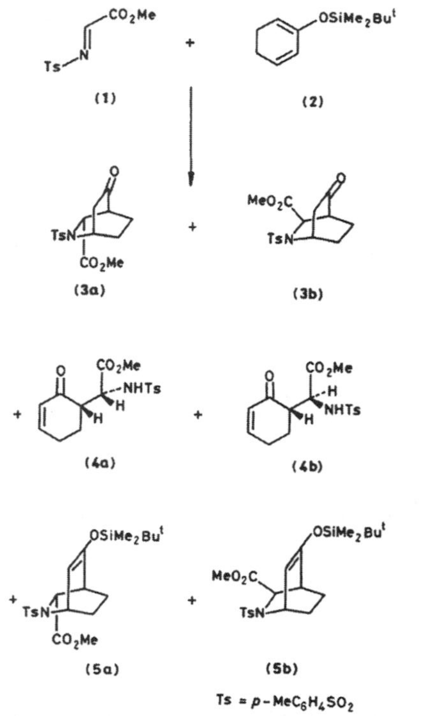

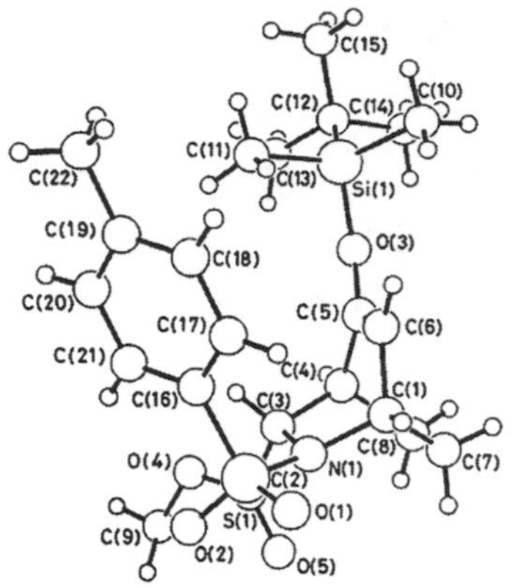

Figure 1. The structure of the *exo* silyl enol ether (5a).

interaction between the *N*-tosyl and t-butyldimethylsilyl groups is minimised,[5] while (4a) is derived from an open transition state in which the bulky silyl substituent avoids the steric clash with the ester group of the imine (1). Thus the proposed mechanism (c) in the previous paper,[1] involving a dual pathway explanation[6] for the origin of bicyclic and monocyclic products, seems to be supported in the present study.

We thank the S.E.R.C. for supporting this work, Merck, Sharp and Dohme, Harlow for a CASE award (A. B. T.), and Clare College for financial support (T. N. B.) and the award of the Denman Baynes Studentship (A. B. T.).

Received, 11th May 1988; Com. 8/01856A

The results of a systematic variation of solvent, temperature, and Lewis acid on the outcome of the reaction of (1) with (2) resemble the pattern reported for the trimethylsilyl analogue of (2), but with several important differences. Firstly, the *exo/endo* ratio of bicyclic adducts derived from (2) is greater than that for the trimethylsilyl analogue, and secondly, there is a corresponding greater preference for (4a) over (4b). Under certain low temperature conditions only cyclohexenone products (4a,b) and no bicyclic products were obtained. Hydrolysis of the hindered bicyclic silyl enol ether (5a) under relatively forcing conditions (AcOH–THF, 50 °C) barely produced any retro-Michael product (4b). These observations are consistent with (3a) and (5a) being derived from a [2 + 4] transition state in which the unfavourable steric

References

1 T. N. Birkinshaw, A. B. Tabor, A. B. Holmes, P. Kaye, P. M. Mayne, and P. R. Raithby, preceding communication.
2 W. G. Dauben and R. A. Bunce, *J. Org. Chem.*, 1982, **47**, 5042.
3 F. Brisse, D. Thoraval, and T. H. Chan, *Can. J. Chem.*, 1986, **64**, 739.
4 A. J. Kirby, 'The Anomeric Effect and Related Stereoelectronic Effects at Oxygen,' Springer-Verlag, Berlin, 1983; K. L. Brown, L. Damm, J. D. Dunitz, A. Eschenmoser, R. Hobi, and C. Kratky, *Helv. Chim. Acta*, 1978, **61**, 3108; L. I. Kruse, C. W. DeBrosse, and C. H. Kruse, *J. Am. Chem. Soc.*, 1985, **107**, 5435 and references cited therein.
5 S. M. Weinreb and R. R. Staib, *Tetrahedron*, 1982, **38**, 3087.
6 S. J. Danishefsky, E. Larson, D. Askin, and N. Kato, *J. Am. Chem. Soc.*, 1985, **107**, 1246.

Figure 14-2 A page (the second of two) from a contemporary article: T. N. Birkinshaw, A. B. Tabor, A. B. Holmes, and P. R. Raithby, *J. Chem. Soc. Chem. Commun.* (1988): 1602. Reproduced by permission of The Royal Society of Chemistry.

The ability to see a conformation behind a constitution, so that the set of symbols 3 is understood by an organic chemist as 4, engages two types of skills. The imagination, the visual ability to perceive a shape in three-dimensional space runs second to *translation*. To go from the first, "flat" set of symbols, to the second, "skewed" set of symbols is equivalent to switching from one language (the word "science") to another (the word "Wissenschaft"). This example drives home how much of a linguistic component there is in the thought processes of chemists, as activated by these small sketches that we call chemical structures. Part of our paper deals explicitly with chemical structures as constituting a language.

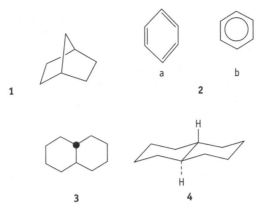

Figure 14-3 Norbornane (**1**), two ways of drawing benzene (**2**), and two ways of drawing trans-decalin (**3** and **4**). See text for discussion.

The Shapes of Molecules and How They Are Communicated

When we wish to probe something which is so ingrained in us that it is second nature (as the structural formula is), it is useful to put oneself outside. Imagine telling someone intelligent, attentive, and sympathetic, but someone who is not a chemist, about structure and its communication in our science. Here is what we might say:[1]

Shape matters in chemistry. Two molecules as subtly different from each other as a left hand is from a right, may have quite different physical, chemical, and biological properties. Thus, the mirror image of carvone, the main component of oil of spearmint, smells of caraway. The arrangement of atoms in space is not just a laboratory curiosity, it can be a matter of life or death. Thalidomide, a sedative of the early sixties, was responsible for thousands of fetal malformations in Europe. The pharmaceutical marketed was a mixture of left- and right-handed mirror image molecules. One form was teratogenic, causing malformation, its mirror image was not. Had this been known at the time, great anguish and human loss could have been prevented.[2]

Molecules are made up of atoms. (Forgive us, chemist friends; we're still continuing our hypothetical monologue directed at the nonchemist.) But molecular structure is not just the identity of the atoms. Or even how they are connected up, those elemental atomic building blocks within the molecule. At the operative level of modern chemistry, structure means the three-dimensional arrangement of atoms in space.[3] It is a graph, at the very least, a three-dimensional set of points connected by lines called bonds.

It is critical that chemists easily communicate this structural information among themselves. Via what's at hand, which are two-dimensional media—paper, a screen. The information is complex—many atoms, many bonds, a richness of geometrical structure. The information is at some important level inherently graphic—it is essentially a shape to be drawn. And now we come to the crux of the matter. The group of professionals to whom this visual, three-dimensional information is essential are not talented (any more, any less) at transmitting such information. Chemists are not selected, do not select themselves, for their profession on the basis of their artistic talents. Nor are they trained in basic art technique. The authors' ability to draw a face so that it looks like a face atrophied at age ten.

So how do they do it, how do we do it? With ease, almost without thinking, but, as we will see, with much more ambiguity than we, the chemists, think there is. The process is *representation,* a symbolic transformation of reality. It is both *graphic* and *linguistic.* It has a *historicity.* It is *artistic* and *scientific.* The representational process in chemistry is a shared code of this subculture.

Let us begin our look at the process by a look at the outcome. This was shown in Figure 14-2, a typical page from a modern chemical article. The substantial amount of graphic content just stares one in the face. There are little pictures here. Lots of them. But the intelligent observer who is not a chemist is likely to be stymied. He finds himself in a situation analogous to that of *Roland Barthes* on his first visit to Japan, beautifully described in his "The Empire of Signs."[4] What do these signs mean? We know that molecules are made of atoms, but what is one to make of a polygon such as structure **5**, here representing a white, waxy medicinal compound with a penetrating aroma, camphor? Only one familiar atomic symbol, O for oxygen, emerges.

Well, it's a shorthand. Just as the military man gets tired of saying Commander in Chief, South Pacific Operations, and writes CICSPO, so the chemist tires of writing all those carbons and hydrogens, ubiquitous elements that they are, and draws the carbon skeleton. Every vertex and endpoint that is not specifically labeled otherwise in structural representation **5** (see Figure 14-4) of camphor is carbon. Since the valence of carbon (the number of bonds it forms) is typically four, chemists privy to the code will know how many hydrogens to put at each carbon. The polygon drawn **5** is in fact a graphic shorthand for structure **6**.

But is **6** the true structure of the molecule of camphor? Yes and no. At some level it is. At another level the chemist wants to see the three-dimensional picture, and so draws **7**. At still another level, he or she wants to see the "real" interatomic distances, that is, the molecule drawn in its correct proportions. Such critical details are available, with a little money, a little work, by a technique called X-ray crystallography. And so we have a drawing **8**, likely to have been produced by a computer.[5]

This is a view of a so-called ball-and-stick model, perhaps the most familiar representation of a molecule in this century. The sizes of the balls representing the carbon, hydrogen, and oxygen atoms are somewhat arbitrary. A more "realistic" representation of the volume that the atoms actually take up is given by the "space-filling" model **9**. Note that in **9** the positions of the atoms, better said of their nuclei, become obscured. And neither **8** or **9** is portable. It cannot be sketched by a chemist

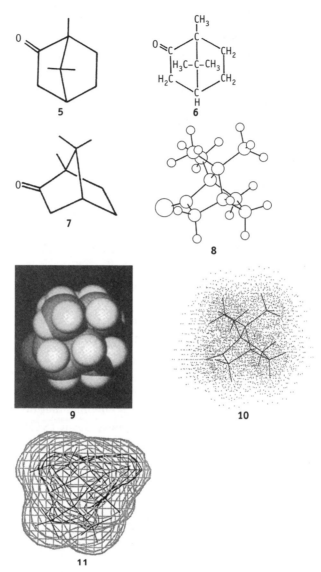

Figure 14-4 Seven representations of camphor.

in the 20 seconds that a slide typically remains on a screen in the rapid-fire presentation of the new and intriguing by a visiting lecturer.

The ascending (descending?) ladder of complexity in representation hardly stops here. Along comes the physical chemist to remind his or her organic colleagues that the atoms are not nailed down in space, but moving in near harmonic motion around those sites. The molecule vibrates; it doesn't have a static structure. Another chemist comes and says: "You've just drawn the positions of the nuclei. But chemistry is in the electrons, you should draw out the chance of finding them at a certain place in space, the electronic distribution." As one tries to do in **10** and **11**.

We could go on. The literature of chemistry does. But let's stop and ask: Which of these representations, 5 through 11, is right? Which *is* the molecule? Well, all are, and none is. Or, to be serious—all of them are models, representations suitable for some purposes, not for others.[6] Sometimes just the name "camphor" will do. Sometimes the formula. $C_{10}H_{16}O$, suffices. Often it's the structure that's desired, and something like 5 or 7 is fine. At other times one requires 8 or 9, or even 10 or 11.[7]

A final story needs to be told of camphor. We picked this molecule as one recognizable to the public, but carrying within it minimal complexities of representation. One of us *(R. H.)*. having forgotten its structure, checked it in a textbook, then specified to some friends the geometry needed to produce the beautiful drawings of camphor in this paper. Every one of them was the mirror image of what you see, which is the naturally occurring, dextrorotatory material *(1R,4R* in configuration)! That we had the wrong absolute configuration was pointed out to us by a careful reader, *Ryoji Noyori,* the 1990 Baker Lecturer at Cornell. A literature search then revealed the wrong configuration disported by many, if not most textbooks, the Merck Index, and numerous literature papers, such as the important one by G. M. *Whitesides* and D. W. *Lewis,*[8a] on the use of an NMR shift reagent to determine enantiomeric purity. The structure is correctly given in the Sigma, Aldrich, and Fluka catalogues; the references to the assignment may be found in the Klyne and Buckingham compendium.[8b]

Naive realism asserts that chemical formulas resemble reality: they do. It is possible to obtain pictures of benzene rings by physical means. They look, sometimes, like the benzene rings of the chemist (2). Sometimes they don't. The scientist who thinks that now, with scanning tunneling microscopy (STM), one can finally see atoms in molecules, has a shock coming when he or she looks at an STM image of graphite. Half the hexagonal lattice atoms are highlighted in the image, half not. For good reasons. Seeing and believing have a complex relationship to each other. These (the benzene rings of the chemist) are rough approximations. They stand not unlike a metaphor to the molecular object represented.

Let us fix on the *typical* level of presentation (of Figures 14-1 and 14-2), that of a polygon (5) or a three-dimensional idealization of it (7). But what *are* these curious constructions, these drawings, filling the pages of a scientific paper? We now ask the question from the point of view of an artist or draftsman. They're not isometric projections, certainly not photographs. Yet they're obviously attempts to represent in two dimensions a three-dimensional object for the purpose of communicating its essence to some remote reader.

The clues to three-dimensionality in these drawings are minimal. Some are conventional: here and there (one example in 7), there is a line "cut" to establish that some piece of the structure is in front of another, that is, 12 instead of 13 (see Figure 14-5). This is hardly a modern invention, something that one must learn at the Ecole des Beaux Artes. Figure 14-6 shows one of the cave paintings from Lascaux.[9] Note the treatment of the legs of the bisons as they go into the body. Now

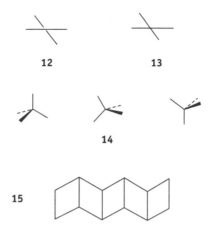

Figure 14-5 Clues to three-dimensionality (**12** vs.**13**), three representations of methane (**14**), and a rhomboid pattern that is difficult to see as flat (**15**).

Figure 14-6 Two bisons from Lascaux Cave.

smart chemists should be able to do what cavemen 15,000 years ago did, shouldn't they? Too often they don't bother.

Scattered about in the drawings of Figures 14-1 and 14-2 are sundry wedges and dashed lines. These are pieces of a visual code, simple in conception: a solid line is in the plane of the paper, a wedge in front, a dashed line in back. Thus, **14** shows several views, all quite recognizable to chemists, of the tetrahedral methane molecule, CH_4. The tetrahedron is the single most important geometrical figure in chemistry. (Be patient, colleagues, think of telling your father-in-law about what you are doing.)

Describing this notation may be enough to make these structures rise from the page for some people, but the neural networks that control representation are effectively etched in, for life, when one handles (in human hands, not in a computer) a ball-and-stick model of the molecule while looking at its picture.

A glance at the more complicated molecules of Figure 14-2 shows that the wedge-dash convention is not applied consistently. Most compounds have more than a single plane of interest; what's behind one plane may be in front of another. So the convention is almost immediately used unsystematically, the author or lecturer choosing to emphasize the plane he or she thinks important. The result is a cubist perspective, a kind of Hockney photocollage.[10] The molecule is certainly seen, but may not be seen as the scientist thinks (in a dogmatic moment) that it is seen. It is represented as he chooses to see it, nicely superimposing a human illogic on top of an equally human logic.

They're floating in space, curiously unframed, these representations. You search in vain for a set of reference planes, a chair, a figure to orient you. But these are not given. The reader must decode the three-dimensionality with little help. Mind you it's not that difficult—there is an innate drive to see things as three-dimensional. Witness the difficulty we have in seeing the pattern of connected rhomboids, **15**, as flat.[11a]

Let us now come out of the didactic monologue. The hypothetical nonchemist, if he or she has persevered, has now partaken of the knowledge (prejudices?) that the chemist has about structures. We can proceed together.

The question remains: what do these chemical structures represent? How are they drawn and read? *Philostratus* tells, much mythologized, the story of *Apollonius of Tyana*, a Pythagorean who lived around the time of Christ. In a dialogue with a disciple *Apollonius* explores what painting is. It's done to make a likeness, to imitate. But what about cloud shapes in the sky, read by us as horses or bulls? Are those also imitations? *Apollonius* and his disciple agree that these are but chanced configurations, that it is we who interpret those shapes, give them meaning. He continues: "But does this not mean that the art of imitation is twofold? One aspect of it is the use of hands and mind in producing imitations, another aspect the producing of likenesses with the mind alone . . . I should say that those who look at works of painting and drawing must have the imitative faculty and that no one could understand the painted horse or bull unless he knew what such creatures are like."[11b]

Knowing is not an unproblematical concept. How does that three-dimensional structure unfold in its full glory in the mind of a chemist? As we said, the direct images produced by contemporary techniques such as scanning tunneling microscopy or electron microscopy (and these are not so "direct" on close examination) are few. Secondary knowledge, through X-ray crystallography, microwave spectroscopy, or electron diffraction, is experienced still by only a small number of specialists. For most of us it is the real, physical handling of models that sets the stage—the analogy to seeing *Apollonius'* bull or horse in the first place. Or looking at many pictures of molecules drawn by others, assimilating thereby the set of conventions shared by chemists. It's much like art, and we will return to this below.

History, Media

How did the chemists' way of seeing and drawing evolve? One can choose a starting point for chemistry almost at will, and ours will be *Antoine Laurent Lavoisier*. In his time, the end of the 18th century, there were no chemical structures of type **5-11**. The idea of an element was just taking form; the symbols for these elements were still not codified. The results of chemical experiments could be perfectly well communicated in words and, especially after *Lavoisier*, with numbers.

Illustrations figure importantly in *Lavoisier's* work, nevertheless. They are often the illustrations of his experimental equipment, of glass and metal containers, gauges, barometers, all exquisitely detailed. A plate from *Lavoisier's* 1789 classic "Traité Elementaire de Chimie" is shown in Figure 14-7.[12a] The appeal of these illustrations is in large part due to their creator: *Mme. Lavoisier, Marie Anne Pierrette Paulze*. Her original sketches, as well as her corrections of these engravings, are available. *Mme. Lavoisier* was an accomplished artist, a student of *David*.[12b]

The media available to the scientific communication process of the time were quite circumscribed. In lecture presentations it was the voice, demonstrations, and blackboard and chalk. Illumination levels were low. In printed presentations one

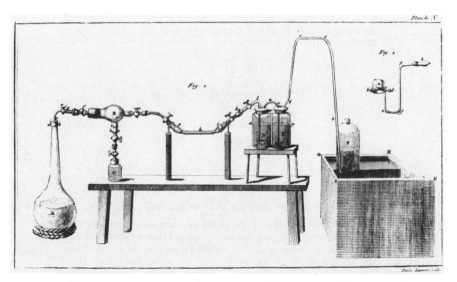

Figure 14-7 Plate X From Lavoisier's classic "Traité Elementaire de Chimie." The drawing is by Mme. Lavoisier. Courtesy of Division of Rare and Manuscript Collections, Cornell University Library.

had, aside from normal typesetting possibilities, woodcuts and engravings, usually copperplate.

When did it become essential to communicate three-dimensional chemical information? By the middle of the 19th century there arose some pressure to represent the ways atoms were linked to each other. Figure 14-8 shows a page from a crucial 1852 paper by *A. W. Williamson*.[13] Note the illustration, presumably engraved, and the way chemists of the time tried to represent, within the linear confines of a printed text, that in ordinary alcohol, ethanol, one has a C_2H_5 group and H atom connected to an oxygen. We see here the beginnings of the tension between the medium and the content. The content, incidentally, is fascinating, in that it reveals a struggle to give structure to and withhold it from the molecular formula.

The study of optical isomerism led *L. Pasteur* to the remarkable insight that the still unseen microscopic molecules must be in some way three-dimensional and left- or right-handed.[14] It seems in retrospect almost strange that it took another quarter of a century for a correct model to take shape. But so it did, in 1874, in the tetrahedral carbon atom of *J. H. van't Hoff and J. A. Le Bel*.[15] Structure **16** in Figure 14-9 reproduces one of *van't Hoff's* original drawings.

A carbon lies at the middle of each tetrahedron. *Van't Hoff* was led to this model some forty years before anyone "saw," even indirectly, the disposition of atoms in

The reaction is easily understood by the following diagram, in which the atoms

are supposed to be capable of changing places by turning round upon the central point A.

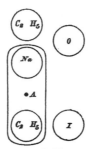

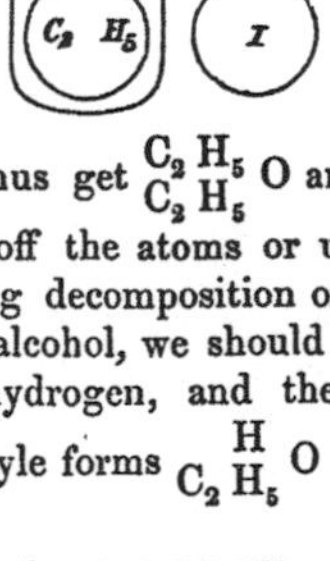

It is clear that we thus get $\frac{C_2H_5}{C_2H_5}$ O and NaI. The circles are merely used to separate off the atoms or units of comparison. To express the corresponding decomposition of iodide of ethyle by hydrate of potash, forming alcohol, we should replace the ethyle of the sodium-compound by hydrogen, and the same change of place between sodium and ethyle forms $C_2H_5{}_{H}$ O (alcohol) and Na I.

Figure 14-8 An excerpt from an article by A. W. Williamson, On Etherification, *J. Chem. Soc. London* 4 (1852): 229. The original runs across two pages.

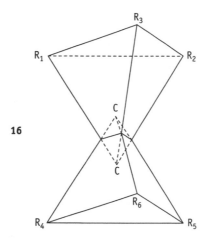

Figure 14-9 An ethane, as drawn by J. H. van't Hoff.

space in a molecule. *N. J. Turro* in a perceptive article addressing some of the same issues discussed here, has aptly called the tetrahedral carbon atom "a triumph of topological thinking."[16]

The increased reliance of chemists upon molecular formulas derived much from the example of another pivotal 19th-century figure, *Kekulé*. As a teenager, *Kekulé* had hesitated between two professions, architecture and chemistry. His strong leanings for the former showed up in his routine building of molecular models, which he would display in his lectures. He would make projections onto planar screens of these models and use these projections as graphic formulas in his publications.

With the contributions of *Kekulé, van't Hoff,* and many others, the structural theory of organic chemistry evolved apace, and by 1890, it was clear that one needed the capability to set in print quasi-three-dimensional drawings of type **7**.

There was no problem in doing so on a blackboard. But the printing media were not up to it, at least not at the budgetary levels appropriate to mass dissemination of a scientific journal. Photography had been around for a good part of the century, but it also had not yet affected the routine printing process. Nor did it *ever* subsequently make an impact on the communication of chemical structures. This is worth a digression relevant to the question of what one is trying to communicate in these drawings, which we will take later.

Engraving still remained the technique of choice in the printing process of a hundred years ago. It was expensive to set lines at an angle. So if a journal had to set the molecule of norbornane, C_7H_{12}, the "skeleton" of camphor, it was represented as **17** rather than **18** (see Figure 14-10). This was done not only in 1890, but until 1950, even when engraving was replaced by photogravure.

The policies of journals, their economic limitations, and the available technology put constraints not only on what is printed, but also on how we *think* about molecules. Now everyone (in chemistry) knew since 1874 that carbon was

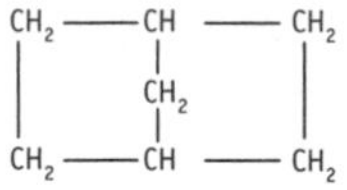

$$
\begin{array}{ccc}
CH_2 \text{---} CH \text{---} CH_2 \\
| \quad CH_2 \quad | \\
CH_2 \text{---} CH \text{---} CH_2
\end{array}
$$

17    18

Figure 14-10 Two representations of norbornane.

"tetrahedral," that its four bonds pointed to the corners of tetrahedron. Molecular models were available or could be relatively easily built. Yet we suspect that the image or icon of norbornane that a typical chemist had in his mind around 1925 was **17** and not **18**. The chemist is conditioned by what he or she sees in a journal or textbook—an image—and might have been moved to act, in synthesizing a derivative of this molecule, for instance, by that unrealistic image.[17] And yet *G. Komppa* did synthesize it![18]

Chemical Representation As Language

Most, if not all, scientists make use of visual imagery for problem-solving, in order to sort out and organize information, to find analogies, to think.[19,20] But chemists are unusual among scientists (though they share this with the architects) in having an iconic vernacular, that of the formulas.

A chemical formula is like a word. It purports to identify, to single out the chemical species it stands for. Chemical formulas embody the ancient dictum (going back, as far as teaching is concerned, to the Czech *Comenius* at the time of the Renaissance): "One thing, one word."[21] Indeed, chemical systematics takes great care to avoid ambiguous situations; we have nomenclatures that deal with every conceivable type of isomerism.

A first problem is easily discarded. When we refer to a familiar object in the outside world by its name, we project a measure of personal experience into the identification. If I say "dog," I cannot help forming a fleeting mental image of a furry or fiery quadruped, wagging its tail or baring its teeth, in between a friend and a threat.[22] The trivial names of chemicals, which often have been with us for centuries, are no different, even when imbued with outdated knowledge that has leached out: sulfuric acid, potash, albumin, *azote*, etc.

This brings up a second, more serious problem. Chemistry is a mature science. It has shed to a large extent its childhood habit of going no further than a phenomenological description of bulk properties, at the macroscopic level (that of sensory perceptions), accompanied by an apt denomination (such as "potash"—because the compound was first found, literally, in pot ash).

Chemistry has become a microscopic science. Explanations nowadays go routinely, paradigmatically from the microscopic scale to the observable: from the way the electrons are distributed in a dye molecule to its color; from the detailed shape of a molecule and of the electrostatic potential around it to its pharmacological activity; from bond energies to the superior tensile strength of Kevlar.

Such a seminal characteristic of chemical sciences was noted already at the time of the Enlightenment. The entry for "chemistry" in the *Encyclopédie* by *Diderot* and *D'Alembert* points out that chemists (but not the Newtonian physicists of the time, who explained everything with central forces acting on material points) are wont to posit invisible and intangible entities or qualities to explain observations. They are forced to do so by the smallness of their particles. We are still under a similar constraint, even though "images" of some molecules and of some atoms have become available very recently. Indeed we know our building blocks, molecules, much better. But it remains a long, long way from the molecular scale to the macroscopic world of the senses. We still have to represent molecules, and chemists have to do so as much as other people. And we tend to represent to ourselves atoms as if they were normal objects in our everyday experience: with a size, with a certain hardness or softness, with measurable attractions to other atoms or to electrons, and so on. This is a little naive, unavoidable, and endearing—not unlike a belief in angels in past centuries.

We made earlier the same point. Humans delude themselves with mental constructs projected upon reality, often deemed as "realism." Chemistry, in like manner, is a mix of a molecular engineering, based on extrapolations from the macroscopic to the microscopic, and a science, coming to grasp directly with the microscopic. The puritans among us wish to get rid of the former, and they push for our minds to handle abstractions only, as in mathematics. Thus, they tend to measure our understanding of the physical world by its secession from everyday experience. While we have a great deal of respect for such an inclination, a constant temptation to most scientists one would think, we are not such purists ourselves. The needs of communication and teaching are one, circumstantial, reason. A second, and more potent, reason is the centrality of semirealistic representations to chemical thought. This is the central theme in our discussion.

What we have is a fascinating philosophical quandary, and an instance of quasi-circular reasoning,[23] too. The above cited quantities—the size of an electronic cloud in an atom, the atom's *electronegativity*—are neither derivable from first principles[24] nor are they observables. They, these Platonic archetypes, have evolved slowly among chemists, in an inductive manner, from experimental determinations; and we attempt to *deduce* from them in turn consequences, for objects in the physical world only very slightly different from those that formed the basis for the formulation of these qualities. This feature is one of the major hurdles for students learning chemistry. Our discipline is a curious mix of empirical observation *and* abstract reasoning. This is not unlike music—but it parts chemistry from the rigor of pure mathematics. Students brought up on deductive logic have a hard time with chemistry!

True, chemical formulas have been severed from subjective life experiences to a considerable extent, much more so than words such as "dog," "automobile," or "contraception." Yet, despite such an excision, chemical structures retain a strong connection with sensory experiences; by contrast with such mundane words, they carry an essential *representational* component.

Let us elaborate some on this interesting paradox. As *Ferdinand de Saussure* pointed out, in one of his fundamental insights, the *word-thing* relationship is that of the *signifier* to the *signified;* and a key notion is that of the *arbitrariness* of the signifier.[25] When we say "dog" (or "Hund" or "cachorro"), each of these terms has been selected more or less randomly in the history of the language. It could have been a "snark" or a "livel" or a "rop" in English, who knows! The word "dog" has settled into its niche (forgive the pun) in the common language by way of the numerous cross-relations it bears with other words in the dictionary (such as "leash," "dog food," "bitch," or "seeing eye"). We flesh out a word because we cannot help loading it with our private empirical experience, just as we do with atoms. But in actuality the word is a total fabrication. It could have been an entirely different choice.[26]

With this reminder, we return to this paradox: formulas forsake reality in a sense, and they aim to stand for reality in another. Indeed, chemical structures differ from words in the normal language because they combine symbolic and representational (iconic) values. Take the case of the molecule of natural gas, methane, drawn in structure **14** in Figure 14-5. The various types of strokes in **14** indicate *both* connection of the carbon with the hydrogens via chemical bonds (the symbolic statement) and whether the bond butts out of the plane of the paper, or recedes from the viewer to the backside (the representational or geometric statement). The chemical formula is trying to signify a lot with the utmost economy in graphics. It aims at portraying accurately the connectivity of atoms—the nearest-neighbor relationships as stemming from chemical bonds—and the geometry.

Thus, we can state the second problem with chemical formulas: they are in between *symbols* and *models*. This hybrid status is an uneasy one. These two poles pull formulas toward opposite and sometimes incompatible requirements.

As models, chemical formulas ought to be reliable and accurate representations of what one might term "molecular reality" (we shall return to this useful figment of the collective psyche of chemists). As symbols, they ought to be arbitrary, to a large extent. This is reflected, to give an all-too-familiar example, in the use of a capital O to represent an oxygen atom: behind this universal convention, there is an assumption of transferability (an oxygen atom *here* is very much like an oxygen atom *there*) which is not that easy to put on a firm theoretical basis. An oxygen atom free of a molecule is indistinguishable from any other oxygen atom. But inside a molecule, a ketone oxygen is very different from an ether one.

Words of a common language are highly ambiguous. Take the word "representation" itself. It can mean either the act of representing or the result of such a mental or physical action. In the latter sense, a representation can be a description, according to this or that framework: representations are theory-laden. In a closely related sense a representation will refer to the worldview, to the *style* of whoever makes it; this is the more subjective meaning of representation. In yet another meaning (that of the poet, for instance), a representation is an epiphany; making the represented object be under our very eyes, whereas it truly is absent. This is representation as illusion. Not to mention yet other meanings, such as in the diplomatic lingo, where representations are a protest and a rebuke. Strings of words, as arranged in a poem, any text for that matter, have also a plurality of meanings. Hermeneutics strives to enrich the reading of a text by associating to the more obvious and mundane meaning a deeper meaning as well.[27]

Is the world of chemistry and chemical structures unambiguous, characterized by one drawing, one molecule? To some extent chemistry is such, because it is "un signe imprimé dans la matière," a sign imprinted in matter, as *J.-M. Lehn* has called it.[28] The instructions for making aspirin work here as well as in Montevideo and Karaganda. The article reporting the synthesis of a new drug perhaps needs translation, but in another sense it doesn't, for it is understood around the world, it is infinitely paraphrasable. Chemical structures, chemical formulas are the signing tools of this language.

But in another way the graphic language of chemistry is quite ambiguous. We've seen clear evidence for this in the plurality of answers given for camphor to the simple question: What is the structure of the molecule? It could be argued that once drawn, no matter how drawn, the single molecule "camphor" enters the mind of the chemist. But as we will see later, how a molecule is thought about and subsequently manipulated in the material world is very much influenced by the way we carry it around in our minds.

A chemical formula is at once a metaphor, a model (in the sense of a technical diagram), and a theoretical construct. A chemical formula is part pure imagination, part inference. It is an attempt to depict the real by manipulation of symbols, just as language enables us to talk about the world and about ourselves by combining arbitrary utterances. The simile cannot be pushed too strongly. In a deep philosophical sense, calling something an "acid" and calling something else "red" are identical mental operations. Likewise, referring to "ethanol" or "reserpine" is akin to talking about "Rockefeller Center" or about the "Eiffel Tower."

"Acid" and "red" are ill-defined but most useful concepts relating to everyday experiences. Conversely, "ethanol" and "Eiffel Tower" are defined unambiguously, but they need not be commonplace objects, they are cultural objects in the widest sense.

One of the points C. P. *Snow* made in his "Two Cultures" lecture was precisely that we should give equal status—that the value judgments had to be either suspended or commensurate—for two cultural objects such as, say, the novels of *Jane Austen* on one hand and the structure of DNA or the second law of thermodynamics on the other.

If "ethanol" is in a similar mental category as "Eiffel Tower," the chemical formula of ethanol stands to it not unlike a dictionary definition for "Eiffel Tower": a
chemical formula is a concise paraphrase, in a half-symbolical, half-iconic language,
of some of the attributes of an object, so that the object can be properly and unambiguously identified, that is, differentiated from like objects (Eiffel Tower as distinguished from the Madeleine or the Centre Pompidou; ethanol as distinct from
ethane or from acetic acid).

Language and chemical representation, besides their joint use of names, have
other similarities. They share use of invariant elements. Many words and most chemical compounds are just that, compounds, put together by association of structural
fragments. There is deep similarity between a word such as "one-upmanship" and a
"chemical word" such as C_6H_5-CH_2-CO-OH. In the latter case, just as in the former,
the structural fragments (phenyl, methylene, carbonyl, hydroxy) are stable semes:
to a first approximation they retain their basic meaning whatever the nature of the
other modules they are connected with.

Chemistry is the science of change, of transformations. Every science starts
with axioms about the integrity of certain of its objects. The mechanical engineer
believes in the integrity of a steel girder. The cell biologist believes in the integrity
of chloroplasts or mitochondria: he or she is convinced that these organelles are
interchangeable, playing identical roles in one cell or in another.

For chemistry, it is crucial to make sense of change by constraining it to occur
between well-defined states. Invariance and its equivalent, transferability, are basic
assumptions to chemistry. At each level of understanding (or complexity), the lower
units are set as invariant: starting with atoms, going on to the simple structural fragments such as the above (C_6H_5, CH_2, CO, OH), on to simple molecules, further on
to chains or polymers, the helices of proteins and nucleic acids, and so forth.

The concept of transferability goes back, beyond *Dalton's* atoms, to the Lavoisier
revolution: "Rien ne se perd, rien ne se crée." By a quirk of scientific history, his use
of the balance made *Lavoisier* discover the physical law of mass conservation; but
the linguistic bent he had inherited from *Condillac* made him give a *linguistic* expression to it. *Lavoisier* founded modern chemical language on the explicit analogy to
natural language. No wonder that chemical formulas, to this day, retain an important linguistic component.

It is possible to set up a formal relationship between chemistry and language; an
intriguing initial step in this direction has already been taken by H. W. Whitlock, Jr.[29]
In the terminology of *Chomsky,* a language is defined by a set of symbols (a vocabulary) from which strings (sentences) may be generated by a set of productions or
transformations (rules for making changes).[30] The identification of symbols with
chemical elements or those simple structural fragments (CH_2, CO, OH) we have
alluded to above, and of productions with chemical reactions, is obvious. *Whitlock*
interestingly shows that a certain problem in organic synthesis may be approached
by analyzing it in the context of formal language.[31]

The shared productivity of language, formal or natural, with chemistry applies also to the realm of what has not yet been said, or written, or synthesized. There exist rules, for instance such that we have the competence to pronounce a word that we have never heard: "a roor," "to roat," "the poot." Likewise, we can write Utopian formulas for (so far) unknown chemical species (e.g., **19-21**, Figure 14-11). This serves quite often as an inducement to try and prepare them.

These structures, waiting impatiently to be made, have for chemists the incongruous look, both attractive and shocking, of a novel object deemed by some as an impossibility: not unlike a first look at a laser printer in action, or a monorail train levitating above its magnetized track, a Stealth bomber, or a town-size space station.

This is a strong analogy, that of language and chemistry. Its tiny area of invalidity is homonyms ("rapt" and "wrapped"). They don't occur in chemistry, or very seldom: *"periodic* table" and *"periodic* acid," linguistic look-alikes, in fact, are easily told apart from the context and pronunciation.

Thus both languages, the natural and the chemical, witness an evolution of meaning, from so-called nonsense to highly significant statements. Even though it does not build upon known ("domesticated") words, and it invents instead new ("wild") words, the "Jabberwocky" poem by *Lewis Carroll* makes sense because of its impeccable syntax, which lets the imagination of the reader both be charmed by the word-play and invests some of the words with meaning, by way of various associations. Chemistry has likewise its wild species (benzyne, tetrahedrane, or **19-21**), besides its more usual bottled samples. In fact most of the molecules of chemistry, wild dreams or not, are invented, synthesized. They were not on earth before.

Chemical representation is also language-like when dealing with the class of transformations known as chemical reactions. In an important sense, chemistry is the skillful study of symbolic transformations applied to graphic objects, the formulas. There are one-to-one correspondences between compounds and formulas. Likewise, it is possible to specify unequivocally the transform from one compound to another. If one thinks of chemical compounds as nodes in an infinite and multidimensional grid, then the connecting lines in such a network are the transforms. Chemistry has thus two facets, structural when the focus is static, on the points in the grid; and dynamic when what is examined is the interconversion along the edges in the network. There are strict rules about *rewriting* formulas to express transforms; not unlike rules of musical, composition.

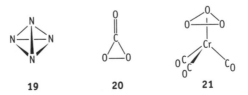

19   **20**   **21**

Figure 14-11 Some so far unknown molecules.

Chemical transforms are the analogues of action sentences in natural language. The parallel is illustrated by the following:

subject	transitive verb	object
substrate	reagent	product

An example of the former is "John saws a log." An example of the latter is "the aldehyde is reduced by sodium borohydride to the alcohol."

Just as action sentences carry subordinates as modifiers, to provide information about time, location, quantity, manner, so the chemical equation is wont to specify the solvent, the reaction temperature, the reaction time, the yield of product, etc.

In another sense, this parallel is too general. The product molecules in a chemical reaction contain the same atoms as the reactants, but reconnected in a different way. The relationship of subject and object is different—except in a category of sentences having a pronominal (or reflexive) verb.

Sentences such as "Jane washes herself" or "John admires himself in the mirror" are somewhat like statements about chemical transforms. Funny that chemical equations come so close to psychological statements!

In *Goethe's* 1809 novel *Die Wahlverwandtschaften* (Elective Affinities), a by-then outdated theory of chemical combination powers a work of fiction.[32] The actions and emotions of the characters of this work embody (and probe critically) the way some people thought molecules behave. One wishes one were able to point today to a similarly inspired literary text. The closest one comes is *Primo Levi's* remarkable memoir *The Periodic Table*, or some chapters in his novel *The Monkey's Wrench*. Nevertheless, it is interesting to reflect on the deep morphological resemblance of chemistry and, if not life, at least language.

In the foregoing, the parallel between chemical formulas and everyday language may have appeared rather artificial and a little forced. We can defend it nevertheless with an anecdote.

One of us was visiting recently the office of Professor C. *Tamm*, at the University of Basel. The conversation turned to a structure of a complex natural product, inscribed on the blackboard and recently elucidated in Professor *Tamm's* laboratory. As an aside, one of us said: "The way I read this structure starts with the six-membered ring, goes on to the *spiro* junction with the five-membered ring, etc." Professor *Tamm* agreed. We discovered in this manner that we shared the same vision of this particular molecule. Both of us were going around this complex network (bonds and atoms) in well-nigh identical manners—obviously a result of our training.

It is thus our contention that chemical formulas are read according to conventional sequences. Recent research in neuropsychology teaches indeed that mental patterns do not spring up whole. They are built up gradually, a part at a time; and the parts, it is found, are visualized in roughly the same order as they are typically drawn. With respect to the two types of tasks (retrieval of archival shapes and coordination of shapes into a mental image), the two brain hemispheres play different roles.[33,34]

The opposite viewpoint is a necessary complement. As *Verbrugge* indicates, discussing the example of *Kekulé's* architectural formulas, his ring structure for benzene, and the oscillation of benzene between the two equivalent ring forms, "scientific understanding develops only when we are prepared to reshape our representation systems in fundamental ways."[35]

We submit that the combined pressures of (1) the learning, early on, of chemical nomenclature, (2) incessant on-the-job confrontations with formulas through seminars, the reading of publications, the handling of molecular models, and (3) the demands of communication with other chemists have built this largely unconscious and stereotyped collective way of seeing. Probably art historians act the same; it is quite possible, even likely that two specialists of Quattrocento painting will both scan a picture in much the same ways.

Let us now explore the opposite viewpoint. Because perception of chemical shapes is so stereotyped, conversely to be able to see a structure in a novel way can be extremely fruitful. Chemistry shares with poetry its notion of elegance, its mission so to say. This is to discover new relations between objects. Very often, the elegance of a key step in a series of chemical transformations is rooted in the perception of a nonobvious connection between parts of the molecular object. More generally, it would repay the student of psychological invention to take a close look at flowsheets, the sequence of molecules made and transformed, in synthesis of natural products. Indeed, the synthetic elaboration of complex chemical structures offers fascinating glimpses into the creative process, in the doing. In the hands of a master craftsman such as *R. B. Woodward*, structural fragments experience what are to the mind—that of the conceiver as well as that of the reader—genuine *Gestalt* shifts, from one part of the synthesis to another. By studying the Woodward syntheses in detail, one can just see him turning a molecule over in this mind, and seeing some of its elements from new angles. The changes of the structures in the course of a Woodward synthesis are metaphors of creativity.[36] They exploit the plurality of meanings embodied in a chemical structure. We submit that it may be difficult to move closer to the creative imagination in action.[37]

A final point, made to us by *H. Hopf.*[38] Chemistry is full of representations other than graphic ones—take the various spectra, IR, NMR, photoelectron, all different (the medium is the message; in German the plotter is even called "Schreiber"). Their music (the tones, overtones and harmonics of spectroscopy) carry incredible dynamic detail far beyond the static structure. Some (NOESY plots) claim to embody the same informational content as conformational drawings. It's a wondrous multitude of representational imagery out there, artists/scientists working in so many media to capture the essence of the real.

But Is It Art...

Art or the reaction to it, the aesthetic response, has never been easy to define. There are so many forms of pleasing human creation, so many constructed objects or patterns evoking emotional reactions. Cognizant of the complexity and venerable history of aesthetics, let us hazard a definition.[39,40] While it is one contestable in all of its parts, perhaps it touches on most of the qualities of what we've chosen to call art. Then we will examine representation in chemistry as it measures up against this definition.

Let's call Art those symbolic acts or creations of human beings which aspire to the extraction from the complex realm of Nature, or the equally involved world of the emotions, of some aspect of the essence of these worlds. Art functions by communication of a symbol, meant to convey information and/or evoke an emotional response.

The essential components of the aesthetic system are (1) the creator—painter, composer, photographer, writer, dancer, (2) the audience—both that perceived in the creator's mind and the real one, the viewers, (3) the set of symbols by which communication takes place—the watercolor, sound waves, and images evolving in time, a text, (4) the act of communication itself—to an audience that is present (watching a dance, listening to a cantata) or absent (reading a novel).

If this definition sounds too rational, too "scientific," devoid of the gut emotional response we should like good art to hit us with, so be it. *Nelson Goodman* argues persuasively that "in aesthetic experience the *emotions function cognitively*," that feeling is knowing. He goes on to examine the usual attributes of art (that the aesthetic is directed to no practical end, that it gives immediate satisfaction, that in art inquiry is a means of obtaining satisfaction, etc., all of these existing as marks of art mainly by contrasting them with an opposite attributed to science), and he finds them wanting. He says: "...the difference between art and science is not that between feeling and fact, intuition and inference, delight and deliberation, synthesis and analysis, sensation and cerebration, concreteness and abstraction, passion and action, mediacy and immediacy, or truth and beauty, but rather a difference in domination of certain specific characteristics of symbols."[41]

To return to the question heading this section: are chemical structures art? It seems clear that they possess all the components of the "aesthetic system." Structural formulas are symbols created by one chemist (or several) to communicate information to others. The drawing of the structure of camphor is certainly a symbolic motion, a communication of an essence—the arrangement in space of the atoms of this molecule. Some might call it just a sketch, an information-reducing stratagem by someone not able or willing to compute and show others the all-important electron density around the nuclei. *That* electron density is the real molecule; the structural formula—well, that's "just a poor representation."

Of course, that simple structural drawing is more. It is the appropriate tool, the model fitting the occasion. Indeed, it's the extraction from the complexity (of something so simple as a molecule!) of one aspect of its essence. It certainly conveys information, that structure. And, in the few careful readers of the paper, the structure may even evoke an emotional response, a wonder that it was made, jealousy of the man or woman who made it first.

Does it matter that the aesthetic response will be provoked instantly only in the minds of those initiated to the chemical structure code? We don't think so. To be sure one has to be taught to understand the chemists' ways of signing. But it's not that difficult. Responses of the population at large to abstract art, or some primitive tribal groups to realistic Western art, have not been, are not likely to be, immediately appreciative.

The chemical structure is an artistic construct because it is a transformation of a model of reality (note the secondhand if not nth-hand relationship to the real) for the purpose of communication. Neither chemistry texts nor anatomy books are much illustrated with photographs. There are some photographs, to be sure. But by and large a photograph (and we certainly don't wish to imply that photography is mere representation; it is far from that) contains too much detail. What one wants to communicate is the essence needed for the moment. One wants to teach, to evoke a response. Drawings, with their artistically selected detail, are much better for that purpose.[42]

The symbolic nature of the chemical formula, the fact that chemists know that a hexagon stands for a ring of carbon atoms that in turn is much, much smaller, the implicit knowledge that that hexagon is *not* an enlarged photograph of the ring, all that symbolic distancing of course enhances the metaphorical nature of the chemical discourse. Structures are not what they stand for; they stand for what they are not.

But is it art? Having argued that chemical representations, such as structural formulas, share all the symptoms of art, let us take the opposite tack, at least for a while.

Just as it is impossible to ignore the artist and his audience, the mental set of both, so it is impossible to put out of mind the context of a picture or a scientific illustration. And chemical structures, in particular, are often really part of the text. Oh, they may have artistic value or expressive power on their own. One could mount a good art exhibit around them. But their function, their organizing referent, is the text.

By way of illustration, Figure 14-12 shows the beginning of a paper written by one of us.[43] Note four little drawings in the first two paragraphs. They are typically floating in midair, typically using the wedge-dash-line notation. The structures are numbered boldface, and referred to specifically in the text. In fact they are part of the text, and that they are referenced to (by their boldface numbers) even in the middle of a sentence, confirms this role. In one chain (labeled **2** in Figure 14-12) the units of five telluriums are related by a so-called screw axis, whereas in another chain

The recent literature contains a number of examples of a square-planar tellurium structural unit which may formally be defined as Te_5^{n-} (1). In some cases such as Rb_2Te_5[1] and Cs_2Te_5[2]

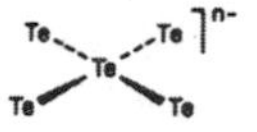

the stoichiometry clearly defines the charge on the unit as 2–. In these instances the unit is the basic building block of a one-dimensional anionic chain. In Cs_2Te_5 the Te_5^{n-} units are screw axis related (2) while in Rb_2Te_5 the Te_5^{n-} units are related by translation (3).

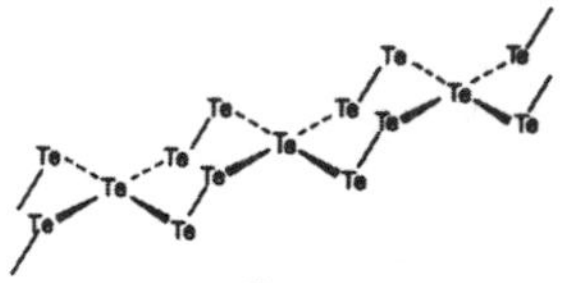

For most of the other cases in which the Te_5^{n-} unit appears, the attribution of a particular charge to the unit is more ambiguous. For instance, on K_2SnTe_5,[3] the unit again appears embedded in a one-dimensional chain (4) but the Te_5^{n-} units alternate

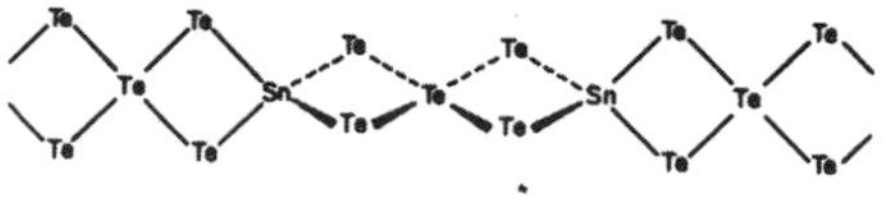

with tetrahedral Sn in making up the chain. Formally, at least, the square-planar units could still be considered Te_5^{2-} if the tin is Sn(0), but the Te–Sn bond length of 2.74 Å is almost identical with the sum of the tetrahedral covalent radii (2.72 Å) given by Pauling.[4] This suggests a formal oxidation state of 4+ for the Sn and a net charge of 6– on the Te_5^{n-} unit.

Figure 14-12 The beginning of a paper by one of the authors. The original runs across three columns in the journal.

(labeled 3 in Figure 14-12) they are simply translated, one relative to the other. All that could have been said in words. But it was easier to draw a picture. So a structure was born.

Actually, the motives of the representation are not so simple here. The authors use the pictures not only to save space. They are also plumbing strategies to capture their audience. These tellurium compounds are not immediately interesting to everyone in chemistry. Specialization is a plague in any field of scholarship. The geometries of the tellurium structures are difficult to see. Since in science, as everywhere else, what we do not understand we are afraid of, what we do not understand we find uninteresting, the authors are using the visual appeal (and density of explan-atory power) of a drawing to inform, to pull in, to attract, to seduce. Still another motive: It is the "style" of one of the authors *(R. H.)* to decorate his papers with such drawings. He is establishing a visual signature.

Still another argument against assigning full artistic value to a chemical struc-ture is that it is not, to use *Goodman's* terminology, "replete."[44] Not every stroke in the representation matters; the lines could be a different color, the molecule readily recognizable as what it is even if drawn from a somewhat different viewpoint. This is in contrast to a Goya etching, which, were similar changes made in it, would be another Goya etching, a different work of art, or at the very least a different "état."[45] To put it in another way, the chemical structure addresses the inherent paraphras-ability of scientific knowledge (paraphrase as one of the [few] differentiations of art and science has been persuasively forwarded by G. Stent[46]). It's the same molecule, intended to be the same, perceived to be the same. But what if the slightly different representation enters the unusual mind, prompting an experiment untried before? The icon powers the iconoclast.

Perhaps another way to approach the artistic content of chemical drawings is to think of their relationship to various visual art genres. For instance, we see a similarity between chemical structures and what has been called, not without con-troversy, primitive art. Tribal art, be it of Australian aborigines or an Eskimo clan, often appears schematic to us, deficient in perspective. That's our problem, for to the native group which shares the culture that informs that art, the representation may be highly accurate and perceptive. So it is with chemical drawings—their per-spective may be inadequate, their representation artistically unsophisticated. But they tell a concise story to the chemical reader. Like primitive images or sculptures, chemical drawings will usually distort a view if the viewers' ability to clearly *classify* an object is enhanced by that distortion. So if it is important to show that someone owns six sheep, the sheep will be posed so that they are distinct, and clearly seen as sheep. A deity will be represented by its symbolic attributes so that we cannot confuse it with any other.

The iconic representation of camphor (see drawing 7) is simplified and distorted (compare with 9 or 11) so as to allow us to identify the molecule, to trigger a con-nection in the mind. If a piece of a molecule, some functionality such as an aldehyde group is essential, even if it is hidden behind another part of the molecule, it will be brought forward without much regard to faithfulness of representation.

There is still another anthropological point of contact. In some cultures knowing the true name of an object or a person forms a special bond, even gives power. So it is in chemistry. Knowing the "name" of a compound, which means its structure, gives the chemist tremendous power over the molecule. A range of its properties, its behavior are implied by that structure.

Guy Ourisson, in a perceptive article that deals with many of the same issues we have discussed, identified the chemical structure as an ideogram or pictogram, a symbol that represents an idea or object directly.[48] This point was also made by R. *Etiemble.* He makes the analogy to Chinese characters, and perhaps one could also do so productively to Egyptian hieroglyphs. Like the character, the chemical sign enters the conscience of a chemist directly. All its meanings are attached, and the chemist manipulates that little picture mentally in a multiplicity of ways. The chemical structure implies not only a molecule but its physical, chemical, even biological properties.

Rudolf Hoppe has prefaced one of his review articles with stimulating reflections on the symbolic language of the chemist; his graphics are admirably chosen; they more than hint at the convergence of *la pensée sauvage* and the language of chemistry.[49]

The drawing of chemical structures also has a kinship with caricature and comic strips. If one examines a successful cartoon or schematic book illustration closely one finds that a wide spectrum of emotion—grief, terrible anger, ecstasy—is communicated in just a few strokes of a pen. Think of the Dr. Seuss books, *Jean de Brunhoff's* Babar, *Hergé's* Tintin books, *Tove Jansson's* Moomintrolls, or *Walt Disney's* numerous cartoon characters. And not just for children. *Gombrich* discusses the effect of caricature perceptively, arguing that we "accept the grotesque and simplified partly because its lack of elaboration guarantees the absence of contradictory clues."[50] In examining a work of art we look at the information in it, and unconsciously seek relationships. It's not the absolute flux of light entering our eye from a painted white spot that makes us see it as brightly lit, it is its differentiation from neighboring patches of paint.

The act of viewing is collaborative (between painter and viewer) and forgiving. We always create space, in our minds, when we see a two-dimensional representation. And we elaborate the information, interpolating our experience to fill out what is omitted. At least providing there are no contradictory clues, no signals to tell us that we are wrong. Caricatures or cartoons (or if those are not "serious" enough, take a Goya or Picasso etching) work by providing the appropriate minimal information.

So do chemical structures. The chemist's mind is very tolerant. It will accept both **22** and **23** as representations of an eclipsed ethane (Figure 14-13). It will read the substituent attached at lower left to the six-membered ring in **24** through **27** as a

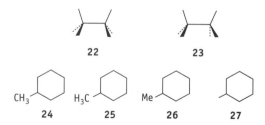

Figure 14-13 Two representations of an eclipsed ethane conformation (top); four representations of methylcyclohexane.

methyl group. It will fill in the missing three-dimensional background required to make these molecules come to life; it will geometrize the floating world of these little symbols.

In any case, the drawing does not exist by itself, but is an integral part of the text. It is as if the chemist invented a new language, part text, part the tactile sense that is behind the model-building that is behind the ability of a chemist to reconstruct a molecule in his or her mind from such minimal information. We are pushed back to the logic of language. And in another way, to viewing the combined text-structure complex as an art form.

Note added in proof (July 1990): The reader's attention is directed to two recent contributions: *Stephen T. Weininger,* Worcester Polytechnic Institute, explores in an important, well-documented lecture-essay a semiotic perspective on chemistry.[51] And *Pier-Luigi Luisi* and *Richard M. Thomas,* ETH Zurich, make a number of comments similar to ours (and many original ones) in an article on pictographic communication in chemistry and biology.[52] There is also a beautiful discussion of the language of chemistry in two recently translated essays by *Primo Levi.*[53]

Acknowledgments

We've benefited much from the critical comments of Kurt Mislow, Hearne Pardee, Noel Carroll, Vivian Torrence, Gilles-Gaston Granger, and Henning Hopf. The assistance of Dennis Underwood, Donald Boyd, and Laura Linke is also gratefully acknowledged, as are the photographs of Stephen R. Singer, the drawings of Jane Jorgensen and Elisabeth Fields, and the typing of Joyce Barrows. A slightly different version of this essay has appeared in *Diogène,* Paris.

Notes

1. For a general introduction to molecules we might direct that hypothetical listener to the beautiful book by P. W. Atkins, *Molecules* (New York: Scientific American Library, 1987).
2. The story is more complicated. It seems that, if one subjects the harmless form to physiological conditions, it transforms into a mixture of mirror images, one harmful, one not. See W. Winter and E. Frankus, Thalidomide Enantiomers. *Lancet* 339 (1992): 365.
3. The precise definition of molecular structure is still a subject of debate in chemistry. See, for instance, R. D. Brown. Kinky Molecules. *Chem. Br.* 24 (1988): 770; R. G. Woolley, Must a Molecule Have a Shape? *J. Amer. Chem. Soc.* 100 (1978): 1073; and the articles by Mislow and Turro cited below.
4. R. Barthes, *L'Empire des Signes* (Milan: Skira, 1980); English: *The Empire of Signs* (New York: Hill and Wang, 1982); German: *Das Reich der Zeichen* (Frankfurt: Suhrkamp, 1981).
5. Not really. What we have is a human being guiding a tool, which in turn was programmed by other human beings, not to speak of it being built by still other ones and their tools. I am grateful to Dennis Underwood and Don Boyd for their help with structures **8-11**.

6. For methodological discussions of how models are used in chemistry, see K. E. Suckling and C. W. Suckling, *Chemistry Through Models* (Cambridge: Cambridge University Press, 1978); C. Trindle, *The Hierarchy of Models in Chemistry. Croat. Chem. Acta* 57 (1984): 1231: J. Tomasi, Models and Modeling in Theoretical Chemistry. *J. Mol. Struct. (Theochem.)* 179 (1988): 273. And for the different meanings of "model," see the amusing comment by N. Goodman in *Languages of Art,* 2nd ed. (Indianapolis, IN: Hackett, 1976), p. 171.

7. That there are many ways to look at a molecule's structure is, of course, well known to the chemical community; we are not saying anything new here. See, for instance, G. Ourisson, Le Langage Universel de la Chimie: Ambiguités et Laxismes. *Actual. Chim.* (1986): 41.

8. (a) G. M. Whitesides and D. W. Lewis, Tris[3-(tert-butylhydroxymethylene)-d-camphorato] europium(III). A Reagent for Determining Enantiomeric Purity. *J. Am. Chem. Soc.* 92 (1970): 6979; (b) W. Klyne and J. Buckingham, *Atlas of Stereochemistry, Vol. 1* (London: Chapman and Hall, 1978), p. 85.

9. J. Vouvé, J. Brunei, P. Vidal, and J. Marsal, *Lascaux en Périgord Noir* (Paris: Pierre Fanlac, 1982), p. 31.

10. For an introduction to David Hockney's neocubist perspective, see D. Hockney, *Camerawork* (New York: A. A. Knopf, 1984); German: D. Hockney, *Cameraworks* (Munich: Kindler, 1984).

11. (a) E. H. Gombrich, *Art and Illusion,* 2nd ed. (Princeton: Bollingen Series, Princeton University Press, 1961), p. 262; German: E. H. Gombrich, *Kunst und Illusion. Zur Psychologie der bildlichen Darstellung* (Stuttgart: Belser, 1986). See also the comments of R. Thorn in *Stabilité Structurelle et Morphogenese,* 2nd ed. (Paris: Intereditions, 1977); English: *Structural Stability and Morphogenesis* (Reading, MA: W. A. Benjamin, 1975) on the inherently geometrical in our minds; (b) F. Philostratus, *Life of Apollonius of Tyana,* Book II, Chap. 22 (New York: Macmillan, 1912), I, p. 175; German: *Appollonius van Tyana* (Artemis, 1983). We owe this story to E. H. Gombrich (*Art and Illusion* [see above]), p. 181.

12. (a) A. Lavoisier, *Traité Elementaire de Chimie* (Paris: Cuchet, 1789); (b) J. Brody, Behind Every Great Scientist—Lavoisier, Antoine Laurent. *New Sci.* 119 (December 24-31, 1987): 19. The originals of many of Mme. Lavoisier's drawings and her annotations on proofs of the etchings are in the collection of the History of Science Library at Cornell University.

13. A. W. Williamson, XXII.—On Etherification. *Q. J. Chem. Soc. London* 4 (1852): 229.

14. For a beautifully reasoned "epistemological note on chirality," the property of molecules existing in such nonsuperimposable mirror image forms, see K. Mislow and P. Bickart, An Epistemological Note on Chirality. *Isr. J. Chem.* 15 (1976/1977): 1. An important point in their discussion is the distinction between geometric figures or models (where the notion of chirality is well-defined) and real molecules, where it is fuzzy, depending upon conditions of measurement.

15. J. H. van't Hoff, *Voorstel tot uitbredning der tegenwoordig...* (Utrecht: J. Greven, 1874); English: *The Arrangement of Atoms in Space,* 2nd ed. (London: Longman, Green and Co., 1898). J. A. Le Bel, Sur des relations qui existent entre les formules atomiques des corps organiques et la pouvoir rotatoire de leur dissolutions. *Bull. Soc. Chim. Fr.* 22 (1874): 337. Le Bel and van't Hoff infused their independent and quasi-simultaneous discovery with quite distinct yet complementary world views. Their scientific approaches differed, and, as far as we can appreciate, so did their *themata* (in the sense of G. Holton) and their impact on what came after. See on this point, of great interest to philosophers of science, J. Weyer, A Hundred Years of Stereochemistry—The Principal Development Phases in Retrospect. *Angew. Chem.* 86 (1974): 604; *Angew. Chem. Int. Ed. Engl.* 13 (1974): 591; F. G. Riddell and M. J. T. Robinson, J. H. van't Hoff and J. A. Le Bel—Their Historical Context. *Tetrahedron* 30 (1974): 2001; R. G. Grossman, Van'tHoff, Le Bel, and the Development of Stereochemistry: A Reassessment. *J. Chem. Ed.* 66 (1989): 30; S. F. Mason, The Foundations of Classical Stereochemistry. *Topics Stereochem.* 9 (1976): 1; O. B. Ramsay, ed., *van't Hoff-Le Bel Centennial* (Washington, D.C.: American Chemical Society, 1975).

16. N. J. Turro, Geometric and Topological Thinking in Organic Chemistry. *Angew. Chem.* 98 (1986): 872; *Angew. Chem. Int. Ed. Engl.* 25 (1986): 882.

17. The gift for seeing structures in space with one's eyes, and the cognate imaginative prowess of seeing them with the mind's eye, are important components of creativity for a chemist. And perhaps not only for chemists. For instance, mathematicians still discuss, since Hadamard's pioneering study of mathematical creativity (K. Hadamard, *The Psychology of Invention in the Mathematical Field* [Princeton: Princeton University Press, 1945]) whether iconic representations are involved in the thought processes of mathematicians (A. Muir, The Psychology of Mathematical Creativity. *The Mathematical Intelligencer* 10 [1988]: 33). If one were to rely on the example of chemists, the answer to this much-debated question would have to be "Yes, of course!"

18. G. Komppa, Über die Totalsynthesis des Camphors. *Ber. Dtsch. Chem. Ces.* 41 (1908): 4470.

19. See R. R. Hoffman in *The Ubiquity of Metaphor,* ed. R. Dirven and W. Paprotte (Amsterdam: John Benjamin, 1985).

20. We take the view that such figures of speech as the metaphor, should be seen as cognitive tools, and not as deviations with respect to the prosaic norm. See D. Sperber and D. S. M. Wilson, *Relevance: Communication and Cognition* (London: Blackwell, Oxford/Harvard University Press, 1986).

21. J. A. Comenius, *Orbis Sensualium Pictus* (Nürnberg: Michael Endler, 1658).

22. We have not succumbed to the nominalist fallacy! We have elected our Fido example because the view, admittedly restricted and restrictive, of language as nomenclature is precisely that embodied in the language of chemistry. Saussure, with his idea of signs as uniting a signifier and a signified, swept off the illusory notion of language as nomenclature for objects. The word "dog" refers, not to a real object, but to an attribute shared by several different entities.

23. Circular reasoning is not all bad. Elsewhere, one of us makes a sort of a case for its utility in real chemistry: R. Hoffmann, Nearly Circular Reasoning. *Amer. Sci.* 76 (1988): 182. Chapter 5 in this book.

24. Though actually some convincing progress has been made in this direction, see R. G. Parr, Density Functional Theory. *Ann. Rev. Phys. Chem.* 34 (1983): 631.

25. J. Culler, *Ferdinand de Saussure,* revised ed. (Ithaca, NY: Cornell University Press, 1986).

26. K. Mislow brought to our attention in this context the story *Feynman* tells of how his father taught him the difference between knowing the name of something and knowing something: R. P. Feyman, *What Do You Care What Other People Think?* (New York: W. W. Norton, 1988), p. 13.

27. P. Ricoeur in *The Conflict of Interpretation,* ed. D. Ihde (Evanston, IL: Northwestern University Press, 1974).

28. J.-M. Lehn, Langue de la Science et Science des Langues: Multilinguisme ou Langue Unique? Le Point de Vue d'un Utilisateur. *Traduire* 116 (1983): 62.

29. H. W. Whitlock, Jr. in *Computer-Assisted Organic Synthesis* (ACS Symp. Ser. 61), ed. W. T. Wipke and W. J. Howe (Washington, D.C.: American Chemical Society, 1977), p. 60. We thank K. Mislow for reminding us of this work.

30. N. Chomsky in *Handbook of Mathematical Psychology, Vol. 2,* ed. R. D. Luce, R. R. Bush, and E. Galanter (New York: Wiley, 1963), p. 323; N. Chomsky, *Cartesian Linguistics* (New York: Harper & Row, 1966).

31. See also D. I. Cooke-Fox, G. H. Kirby, and J. D. Rayner, Computer Translation of IUPAC Systematic Organic Chemical Nomenclature. 1. Introduction and Background to a Grammar-Based Approach. *J. Chem. Inf. Comput. Sci.* 29 (1989): 101; D. I. Cooke-Fox, G. H. Kirby, and J. D. Rayner, Computer Translation of IUPAC Systematic Organic Chemical Nomenclature. 2. Development of a Formal Grammar. *J. Chem. Inf. Comput. Sci.* 29 (1989): 106; D. I. Cooke-Fox, G. H. Kirby, and J. D. Rayner, Computer Translation of IUPAC Systematic Organic Chemical Nomenclature. 3. Syntax Analysis and Semantic Processing. *J. Chem. Inf. Comput. Sci.* 29 (1989): 112.

32. J.-W. von Goethe, *Die Wahlverwandtschaften* (Frankfurt: Insel Taschenbucher, 1972); English: *Elective Affinities* (New York: F. Ungar Publ., 1962).

33. See S. M. Kosslyn, Aspects of a Cognitive Neuroscience of Mental Imagery. *Science (Washington D.C.)* 240 (1988): 1621.

34. For an illuminating series of articles on biological aspects of aesthetics, see I. Rentscher, B. Herzberger, and D. Epstein, eds., *Beauty and the Brain* (Basel: Birkhauser, 1988).

35. R. R. Verbrugge, The Role of Metaphor in Our Perception of Language. *Ann. N. Y. Acad. Sci.* 433 (1984): 167.

36. We have had the privilege of reading in manuscript a remarkable psychobiography of R. B. Woodward, written by his daughter Crystal Woodward. Some of the observations on science, creativity, and art that Ms. Woodward makes in her study are similar to ours. One of us *(P.L.)* alludes elsewhere to the relationship, which C. Woodward explores in detail, that operates between molecular models and molecules, on one hand, and the "transitional objects" of D. W. Winnicott, on the other hand, in his book, *La Parole des Choses* (Paris: Hermann, 1993).

37. For a presentation of the marvels of organic synthesis, see N. Anand, J. S. Bindra, and S. Ranganathan, *Art in Organic Synthesis,* 2nd ed. (New York: Wiley, 1988); E. J. Corey and X.-M. Cheng, *The Logic of Chemical Synthesis* (New York: Wiley, 1989).

38. H. Hopf, Universität Braunschweig, private communication.

39. To get a feeling for the complexity of definitions and the range of opinions in this field, see M. C. Beardsley, *Aesthetics,* 2nd ed. (Indianapolis, IN: Hackett, 1981).

40. Related to this section are a series of articles one of us has written on "Molecular Beauty." This is a kind of anthropological study of the objects chemists admit as possessing aesthetic value: R. Hoffmann, Molecular Beauty. *Amer. Sci.* 76 (1988): 389; R. Hoffmann, Molecular Beauty II. Frogs About to Be Kissed. *Amer. Sci.* 76 (1988): 604; R. Hoffmann, Molecular Beauty III. As Rich As Need Be. *Am. Sci.* 77 (1989): 177; R. Hoffmann, Molecular Beauty IV: Toward an Aesthetic Theory of Six-Coordinate Carbon. *Am. Sci.* 77 (1989): 330.

41. N. Goodman, *Languages of Art,* 2nd ed. (Indianapolis, IN: Hackett, 1976), p. 264.

42. See, in this context, the article by C. Rose-Innes, Where to Draw the Line. *New Sci.* (January 7, 1989): p. 42.

43. J. Bernstein and R. Hoffmann, Hypervalent Tellurium in One-Dimensional Extended Structures Containing Te_5^{n-} Units. *Inorg. Chem.* 24 (1985): 4100.

44. N. Goodman, *Languages of Art,* 2nd ed. (Indianapolis, IN: Hackett, 1976), p. 229.

45. This point was made to us by H. Pardee, for whose comments on this paper we're grateful.

46. G. Stent, *Engineering and Science* (Pasadena, CA: California Institute of Technology, September 1985), p. 9; Nobel Symposium, Royal Swedish Academy of Sciences, 1986, private communication.

47. For an insightful, beautifully presented account of Eskimo ways of seeing and representing, see E. Carpenter, *Eskimo Realities* (New York: Holt, Rinehart and Winston, 1973). We're grateful to Vivian Torrence for bringing this book to our attention.

48. G. Ourisson, Le Langage Universel de la Chimie: Ambiguités et Laximes. *Actual. Chim.* (1986): 41.

49. R. Hoppe, On the Symbolic Language of the Chemist. *Angew. Chem.* 92 (1980): 106; *Angew. Chem. Int. Ed. Engl.* 19 (1980): 110.

50. E. H. Gombrich, *Art and Illusion* (see endnote 11, Chap. 10; E. Kris and E. H. Gombrich in *Psychoanalytic Explorations in Art,* ed. E. Kris (New York: International University Press, 1952), Chap. 7.

51. S. I. Weininger, "Contemplating the Finger: A Semiotic Perspective on Chemistry," Society for Literature and Science, 3rd Annual Conference (September 24, 1989), University of Michigan.

52. P.-L. Luisi and R. M. Thomas, The Pictographic Molecular Paradigm: Pictoral Communication in the Chemical and Biological Sciences. *Naturwissenschaften* 77 (1990): 67.

53. P. Levi, *Other People's Trades* (New York: Summit Books, 1989); Italian: *L'Altrui Mestiere* (Turin: Giulo Einaudi Editore, 1985).

15

The Say of Things

ROALD HOFFMANN AND PIERRE LASZLO

In search of a chemical conversation, we are on a farm in Uniow, a little Ukrainian village in Austro-Hungarian Galicia, just before the onset of World War I. In the farm yard we see a big, steaming, lead-lined iron pot. The men have mixed some potash in it (no, not the pure chemical with composition KOH from a chemical supply company, but the real ash from burning good poplar) and quicklime, to a thickness that an egg—plenty of eggs here, judging from the roaming chickens—floats on it.

Elsewhere in the yard, women are straining kitchen grease, suet, pig bones, rancid butter, the poor parts skimmed off the goose fat (the best of which had been set to cool, cracklings and all). This mix doesn't smell good; they would rather toss the kitchen leavings and bones into the great iron pot, but the fat must be free of meat, bones, and solids for the process to work.

They are making soap. Not that we had to go that far, near where one of us was born, for soap was prepared in this way on farms from medieval times until the twentieth century. Fat was boiled up with lye (what the potash and quicklime made). The reaction was slow—days of heating and stirring until the lye was used up, and a chicken feather would no longer dissolve in the brew. One learned not to get the lye on one's hands. The product of a simple chemical reaction was then left in the sun for a week, stirred until a paste formed. Then it was shaped into blocks and set out on wood to dry.

And inside the steaming pot, deep inside, where the fat and the lye are reacting? *There* is the conversation we are after, a hellishly animated molecular conversation. The lye that formed was an alkaline mixture of KOH, $Ca(OH)_2$, and NaOH. In the vat one had hydroxide (OH^-) ions, and K^+, Ca^{2+}, Na^+ all surrounded in dynamic array and disarray by water molecules. Contaminants aside, the fat molecules are compounds called esters, in which an organic base, glycerol, combines with three long-chain hydrocarbon chains. A typical chain is stearate (Figure 15-1 top, left).

If we call just this ion R^-, then the formula for a fat is roughly that shown on the top right side of Figure 15-1. The reason we say "roughly" is that animal and vegetable fats

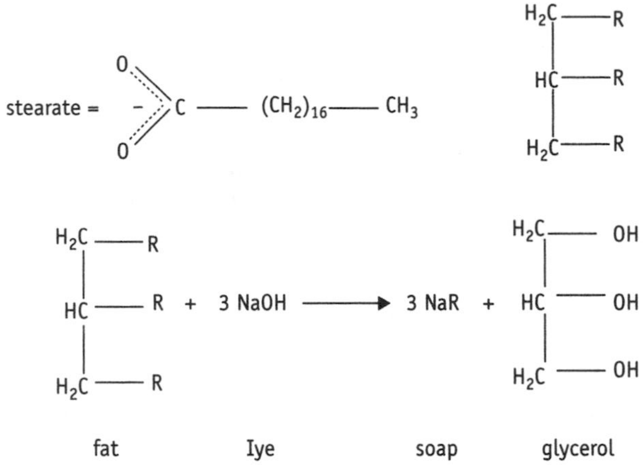

Figure 15-1 Stereate, a component of a typical fatty acid (top left); a typical fat, a triglyceride (top right); the reaction making soap (bottom).

are not just made of the esters of stearic acid, but also of other long chains containing fourteen to eighteen carbon atoms and associated hydrogens. Hardly anything in this world is simple (only political advertisements and the aesthetic prejudices of people who believe that beautiful equations must be true), least of all the products of evolution, which include fats and the human beings who invented the craft of making soap without waiting for professional chemists to tell them how to do it.

And what is soap? A typical soap is sodium (or potassium) stearate, NaR, where R is the stearate group. The reaction in the pot is that written at the bottom of Figure 15-1.

It's a mad dance floor inside the pot. Some 10^{25} molecules of fat are jiggling around in the viscous solution, moving much quicker (if tortuously) than we may imagine. The molecules collide with each other very frequently, as well as with the OH⁻, Na⁺, K⁺, Ca²⁺ ions and waters. Once in a while a hydroxide nears one of the three central carbon atoms of a fat molecule, the knock is just right (men and women are not that different from molecules as they think) and a C—OH bond forms, while the C—R bond loosens. An R⁻ ion slides into the murk, picks up some surrounding waters, and is off onto the dance floor, picking up a positive ion partner.

One of the authors [R. H.] has a fond remembrance of the closest model he has seen for molecular collisions and reaction kinetics. It was outside of Havana, an immense crowd densely dancing as the greatest Cuban band of them all, Los Van Van, played "Muevete."

Lye and fat talk, the triglyceride and hydroxide ions sing this wild riff, entangling, reacting... in the dark of the deep, except that sunlight comes in, and other energy in the form of heat, more energy to be released when nearby bonds are productively broken. The conversation becomes more heated, old bonds are loosened, new ones formed.[1] Eventually, the conversation quiets, and we have... soap.

Is this an excess of anthropomorphism? Molecules, even though they respond to energy and collisions, do not talk. Human beings do. What business do we have, really, to talk of a molecular conversation? Indulge us for a while, and we shall see. Or hear.

Spin to Spin Talk

Scientists have instruments for eavesdropping on conversations of an ensemble of molecules at the microscopic level. These are totally factual chats, as when we book an airline ticket over the phone and supply the clerk with a credit card number. One particular example is provided by nuclei (or electrons) of atoms informing each other of their spin state.

Hydrogen nuclei, for instance, are allowed by the rules of quantum mechanics two spin states, which are often called "up" and "down," but which, for convenience here, we shall term the blue and the red. Such nuclei can be induced to put up either a blue flag or a red flag (so to say) to signal to us their spin state. The inducement is application of one magnetic field and tickling by another.

Now imagine two such nuclei (call them *A* and *B*) not too far from each other. There are four combinations of spins possible (flags they can wave): (red *A*, red *B*), (blue *A*, red *B*), (red *A*, blue *B*), and (blue *A*, blue *B*). If those nuclei are ignorant of each other's presence, the four sets would have equal energy. But the nuclear spins do feel each other, just a little, and with the help of a strong magnet we can translate that feeling into a difference in energy between those four sets, and eventually into lines in a so-called spectrum. These lines speak to us, they tell us that there are two nuclei there, sensitive to each other. And not three or five, for those would give rise to a different number of peaks and plateaus. Precious knowledge, and we have gotten it by tapping in, nondestructively, on an atomic conversation.

A version of the technique we have just described is used for noninvasive mapping of the inner parts of the body. Once called nuclear magnetic resonance (NMR), it got rechristened in the age of fear and advertising, ours, as magnetic resonance imaging (MRI).

Spins talking to each other is a productive metaphor within the chemical community. But is it just a metaphor? Real talk is sequential, even if frequently overlapping. At what speed does spin communication take place—is it instantaneous, or transmitted at the speed of light? We don't want to get into the fascinating, active field of decoherence and quantum locality, the ways in which contemporary physicists have made Schrödinger's cat meow.[2] The only way the limited human intellect has of getting a handle on what actually happens in microscopic interactions is to divide the process into sequential steps. In a sense, Cartesian analysis forces a conversation between spins to take place.

There is still another kind of conversation between spins: electrons have spin, just as some nuclei do. If there is an electron on one carbon atom in a molecule with a spin of one type (say, a red flag), then the physics works out so that on the carbon next to it the spin of the electron on the average must be of the other type (a blue flag). Red and blue don't matter—you could switch them here (the first could be blue, its neighbor then red). Alterity, being the other, does matter.

Electrons, detected through their spins, are talking all the time. Imagine a molecule with two metal atoms, as the copper acetate drawn in Figure 15-2:

On each copper there is an odd electron. Do these two solitary electrons know of each other? If they do, will they line up with both red (blue) flags aligned (in the trade we call this a high-spin or triplet configuration) or one red, one blue (low spin, singlet)? It turns out that the latter is preferred, by just a little.

Enzymes often do their catalytic magic by shuttling an electron from one part of the protein to another—say from the outside of the protein, where an electron donor docks, to a metal ion in a cleft where the enzyme's appointed action takes place. We think of the conversation between the sites—its speed, for instance. How do these pieces of a large molecule talk to each other? How—through space, through bonds? (See Figure 15-3.) We tweak the molecules in various ways, through transient perturbations of colored lights, or magnets, and listen, with those marvelous spectroscopies we've invented, to the chatter (peaks, valleys, more peaks) that emerges. We recognize a molecule by its speech in the conversation we have with it.

Maya-Spectra

In the *Popol Vuh*, the Council Book of the Quiché Maya, Hunahpu and Xbalanque are the conquering and playful twin heroes. And they are players of the Mesoamerican ballgame, in which a rubber ball is hit with a yoke that rides on the hips. The twins are challenged to a lethal ballgame by the twelve lords of Xibalba, the death-dealing rulers of the underworld, who can be vanquished by the

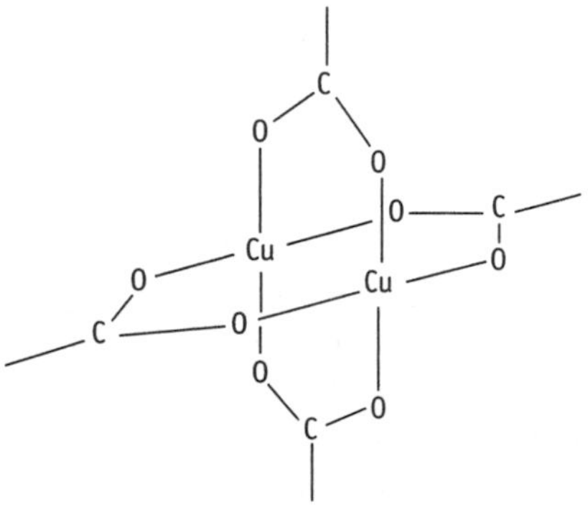

Figure 15-2 The molecular structure of copper acetate.

Figure 15-3 A drawing by Rick Stromoski, reproduced with permission of the artist.

utterance of their real names. The twins are up to extracting those secret names, by stimulating a conversation between the foul gods.[3] This scenario has much to do with the way spectroscopy gives chemists a way to listen in to the language of molecules. A poem [by R. H.] tells the story:

> The bright beam, sent caroming
> off four mirrors of the optical
> bench, into the monochromator,
>
> penetrates, invisible but intent; like
> the mosquito off on his spying
> errand for Hunahpu and Xbalanque,
>
> sly heavenly twins of the Popol
> Vuh. For that light means to sting
> too, inciting the electron clouds'
>
> harmony with a ball, a wave,
> to a state-to-state dance; while
> the mosquito flies—in dark rain,

the sun yet unformed—down the Black
Road to Xibalba, bites the false
wooden idols, registering their blank

of an answer, on to the first, who,
god-flesh-bit, cries out, jumps
and the next dark lord calls

"One Death, what is it, One Death?"
which in turn the mosquito records;
from the light is drawn energy,

like blood, leaving on a plotter
a limp signature of H bonded to C;
sampling down the row of heart-

reeking gods: Pus Master, Seven Death,
Bone Scepter, Bloody Claws. The row,
stung, name each other, as do

carbonyl, methyl, aldehyde, amine
prodded by the beam, caught in the end,
like the ball in Xbalanque's yoke.

The losers are sacrificed, the twins win
and life is made clear by signals from within.

Personalization of Nature

The anthropomorphic turn is *so* natural when we speak of molecules. Why? Personalization of nature is like falling in love: our mind endows the Other with a set of imagined qualities that build on the observed, existing features. Scientists do refer often to nature affectionately. They see it as a good friend, a little bit of a tease on occasion, as someone to respect and certainly not to try and assault, as some fancy us doing routinely!

"As someone to respect," we wrote; this requires a little more elaboration. We respect nature for a number of reasons. We like its good looks; we are awed at all the wonders with which our profligate nature bedazzles us (for instance, when we witness a comet suddenly up in the sky for several months and visible even to the naked eye). Second, in like manner to the strange and difficult Hungarian tongue one of us [P. L.] was hearing as a child and slowly learning how to decrypt, we project an intelligibility onto nature; it gratifies us by seemingly conforming to rather simple behavioral patterns (or "natural laws" such as the Newtonian mechanics) that our feeble brains conjure up.

This second feature, intelligibility, makes nature personable. Human beings both strive and like to understand things. There is a basic harmony, an almost miraculous consonance between the quickness of our brain in deciphering the say of nature, and the goodwill of nature, who is willing to tell any careful listener what it is made of and what it went through.

Thus, personalizing nature tends to provide it with an intellect. We get to believe that the laws of nature form a language, that only some intelligent being (or Being) could ever have so ingeniously contrived. For instance, Maxwell's laws of electromagnetism appear (and appeal) to us in their splendid simplicity as a monumental architecture built by a genius. Such a personalization of nature may account for the crazy mix we perceive in the naive philosophy of scientists (none but ourselves): deep-seated realism about the existence of a complex outside world, together with an intellectualizing and idealistic bent, hell-bent on simplicity.

Meaning and Nonsense

One might start with the following proposition: "our brain learns using language acquisition as model." This assertion would apply, not only to learning a second language, but more generally yet to our learning about the world.

A scientist puts himself in a position quite a bit similar to that of an infant learning a language (Hungarian, say). Hearing speech, the young child starts identifying, in what sounds very much like random noise, some recurring features. Its brain attaches itself to those signals and yearns for their reappearance, for the attendant comforting sense of a coherence. These isolated signals, that the child strives to hear again, start to regroup themselves and to form patterns in its mind. Children very early on start connecting those patterns with their context: the facial expression of the speaker, intonation, speech volume, body language, and so on.

The scientist, in his effort to understand a phenomenon, may feel at times very much like the infant, too dumb at first to understand what nature is trying to tell it … the say of nature.

So, one way to think about science is as a conversation with nature: not only do we listen, as if nature were talking to us, explaining very patiently what s/he does and how she does it; we also most definitely ask experimental questions of nature, and we try very carefully to pick up the meaning from the responses that our questioning elicits. Just as remarkable as the fact that two people conversing about the last World Cup game in sentences that are imprecise, unfinished, and, overlapping understand each other, is that scientists can make consensual sense of a few poorly defined bumps in a spectrum.

Such conversations with nature go quite a way toward explaining why scientists tend to personalize nature, to the extent that they refer very often to it, both in speech and in writing.

Aren't there differences between learning a language and doing science? On the surface, the distinguishing features could not be more obvious. A language is a means by which people communicate, while doing science is an attempt to gain reliable knowledge about the world and to raise new questions about it (why is the sky blue, how can we test our explanation, how long has it been blue, does it need us to see it as blue, and so on ...).

Yet, there is a deep-seated similarity between learning a language and doing science. Both activities rely heavily on interpretation. Crucial to our growing understanding of any new language is parsing, that is, the ability to segment a train of audible frequencies into discrete units (sentences, words, syllables, phonemes). Conversely, the ability to link together discontinuous utterances, the related aptitude to translate a "poorly" uttered phoneme (the speaker has a foreign accent, perhaps) mark competence in a language.[4]

Likewise, it is crucial to the scientific enterprise that the research worker be able to parse the physical world into pieces simple enough for examination, thus restricting his attention to the "system under study," to resort to a phrase commonly used by scientists. Conversely, it is equally important for the scientist to convert by interpolation a set of discrete data points into a continuous curve, which at a later stage he or she may try to express with a mathematical equation.

Indeed, the interpretative skill of a scientist is one of the reasons why science— by contrast to what some historians and philosophers appear to believe—goes beyond, way beyond merely the following of a prescribed procedure, that would lead anyone well-versed in the "scientific method" from observations to conclusions. Leaps of the imagination do occur, and they are as important to the scientist as they are to the artist.

"My Nature in Me Is Your Nature Singing"

Not only scientists have conversations with nature. The trope is well established in poetry; a striking exemplification is given by the greatest contemporary natural philosopher poet, A. R. Ammons, in his 1986 poem, *Classic:*[5]

> I sat by a stream in a
> perfect—except for willows—
> emptiness
> and the mountain that
> was around,
> scraggly with brush &

rock
said
I see you're scribbling again:

accustomed to mountains
their cumbersome intrusions,
I said

well, yes, but in a fashion very
like the water here
uncapturable and vanishing:

but that
said the mountain does not
excuse the stance
or diction

and next if you're not careful
you'll be
arriving at ways
water survives its motions.

Now, Live from the RSC Congress in Durham

Roald Hoffmann Reports:

I am at the annual meeting of the Royal Society of Chemistry in Durham, England. The cathedral, visible from every angle, dominates the town. At breakfast in the college dining hall, I see an old friend who is also an invited speaker there, Arndt Simon from the Max Planck Institute for Solid State Research in Stuttgart. Simon is one of the world's foremost solid state chemists, his clarity of mind is unparalleled, and he has an ability to bridge (as the name of his workplace implies) chemistry and physics. He also publishes on old watches, and is very proud of having discovered one of the few one-minute "repeaters" ever made.

I steer my way to sit with Simon. We both decide to try some dry oatmeal cakes, with mixed results. We talk of a brilliant young Russian who just finished his Ph.D. in my group, now a postdoctoral research associate with Simon (the many ties that bind...). Arndt says he is very pleased with Grisha. I say that was to be expected. We speak of Simon's lecture—he asks me how I found it, because he knows I know the field and the audience and care about presentations. I say it was superb, but perhaps had too much material in it for the non-solid-state-chemist audience. Arndt mentions some new solid state compounds he had made, containing nitrogen, called nitrides. They are related to the new inorganic superconductors, copper oxides. He looks around for a napkin to draw the structure of the molecule; I grope in my back pocket for a folded sheet of paper that I usually carry around for just such purposes; I can't find it, the napkin will have to do.

Arndt draws a picture of the molecular lattice he has made, in a few strokes. I see it, I am used to seeing these; I reconstruct the three-dimensionality of the molecule from his suggestive strokes. We share a semiotics honed by years of practice.

I mention to Arndt an idea I once had, of forming nitrogen-nitrogen bonds in the solid. We know of nitrides (compounds with isolated nitrogen ions) and azides (very explosive, three nitrogens in a row bonded to each other), why then not other extended structures with nitrogen-nitrogen bonds? I say (certainly not as grammatically as this): "The way to make them is to look at some existing nitrides, find those with the closest N...N contacts and with very electronegative transition metals around, and then apply pressure to them, maybe along the preferred bond-forming directions of the crystal." Simon notes this mentally, we go on to talk about his theory of superconductivity in carbides, a class of materials he has been studying; we talk of the new appointments at the Max Planck Institute, or at Cornell, of some research projects of mine. There's a lot to talk about between friends; we are late for the first lecture.

A week later Simon calls with a query, and mentions in a verbal postscript: "I found our conversation very interesting; we are going to try to do something about those nitrides." I say "I hope they don't explode." We laugh. This is chemistry.

Later in the day, I meet Norman Goldberg, a former postdoctoral associate, now carrying out research (for his habilitation) at the University of Braunschweig. Norman, a former graduate student, Greg Landrum, and I had done several papers on a kind of unusual bonding called electron-rich three-center bonding, or sometimes hypervalent bonding (perhaps an intrusion of hyper this or that, "postmodernist" hype, into chemistry?) These young guys should be left alone to work on problems of their own choosing, but here I had just been at Arizona State University, and Omar Yaghi there had told me in the course of conversation (and had shown me a model) of a new kind of extended structure or polymer he had made with T-shaped junctures at a copper or silver atom; the top of the T was formed by two organic units going off linearly, while the vertical part of the T was a bond to another copper atom. Now that's pretty interesting by itself, I told Norman (actually I had held that part of the conversation in a preceding letter and now I was reminding him of the story), as I sketched the Yaghi network (Figure 15-4):

"What it made me think of," I said, "is an infinite network of T-shaped hypervalent atoms." Why? Because the T shape is typical of such molecules; it occurs in BrF_3. "Could we do a calculation on this?" In the usual way of bosses in research groups, by "we," I meant "you."

Norman said "I like the problem; I think we can do it." He continued "But what would be the best system to try?" He meant to try calculations on..."Iodides?"

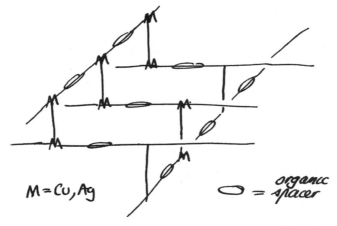

Figure 15-4 A sketch by Roald Hoffmann of an extended system made by Omar Yaghi and coworkers.

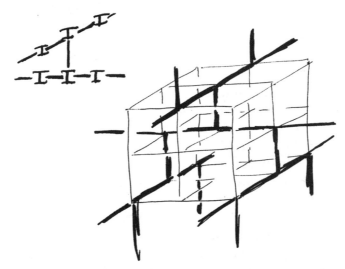

Figure 15-5 A hypothetical infinite network of T-shaped atoms, sketched by Roald Hoffmann.

I wondered, and drew this chain of iodides, with T junctions. As I did, I began to wonder if that was realistic, or if the atoms in the chains were too close to each other (not a good thing for a realistic chemical structure). "I'm worried, Norman," I said, as I began to draw a simple cubic structure, Escher cubes wandering off to infinity, and thickened some of the lines (I didn't have a colored pencil at hand), "if this is just a coloring of a cubic lattice; that won't work, we need a bigger spacer." But then I drew the structure more carefully, as Figure 15-5.

It turned out the structure was fine; only every second cubical site was occupied and the atoms well spaced. We ran out of paper, but I don't think either Norman or I will forget.

The Chemical Place Setting

That so much human conversation takes place around eating might surprise a visitor from another planet—after all, isn't the primary function of visiting restaurants to replenish the chemical feedstocks of this factory of ten thousand chemical ways of breaking down and building up? The mouth should stick to its primary task. Being thus otherwise engaged, there shouldn't be much talk.

But eating is a social activity for humans, as well as a biological one. We eat, and we relax, and we talk. And, especially if we are chemists, we *draw* as we talk. Notice has been taken of the importance of Chinese restaurants, especially a small one near Columbia University, in the development of modern physics. One of the coauthors [R. H.] can introduce a further piece of evidence of the magic of Columbia University, and one illustrative of the sheer quantity of visual information that is communicated by chemists, by showing two sides of a paper place setting he saved from a visit to Columbia (Figure 15-6). The dating of this artifact is not easy (who dates napkins?), but there is circumstantial evidence in the gendered faculty club name, the chemistry on the napkins, and the reasonably good trip files of the coauthor. It is likely that this conversation took place a long time ago, in March 1968.

Do chemists draw more than other people? And if so, why? The answer to the first question is clearly "yes." You just have to take a look at any page of a chemical journal. On the average, 40 percent of the printed page is covered with visual material, and that does not include chemical or mathematical equations. To be sure, there are graphic representations of the results of experimental measurements, but the greatest part of that 40 percent consists of iconic representations of molecular structures. Chemistry was and is the art, craft, business, and now science of substances and their transformations, but in the last two centuries we have learned to look inside the innards of the beast, and to think of the persistent groupings of atoms that are molecules, and of their transformations. Chemists navigate in macroscopic and microscopic worlds, using a necessary mix of symbolic and iconic languages.[6] Chemical structures are the latter, and they just flow across the page.

And from the pens and pencils of chemists engaged in conversation, for it is impossible to talk of molecules without drawing them. So chemists talking to each other immediately gravitate to a blackboard, or place a pad of paper between them, or, in places not amenable to either of these, go through a familiar shuffle of probing all of their pockets for a paper scrap or a Paris Metro ticket. *Ergo* that Columbia Men's Faculty Club place setting.

We have written elsewhere of representation in chemistry.[7] Three-dimensional information of the shape of molecules is critical, often literally a matter of life or death in the activity of a drug molecule. Communication of that information takes place just as you see, through little drawings on two-dimensional surfaces. But the group of people to whom this task devolves (chemists) are not talented at drawing!

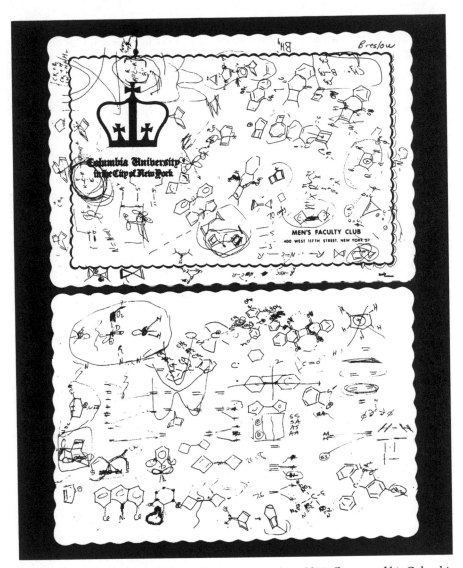

Figure 15-6 A place setting, front and back that served Roald Hoffmann and his Columbia University colleagues well in the course of a meal at the Men's Faculty Club of Columbia University in March 1968.

Now comes the quotidian miracle: untalented as we are, we not only cope, we communicate three-dimensional information effectively and creatively.

Chemistry is not just still substances or molecules, but molecules in transformation. How does one communicate essential change? In the iconographies of Klee-like arrows and bond-sundering wavy lines. And chemistry is microscopic, and at the level of atoms and electrons there are limitations to architectonic principles derived from the macroscopic structures (or the billiard-ball dynamics) of the macroscopic world. Down there, as mentioned already, quantum mechanics rules. How is that shown? Through another set of icons, little figure-eight shapes indicating in this case where the wave/particle electrons in the molecule are. See if you can spot these "orbitals" in Figure 15-6. One of us has gotten a lot of mileage out of them ·[I'll vouch for that, P. L.].

So much less would have been said in the absence of that place setting.

Group Meetings, Posters, and Making the Rounds

There are times when we are quiet, think, eat, read, sleep, watch Zinedine Zidane as he ties a spiritual bond between the ball and his head. There are times when we speak to ourselves. But most often we speak when we and others meet, in institutionalized settings that facilitate or even demand that a conversation take place.

Science is replete with such settings. Some, such as

- the coffee hour (tea time in the United Kingdom),
- the discussion after a seminar,
- the talk with a prospective graduate student,
- a topical conference, and the discussion at it,

are not peculiar to science. But some are; among them we pick

- research group meetings,
- poster sessions,
- visits to lecture at a university; appointments during such a visit.

Let's look at what makes these settings special.

Research Group Meetings

At the end of college, the American twenty-two-year-old chemistry major with a bachelor's degree is probably two years behind the corresponding European university graduate—in chemistry. Five years later, at the Ph.D. level, the two are competitive. Obviously, we must be doing something right in our American graduate education.[8] Among several factors, we would point to the social structure of the research group and its meetings. The research group is often family-like, at times even more strongly bonded than the real nuclear family. An incredible work ethic infects American graduate students. The time spent in the laboratory is great—the late F. A. Cotton, America's leading inorganic chemist, said in 1998:

> I tell the same thing to all my students, "If you're not willing to work 60 hours a week on chemistry even when you're a graduate student or a postdoc, I don't hold out great hope for you."[9]

Many graduate students work still longer hours—you can recognize the chemistry buildings on campus, because in them the lights burn later than anywhere else on campus, with perhaps the exception of architecture drawing rooms.

What goes on at group meetings? Here is how R. H. tells it, for his group:

> *We meet twice a week, for two to three hours each time. After some banter I ask if someone has something from the literature. Anything interesting is game for us, whether we can do a calculation on the molecule or not. It's just a matter of making a couple of transparencies, and then one talks around them. I take the opportunity to provide some background for the problem, if I have some experience with it. Since the people in the group are not only students but also postdoctoral associates and some visiting scholars who are older (we are six to nine people in all, what would be a medium-sized group; some research groups may have as many as thirty people in them) they will chime in.*
>
> *We digress, all the time: In what order are Vietnamese names written? What is the significance of single author papers of German origin? We stop when someone uses slang, and look at every sports metaphor, for there is only one American in this group. When a graduate student is months away from a critical Admission to Candidacy oral exam I pick on (torture?) him or her by calling them to the board to explain something.*
>
> *On other days someone presents their research at an intermediate stage. My own strategy is actually to defer personal discussions on science with people in my group, and say "Why don't you talk about this to the group?" I push for such presentations, for they give desperately needed rhetorical experience, teach presentation methods, and engender planning that is needed for writing papers.*
>
> *While there is substantial variation in the success of what happens, much depending on the interpersonal dynamics of the group (and how much sleep they got the previous night), what transpires is much of the time conversation. Yes, there are also formal presentations, my minilectures as well. But most of the time it is talk, free and unfettered in the way that a family discussion can be. People are not afraid to say foolish things. They know I will jump on them if they are unclear, but if I am lucky I make them feel that their understanding is what I care about.*

Some conversations are better than others.

Poster Sessions

This is a relatively recent way to present material at a national or regional scientific medium, and it has spread like wildfire. It is also a means of presentation that engenders scientifically pregnant conversation, in ways that surpass all other modes of presentation.

In a poster session there is a space with, say, typically thirty to one hundred often rickety poster boards. Each poster presenter is given about a meter by a meter and a half, and can paste or pin up anything in that space. There is tremendous variation in the quality of the visual displays—from a dismal pinning up of pages of tightly packed numbers, to colorful computer-produced integrated posters. Often there is a personal touch—a ribbon, a flag of the country, a photograph of the laboratory.

The posters usually stay up for a day, and the poster presenter is supposed to be at his or her poster at a specified hour, typically for an hour or two. Sometimes refreshments are served in the room. At the appointed hour, the room fills up, and people begin their promenade. The psychology of poster presentations deserves a separate essay, or better still, a play. The viewer tries to keep a distance, so as not to be sucked into a windy presentation of a boring subject; the graphics of the poster should be such as to lure that bashful-to-recalcitrant viewer in to talking distance. Flexibility is important—here the presenter might be in the middle of telling the story to some person Y, when there swims into view well-known Professor X, who might be the source of a postdoctoral position next year—how can the presenter begin the whole story for X without being impolite to Y?

Curious—a poster is designed to be *read*. But the measure of its success is its ability to engender *conversation*.

Much, much more conversation takes place in the poster setting than in the lecture format. First, more people come by; viewing a poster is less intimidating than raising your hand to ask a question. The format invites one person to ask, the other to answer. You can look at nametags of people. There are many presenters, and a natural empathy among them ("no one is coming around to talk to us!"). They begin to talk to each other. The presenters are often at the same stage of their careers—graduate students or postdoctoral associates—yet they come from different places. In the darkened lecture hall, they do not see each other; here, the bright light necessary for the presentation draws them to cluster into groups, to gossip, to break the social barriers that stand in the way of meaningful communication.

Lewis Thomas describes hauntingly the sound of scientists talking: "it is the most extraordinary noise, half-shout, half-song, made by confluent, simultaneously raised human voices, explaining things to each other."[10] We have often been able, as we wander down a corridor at a meeting, to find the room where the poster sessions are, simply from the hubbub of the talk.

Making the Rounds

Like other human groups, scientists tend to ritualize their interactions. If a guest seminar speaker is arriving from another campus, there is a set procedure so that the visitor may become acquainted with some of his colleagues during his stay. Upon arrival, the speaker is given a schedule on which, besides the time and location of the lecture, are shown time slots for interviews (set at intervals typically of a half hour to an hour) with various professors in their individual offices. Each such conversation thus conforms to the Aristotelian Rule of the Three Unities, which was deemed unassailable by French drama writers of the seventeenth century: unity of location, unity of time, unity of action.

As soon as the visitor has settled, and after a few gestures and grunts of hospitality from the host (a cup of coffee, a Coke, and so on), the conversation can start. The worst-case scenario, one all too familiar to us, has the host delivering a well-memorized monologue, that tells the guest in some detail a recent study that is about to be published. In so doing, the host may show a set of slides on a computer serving as visual aids in a standard sell: salesmanship of the most ordinary kind, no different from someone offering insurance or vacuum cleaners.

Usually, however, the interview starts with some effort at communication, at consonance even: "Which story shall I tell you?" says the host rhetorically, fishing for a topic that will overlap a little with the visitor's interests, which are indicated by the title and the summary of the seminar lecture to be held on the same day—unfortunately, often after the meeting.

Such an interactive scenario—by contrast to the first, the commercial—can lead to genuine discussion. The host will then benefit from points raised or from suggestions offered by the guest, perhaps even encouraging the visitor in such a direction at the outset with something like: "I'd like to have your reaction to these results and ideas," or in the catch-all phrase, "I'd like to pick your brain on this."

Making the rounds in a building and meeting with colleagues in the manner described has at least five virtues: educational—in our age of overspecialization we are thus able to broaden our outlook by learning of advances in other areas; olfactory—if our antennas are delicate enough to sense the prevailing atmosphere in the department visited, whiffs of internal warfare included; seminal—if and when the discussion provides mutually beneficial ideas; congenial—in those not infrequent occasions when host and guest take to one another, and their meeting could conceivably even start a friendship; and, last but not least, the virtue of friendliness—indeed if the two know already each other and like one another's company, their reunion may be pleasant and even festive.

Listening to a colleague talking about his or her work, by contrast with later reading the published paper—often one leaves the office clutching a bunch of "preprints"—has the merits of a focus on the essentials and of learning how the author values his or her contribution, where he or she puts it within the development of a field.

During the presentation, the visitor acquires extraneous information, too. There are all the nonverbal aspects: the seating arrangement, on a scale from distant formality to close informality and geniality; the titles of books on the shelves; little mottos and cartoons up on the walls; the degree of messiness in the office; the personal pictures of family and of hobbies outside work—from skiing or mountain climbing to sailing and scuba diving, not to mention playing the cello, square dancing, or gourmet dining—that hint at more dimensions in the person than is let on explicitly during the meeting.

All in all, such behavior as just described goes back a long time, much before the advent of modern science. Homer's *Odyssey* repeatedly shows us Ulysses being greeted in a city and palace. He tells the story of his peregrinations, and he is treated to some narrative in turn.

Famous Conversations in Chemistry

If Mephistopheles were still offering bargains, what would tempt us would be time travel—to hear and see a Greek chorus in a Sophocles tragedy performed at Epidaurus or Segesta. Or to sit in on that Paris dinner in October 1774, given by M. and Mme. Lavoisier, and attended by that remarkable Unitarian clergyman and scientist, Joseph Priestley. Priestley later was hounded out of his home in England for praising the revolution whose excesses took Lavoisier's life, but in 1774 what brought them together was good science.

Priestley had earlier that year made oxygen by forming *mercurius calcinatus per se* (which we would now call mercuric oxide, HgO) by heating mercury in the presence of air. On decomposition (people then spoke of the compound "reducing itself") it gave mercury and a previously unknown gas that supported combustion. Priestley, thinking in the older chemical framework that Lavoisier's experiments eventually demolished, called that gas "dephlogisticated air." Meanwhile Lavoisier had embarked on a careful series of studies of metal-gas compounds *(calxes)* and the processes of combustion and reduction. He was close to discovering oxygen—the gas was in the air in more than one sense. But until Priestley showed up in Paris, it is the considered opinion of most that Lavoisier did not know of the essential piece of the puzzle—oxygen. Here is how Priestley describes the dinner we wish we could relive:

> *Having made the discovery some time before I was in Paris, in 1774, I mentioned it at the table of Mr. Lavoisier, when most of the philosophical people of the city were present, saying that it was a kind of air in which a candle burned much better than in common air, but that I had not then given it a name. At this, all the company, and Mr. and Mrs. Lavoisier as much as any, expressed great surprise. I told them I had gotten it from precipitate per se and also from red lead. Speaking French very imperfectly and being little acquainted with the terms of chemistry, I said plomb rouge, which was not understood until Mr. Macquer said I must mean minium.[11]*

Actually the gas had first been discovered (but remained unpublished—an interesting story), more than two years before, by a modest Swedish apothecary, Carl Wilhelm Scheele. By heating manganese oxide, Scheele got a gas he called descriptively *eldsluft*, or "fire air." He wrote of it to Lavoisier. Around October 15, 1774, Lavoisier received Scheele's letter telling him in substantial detail of the synthesis and properties of oxygen. The letter was found, unanswered, in Mme. Lavoisier's files 115 years later.

October was not an easy month for Antoine Laurent Lavoisier. In the course of his epochal subsequent work on combustion, Lavoisier failed to give proper credit to Priestley for the crucial information in their October dinner conversation. There may have been a flaw of character in the great French chemist.

Let us recount another conversation. One of the great success stories of twentieth-century chemistry is the renaissance of inorganic chemistry, and the development of organometallic chemistry, a borderland between inorganic and organic chemistry. There were hints of the existence of this fertile region, but that's all there were—hints—until the beginning of an explosive growth of the field in the 1950s. This beginning is very clearly defined; it is the report in 1952 of a very simple compound, $(C_5H_5)_2Fe$, or ferrocene. The compound has two five-membered carbon rings, each carbon carrying a hydrogen. Each of the two groups reporting (independently) the synthesis of this beautiful orange compound had the connectivity of these parts of the molecule right, but guessed quite incorrectly about the way the cyclopentadienyl rings (as they are called) are linked up to the iron. They postulated the molecule at left in Figure 15-7, but the structure is the much more interesting (and simpler) one at right. This is the first of a multitude of organometallic "sandwich" compounds.

Two chemists at Harvard, both good at keeping up with the literature, saw the initial English reports on pretty much the day they came into the library. One of them was a new assistant professor from England, Geoffrey Wilkinson. He had been less than a year at Harvard at the time, just getting his laboratory set up. Wilkinson was still to find his first graduate student. The other Harvard chemist to see the report of the still to be named ferrocene was also young, but already recognized as one of the outstanding organic chemists of the world, Robert B. Woodward. Woodward was that rare person in chemistry, a child prodigy. And while the world thought of him as the shaper of paradigms in the art of organic synthesis, his awesome intellect roamed all over chemistry.

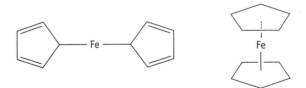

Figure 15-7 Two structures for ferrocene: (left) the incorrect one, first suggested; (right) the correct structure.

Woodward and Wilkinson saw the English papers with the incorrect structure independently on January 30, 1952.[12,13] Both felt intuitively the structure suggested was wrong. Woodward had a sizable research group, and in it was a graduate student, Mike Rosenblum (since that time a distinguished professor of chemistry at Brandeis University). Apparently, Woodward told Rosenblum of the problem, and set him to find some ruthenium, so that a ferrocene analogue (Ru is right below Fe in the Periodic Table, and so should have similar properties) might be made. Here is how Wilkinson describes what happened next:

> *The first I knew of this [of Woodward's interest in the problem] was on Saturday when Mike Rosenblum came into my laboratory asking if I had got some ruthenium. I can't remember what I said, though I remember being more than considerably annoyed, except that I think it was along the lines of "let me tell you what you want that for." However the upshot was that Woodward and I had lunch at the Harvard Faculty Club on Monday and sorted things out. The possibility that the C_5H_5 ring in the iron compound could possibly undergo Friedel-Crafts or other aromatic reactions simply had not dawned on me, but after the structure, this seemed to be Bob's main interest, whereas mine was to go to other transition metals.*[12]

The last comment is actually telling; it cuts to one of the very, very few failures of judgment in Woodward's remarkable career. Wilkinson took off in the right direction, for the explosive (productively, not literally) path of organometallic chemistry led to other metal and other "ligands" (organic groups attached). Ferrocene was just the opening to a new universe. We suspect that Woodward, more than Wilkinson, understood the essence and necessity of the sandwich structure. But this greatest of the century's organic chemists saw ferrocene as an aromatic molecule (he was right), and then got caught up in finding in its chemistry the telltale markings of other aromatic molecules. We think that to Woodward what was interesting was not so much to go on to other metals as to find out how ferrocene was the same and not the same as benzene, the archetypical aromatic molecule.

Simulated Dialogues

Our libraries hold precious few transcripts of actual oral exchanges between scientists, but they hoard a wealth of fictionalized conversations about science. Use of such books, much in fashion at certain periods in history, can be didactic. The two conversants stand for a teacher and a student.

While it is true that science builds up a body of knowledge and that scientists should and do share the specialized knowledge that they have acquired, their forte is not so much to parcel out what they know, but rather to ask questions: unusual, incisive, and radical. The scientific enterprise is a quest for (provisional) truths about nature. It proceeds by raising questions, about anything and everything, including the revision of well-accepted notions. Cavaliere Ripa's early seventeenth-century icon for *Investigatione* (Figure 15-8)[14,15] is a pretty good one to this day.

George Richardson (*Iconology*, London, G. Scott, 1779, pp. 85, 86) writes of this emblematic image: "with wings at her temples, to signify elevation of the understanding, by which this faculty should always be directed. Her garments are overspread with emmets [ants], which, by the Egyptians, were assigned as the hieroglyphick to investigation, they being, of all other animals, the most diligent searchers after every thing necessary for their support. The figure points to a crane in the air with her right hand, and with the left she points to a dog, who is in the action of searching after his prey. The Egyptians understood a crane to be the sign of an inquisitive man."

The dialogue form is near to being ideal for conveying an *impression* of this questioning, which is so basic to science, a skill in which scientists train themselves for years. To quote Leonard Woolf, "Journey, not the arrival matters."[16,17] The dialogue format is thus oftentimes couched in questions from one actor, "the pupil," to the other, "the teacher."

Jane Marcet, in a time (early nineteenth century) when women were considered as unsuited to intellectual pursuits, published such a book. Her *Conversations*

Figure 15-8 A representation of Investigatione, from the 1618 edition of Cesare Ripa's *Iconologia* (Ref. 14).

in Chemistry (1806)[18] introduces to her readership the still-pristine science of chemistry: luminaries with the stature of Lavoisier and Humphry Davy had set it on its course during the previous few decades. She presents both facts and their interpretation. Compared with the contemporary textbooks of chemistry, Marcet's dialogue format endows her text with superior readability.

Such a didactic use of a dialogue goes back to Antiquity. The Book of Job and the Platonic dialogues are the archetypical example of the use of a written dialogue to present ideas. Moving to the seventeenth century, to the time when philosophy and physics parted company, the lively style of Galileo's *Dialogo sui Massimi Sistemi* (1632),[19] together with it being written in the vernacular, a major innovation ensuring a larger and different readership, was responsible both for the widespread diffusion of this dialogue and for its author's well-known troubles with the Church. Galileo could not resist putting a pope's opinion into the mouth of Simplicio.

Bernard le Bovier de Fontenelle was a playwright in Paris when he published in 1686 his *Entretiens sur la Pluralité des Mondes*.[20] This was a time of transition in astronomy, many new discoveries were being made and there was rapid growth of public interest in science. Some necks must still have been strained from the viewing of Halley's comet, an astounding sight, just a few years before (1680), and Fontenelle was able to build on this interest.

Fontenelle was not a scientist, but he made it a point to revise his book and to keep it up to date with contemporary science, especially after his election as perpetual secretary to the Academy of Sciences in 1697. Thus the *Entretiens* well deserved their extraordinary success, they went through dozens of successive editions. They defend Copernican astronomy and also Cartesian physics against Newton. Fontenelle wished to appeal to women particularly. Hence, the *Entretiens* give us dialogues between the author and a lady of the aristocracy, as the pupil intent upon learning astronomy; this marquise has personality and poise.

Some idea of the sheer verve of this dialogue might be obtained from this fragment. Fontenelle begins by enumerating the planets as one goes out from the sun:

> *"Finally, Mars, Jupiter and Saturn follow, in the order in which I've named them for you, and you can see that Saturn makes the largest circle of all around the Sun, and takes more time than any other planet to make each complete turn."*
>
> *"You've forgotten the Moon," said the Marquise.*
>
> *"I'll find her again," said I. "The Moon turns around the Earth and never leaves her in the circle the Earth makes around the Sun. If she moves around the Sun it's only because she won't leave the Earth."*
>
> *"I understand," she said, "and I love the Moon for staying with us when all the other planets abandoned us. Admit that if your German [Fontenelle has mistakenly called Copernicus a German] could, he'd make us lose her too, for I can tell that in all his actions he had it in for the Earth."*

"He did well," I answered, "to have put down the vanity of men, who had giv-en themselves the greatest place in the universe, and I'm pleased to see Earth pushed back into the crowd of planets."

"Surely you don't believe," she cried, "that the vanity of men extends all the way to astronomy."

During the Enlightenment, Diderot wrote *Le neveu de Rameau* [1761–74].[21] This extraordinary piece presents a conversation between intellectuals, of which the title character, the composer Rameau's nephew, runs the gamut of the emotions (at turns sarcastic, witty, incisive, argumentative), and of the modes of expression, from the confessional to the declamatory; and is brilliant throughout. Diderot, in this philosophical essay purporting to report a conversation, refocussed the published dialogue on the personality of one of its participants. This innovation was not lost on subsequent writers: quite a few books by Richard P. Feynman (or by his friends and students reconstructing Feynman), with their unique mix of entertaining fun and artful presentation of scientific concepts, are very much in the same vein.

Today, didactic fictional dialogues continue to serve their useful function of presenting difficult scientific concepts to the layman. The "Mr. Tompkins" books by George Gamow are small gems in this vein. For instance, the first chapter of *Mr. Tompkins Explores the Atom*, first published in 1944,[22] opens on a discussion by two main characters, Mr. Tompkins and Maud, of a gambling martingale, which had been published in the January 1940 issue of the magazine *Esquire:* In this first chapter, Gamow brilliantly leads his reader from an elementary presentation of probabilities into one of statistical physics.

We have addressed so far two functions of a written, made-up dialogue, the didactic and the admiring. There is a third as well, as an outlet for an unorthodox writer who wishes to prevail nevertheless (the power of the pen) and to take on the rest of the world, if need be. George Berkeley wrote such a controversial text in 1713.[23] Bishop Berkeley, in this dialogue, has Hylas as a representative of mate-rialists while Philonous is a mouthpiece for his own views. Philonous undertakes to convince his interlocutor that matter does not exist, only representations in the mind.

An equally important fictional dialogue is that imagined by Alan Turing in 1950,[24] in one of the few papers he ever published. The dialogue sections form a *Gedanken experiment* intended to demonstrate, by way of a questions and answers exchanged by a human being and a machine locked in a closed room, the seeming absence of any distinction between the intelligence of man and that of a computer. This seminal paper by Turing is entitled "Computing Machinery and Intelligence," and lucidly addresses, for the first time we think, the controversial philosophical question of artificial intelligence.

Dialogues persist to this day. The late Jeremy Burdett was one of the most original applied theoreticians of our time. One of his books, published in 1997, is a wonderfully perceptive *Chemical Bonds: A Dialog*.[25]

Taboos

What cannot enter a scientific conversation? There are certainly things that are excruciatingly difficult to say—to tell a graduate student who announces that he has applied for a job at Cal Tech that you do not think he is good enough for the position, or signal a colleague that he has written a paper that just has too much hype in it, for example. But these are problems faced by all people, not just scientists. Could it be more difficult for some scientists to deal with such problems because of shyness of dealing with human beings? Fortunately, real life is a magnificent corrective, and in the end most everyone, scientists included, learns how to deal with a boss, with love, with a bank account.

Perhaps more interesting are the areas of forbidden discourse special to the scientific enterprise:

- We avoid telling a colleague that we have reviewed his research paper or grant proposal. Research papers are our stock in science, rather than books. These papers typically get sent out to two anonymous reviewers, who (often without coercion) respond in a month or so. Research grants are judged in two ways—either with a large set (seven to ten) of anonymous reviewers, or by a review panel, with one person taking the lead in summarizing and evaluating the work for the other members of the panel. A difficult situation is to hear in a conversation a colleague tell excitedly of his work—when one has already reviewed it, and therefore knows it in unusual detail. No problem if one is equally ecstatic about the work; but what if one has doubts? The conversational setting elicits a response; and the situation provokes inauthenticity at the least, the uneasy feeling that one has not quite said what one really thought.

- To reveal, or to conceal? The first reaction of scientists is, "How can you even ask this question? Science depends on a free and open communication system!" Indeed. But the scientist who responds in this way has simply not been fortunate enough (or misfortunate enough) to be faced with a discovery that has real commercial value. Our societies have evolved a complex system of patents for protecting the economic fruit of invention. Much more can be said about patents than we have space for here—in principle, one has a system of exchanging protection for disclosure. But in practice the art is to claim as widely and to reveal as little as one can get away with.

Chemists in industry are faced with extraordinary constraints on what they can or cannot say, especially when their companies are in the midst of patenting a process. Sometimes our heart goes out to them, for underneath the constraints we see a scientist dying to tell us something new. Sometimes, we have no sympathy, when we see them trying to worm out of fellow chemists (working for a rival company) their precious art.

There is a story to tell here, one that regretfully must be told without specifics, and with some minor changes. Once, in Cold War days, the U.S. Department of Defense supported a major program of classified research in a certain field of chemistry, hoping that it would lead to a new explosive. The program was based, as it turned out to be, on some faulty chemical assumptions. It led to no explosive at all, but instead to some wonderful chemistry, indeed to the synthesis of an entirely new class of molecules. No wonder, for the chemists employed in the program were, as it happened, very creative. Now, it is in the nature of secret programs and the stodgy bureaucracies that surround them that, even when it is perfectly apparent to everyone that on technical grounds there is nothing to hide, nevertheless the secrecy is maintained, and "classification" is not lifted.

Imagine the agony of the chemists who for five years could not publish their work, work that they knew would make a stir. The only thing that would make the stupid bureaucrats declassify the work would be if the Russians published the same research in the open literature! Or if not the Russians, at least another American. Some of the scientists got so desperate that they went around to American academic research groups working in related areas, and dropped broad hints that this area of chemistry was worth pursuing. One of us sat in on such a conversation.

Here is a question we believe probably one should not ask, especially a young scientist should not ask in a scientific conversation: "Do you understand?" On the face of it, what could be more honest and straightforward? The speaker, who may have just presented a difficult concept, or spoken too quickly, has sensed a non-verbal response on the part of his audience/listener, and is stating that he or she is willing to explain things again. But the question, unless asked in just the right tone, and between people of equal status or confidence, may be just as problematic as the question "Do you love me?" If it has to be asked, it may be too late.

Written Thoughts and Spoken Words

Conversation may be bad because it can make a human being break a covenant, forget a promise, be lured into an action showing poor judgment, become sneaky and duplicitous. The Bible, as usual, provides us with some pivotal examples. For a comparison between speech and writing, we go to the fascinating parable of the adulteress:

> *The scribes and the Pharisees brought a woman who had been caught in adultery, and placing her in the midst they said to him "Teacher, this woman has*

been caught in the act of adultery. Now in the law Moses commanded us to stone such. What do you say about her?"

This they said to test him, that they might have some charge to bring against him. Jesus bent down and wrote with his finger on the ground. And as they continued to ask him, he stood up and said to them "Let him who is without sin among you be the first to throw a stone at her." And once more he bent down and wrote with his finger on the ground. But when they heard it, they went away, one by one, beginning with the eldest.

(*John 8:3-11*)

Meaning of Jesus writing on the ground—the Apostle does not give us a clue as to what he may have been scribbling—has intrigued theologians for centuries. The dominant interpretation has Jesus making reference by his gesture to Jeremiah 17:13:

Those who turn away from thee shall be written in the earth for they have forsaken the Lord, the fountain of living water.

The reference to Jeremiah has the merit of mirroring—also in the *testing* and in the attendant judicial-like *charge* as well—the first conversation in the Torah, that of Eve and the serpent "the shrewdest of all the wild beasts that the Lord God had made" (Genesis 3:1-5). The well-known consequence was for Adam and Eve to know henceforth the distinction between good and evil, to be cursed in no uncertain language and cast out of the Garden of Eden.[26]

The Bible thus endows speech with the riches and risks of seduction and treachery, while writing records (as a scribe does) such transgressions. Commenting upon Jesus's enigmatic action in John 8:3-11, Schnackenburg, summarizing centuries of exegesis, writes

Jesus refers (the questioners) to the judgment of God, before whom all men are sinners. They are all fit to be "written in the earth" ... a sentence upon the guilty who know their guilt.[27]

Thus writing inscribes names of the sinners, while speech is what incites a person to sin. What a pessimistic appraisal of those two modes of human expression, oral and written, that have given us masterpieces of narration and of lyricism!

The opposition we might put between the two, for the purpose of this paper, has speech being the medium for reports to others, and writing being the tool for thought; it separates the collective and the social, on one hand, from the personal and the private on the other. Such a tension between the personal and the social dimensions animates the opening line of one of Dylan Thomas's poems:

The conversations of prayers to be said
By the child going to bed and the man on the stairs,
Who climbs to his dying love in her high room,
The one not caring to whom in his sleep he will move
And the other full of tears that she will be dead.[28]

Fecundity out of Repression

The scientific article ossified in the mid-nineteenth century. Take away Asian coauthors and computer graphics, and what you have today is pretty much the same format, governed by similar conventions, to what was there one hundred and forty years ago—an ordered layout, the scholarly apparatus of endnotes and/or footnotes, a neutered third-person diction. "Please, no emotions allowed, we're scientists." Underneath the surface (now as before)—we have impassioned, curious, and fallible human beings engaged in the search for reliable knowledge—and forced to at least express themselves like dispassionate "gentlemen."[29]

The mechanisms of repression are certainly institutionalized here. But the id will out; if not in the most tangible product of scientific activity, the article, then maybe in the dark, hidden places—in the referee's reports, the reviews of proposals. Also in those nooks of society where repression is loosened by licit drugs or by informality. In these settings, often in speech—uncontrolled, unrecorded, in plain talk, not subject to strictures of order and good behavior—in the *Nachsitzung* after a seminar at a German university; in the long drinking evenings of a Gordon Research Conference...

So what's new about this? People are people—they use informal conversation for gossip, innocent or malevolent, for *Schadenfreude,* for eliciting pity, claiming power, stoking the insatiable demands of some guilt. Is there anything in the free talk of scientists that is of *value*, over and beyond normal letting go?

Thinking about real value, if conversation is compensatory of repression—more open just because the written product of scientific work is so constrained—could it be that much more real discovery and creation takes place in conversations? We think so! It is the first place where one expresses understanding outside of the private confines of one's mind. The research group presentation is probably next, the writing of the paper last, very important, place. The conversation—with a colleague, student to student—is where the ideas get expressed. And until they are expressed, in some way they are not real. The conversation reifies the idea; it selects in the mind of the researcher one possibility of many, it is *the first existential act in science*. All the stronger because the talk is free.

Acknowledgments

Pierre Laszlo is grateful to the Department of Chemistry at Cornell University for a visiting professorship conducive to the writing of this article. We appreciate the help of William Ashworth in supplying a crucial illustration, and the Hoffmann research group for its assistance in computer matters.

Notes

1. For a good introduction to chemical change, see P. W. Atkins, *Atoms, Electrons, and Change* (New York: Scientific American Books, 1991). See also R. Hoffmann, *The Same and Not the Same* (New York: Columbia University Press, 1995), Chapters 29–36.
2. D. N. Mermin, The (Non)World (Non)View of Quantum Mechanics. *New Literary History* 13 (1992): 855.
3. *Popol Vuh,* tr. D. Tedlock (New York: Simon and Schuster, 1986).
4. P. Laszlo, *La Parole des Choses* (Paris: Hermann, 1993).
5. A. R. Ammons, *The Selected Poems,* Expanded Edition (New York: Norton, 1986), p. 692.
6. E. Grosholz, and R. Hoffmann in *Philosophy of Chemistry,* ed. N. Bhushan and S. Rosenfeld (Oxford: Oxford University Press, 1998).
7. R. Hoffmann and P. Laszlo, Representation in Chemistry. *Angew. Chem. Int. Ed. Engl.* 30(1991): 1; also in *Diogenes* 147 (1989): 23.
8. P. Laszlo, *Les Universites Americaines* (Paris: Flammarion, 1996).
9. R. Dagani, Home on the Range with Al Cotton. *Chemical and Engineering News* 7 (March 30, 1998): 39.
10. L. Thomas, *The Lives of a Cell: Notes of a Biology Watcher* (New York: Viking Press, 1974).
11. J.-P. Poirier, *Lavoisier: Chemist, Biologist, Economist,* tr. Rebecca Balinski (Philadelphia: University of Pennsylvania Press, 1996).
12. Wilkinson tells his version of the story in a published paper: G. Wilkinson, The Iron Sandwich: A Recollection of the First Four Months. *Journal of Organometallic Chemistry* 100 (1975): 273. Woodward (deceased 1979) never wrote of the matter. See, however, the fascinating story of Woodward objecting to the Nobel Prize in Chemistry awarded to Wilkinson and Fischer: T. M. Zydowsky, Of Sandwiches and Nobel Prizes: Robert Burns Woodward—This Is a Story of the Emotions Behind Scientific Advances and Discoveries, and Receiving Recognition for Them. *Chemical Intelligencer* (January 2000): 29. The sandwich structure for ferrocene was independently suggested by E. O. Fischer and W. Pfab, Zur Kristallstruktur der Di-Cyclopentadienyl-Verbindungen des zweiwertigen Eisens, Kobalts und Nickels. *Z. Naturforschung* 7B(1953): 377.
13. P. Laszlo and R. Hoffmann, Ferrocene: Ironclad History or Rashomon Tale? *Angew. Chem. Int. Ed. Engl.* 39 (2000): 123.
14. C. Ripa, *Iconologia…* (Padua: Pasquati, 1618).
15. W. B. Ashworth, Jr., "Visual Arguments in the Scientific Revolution," 1990 lecture at the University of Oklahoma Centennial Conference on the History of Science, unpublished.
16. L. Woolf, *The Journey, Not the Arrival Matters: An Autobiography of the Years 1939–1969* (London: Hogarth Press, 1969).
17. See also Cavafy's wonderful 1911 poem, "Ithaca": K. P. Cavafy (Kabaphes), *The Complete Poems of Cavafy,* tr. Rae Dalven (San Diego, CA: Harcourt Brace Jovanovich, 1976).
18. J. Marcet, *Conversations on Chemistry, In Which the Elements of the Science Are Familiarly Explained and Illustrated by Experiments* (London: Longman, Hurst, Rees, and Orme, 1806).
19. G. Galileo, *Dialogue Concerning the Two Chief World Systems, Ptolemaic and Copernican;* foreword by Albert Einstein, tr. Stillman Drake (Berkeley, CA: University of California Press, 1962).

20. M. de (Bernard Le Bovier) Fontenelle, *Entretiens sur la pluralite des mondes. Digression sur les anciens et les moderns,* ed. R. Shackleton (London: Clarendon Press, 1955).

21. D. Diderot, *Le neveu de Rameau,* ed. E. J. Geary (Cambridge, MA: Integral Editions, 1959).

22. G. Gamow, *Mr. Tompkins Explores the Atom,* 4th ed. (Cambridge: Cambridge University Press, 1958).

23. G. Berkeley, *Three Dialogues Between Hylas and Philonous,* ed. D. Hilbert and J. Perry (Claremont, CA: Arete Press, 1994).

24. A. Turing, Computing Machinery and Intelligence. *Mind* 59 (1950): 433.

25. J. K. Burdett, *Chemical Bonds: A Dialog* (New York: John Wiley and Sons, 1997).

26. In a recent book (P. Quignard, *Vie secrete* [Paris: Gallimard, 1998]), Quignard (our translation) writes:

    "Tertullian said 'Even in Paradise one needs to dissimulate. Even in Eden, it would have been better for the first woman to have been secretive. Even God is secretive: he is inscrutable to our sight. He is impenetrable in his designs. He is forever silent to Himself.' Eve should have shut up. This was the thesis to which the schismatic theologian from Carthage kept returning. What the serpent had whispered to her in the shadow of the tree, she ought to have kept locked away in her heart."

27. R. Schnackenburg, *The Gospel According to St. John,* vol. 2 (London: Burns and Gates, 1980), pp. 165–166.

28. D. Thomas, *Deaths and Entrances* (London: Dent, 1946), p. 7.

29. R. Hoffmann, Under the Surface of the Chemical Article. *Angew. Chem. Int. Ed. Engl.* 27 (1988): 1593. Chapter 13 in this book.

16

How Symbolic and Iconic Languages Bridge the Two Worlds of the Chemist

A Case Study from Contemporary Bioorganic Chemistry

EMILY R. GROSHOLZ AND ROALD HOFFMANN

Chemists move habitually and with credible success—if sometimes unreflectively—between two worlds. One is the laboratory, with its macroscopic powders, crystals, solutions, intractable sludge, things which are smelly or odorless, toxic or beneficial, pure or impure, colored or white. The other is the invisible world of molecules, each with its characteristic composition and structure, its internal dynamics and its ways of reacting with the other molecules around it. Perhaps because they are so used to it, chemists rarely explain how they are able to hold two seemingly disparate worlds together in thought and practice. And contemporary philosophy of science has had little to say about how chemists are able to pose and solve problems, and in particular to posit and construct molecules, while simultaneously entertaining two apparently incompatible strata of reality. Yet chemistry continues to generate highly reliable knowledge, and indeed to add to the furniture of the universe, with a registry of over ten million well-characterized new compounds.

The philosophy of science has long been dominated by logical positivism, and the assumptions attendant upon its use of predicate logic to examine science, as well as its choice of physics as the archetype of a science. Positivism thus tends to think of science in terms of an axiomatized theory describing an already given reality and cast in a uniform symbolic language, the language of predicate logic. (See especially the locus classicus of this position, Carnap's book.)[1]

The authors of this paper wish to question certain positivist assumptions about scientific rationality, based on an alternative view brought into focus by the reflective examination of a case study drawn from contemporary chemistry. Our reflections owe something to Leibniz,[2a] Husserl,[2b] Kuhn,[2c] and Polanyi,[2d] and draw upon the earlier writings of both authors, Hoffmann[3a] and Grosholz.[3b] We will offer a nonreductionist account of methods of analysis and synthesis in chemistry. In our view, reality is allowed to include different kinds of things existing in different kinds of ways, levels held in intelligible relation by both theory and experiment, and couched in a multiplicity of languages, both symbolic and iconic.

We will argue that there is no single correct analysis of the complex entities of chemistry expressed in a single adequate language, as various reductionist scripts require; and yet the multiplicity and multivocality of the sciences, and their complex "horizontal" interrelations, do not preclude but in many ways enhance their reasonableness and success. Nor is this view at odds with the authors' realism; we want to distinguish ourselves quite strongly from philosophers engaged in the social construction of reality.[4] We understand the reality whose independence we honor as requiring scientific methods which are not univocal and reductionist precisely because reality is multifarious, surprising, and infinitely rich.

I. Formulating the Problem

The paper drawn from the current literature in chemistry that we shall consider is "A Calixarene with Four Peptide Loops: An Antibody Mimic for Recognition of Protein Surfaces," authored by Andrew Hamilton, with Yoshitomo Hamuro, Mercedes Crego Calama, and Hyung Soon Park and published in December 1997 in the international journal *Angewandte Chemie*.[5] The subfield of the paper could be called bioorganic chemistry. One way to look at biology is to examine its underlying chemistry, in a well-developed program that is both one of the most successful intellectual achievements of the twentieth century, and a locus of dispute for biologists. For many years, organic chemists had let molecular and biochemistry "get away" from chemistry; recently, there has been a definite movement to break down the imagined fences and reintegrate modern organic chemistry and biology. The paper we examine is part of such an enterprise.

We have learned something about the structure of the large, enigmatic, selectively potent molecules of biology. But describing their structure and measuring their functions do not really answer the question of how or why these molecules act as they do. Here organic chemistry can play an important role by constructing and studying molecules smaller than the biological ones, but which model or mimic the activities of the speedy molecular behemoths of the biological world.

The paper opens by stating one such problem of mimicry, important to medical science and any person who has ever caught a cold. The human immune system has flexible molecules called antibodies, proteins of some complexity which recognize a wide variety of molecules including other proteins.

> *The design of synthetic hosts that can recognize protein surfaces and disrupt biologically important protein-protein interactions remains a major unsolved problem in bioorganic chemistry. In contrast, the immune system offers*

numerous antibodies that show high sequence and structural selectivity in binding to a wide range of protein surfaces.[6]

The problem is thus to mimic the structure and action of an antibody; but antibodies in general are very large and complicated. Hamilton et al. ask the question, can we assemble a molecule with some of the structural features of an antibody, simplified and scaled down, and if so will it act like an antibody? But what are the essential structural features in this case?

Prior investigation has revealed that an antibody at the microscopic level is a protein molecule that typically has a common central region with six "hypervariable" loops that exploit the flexibility and versatility of the amino acids that make up the loops to recognize (on the molecular level) the near infinity of molecules that wander about a human body. The paper remarks,

> *This diversity of recognition is even more remarkable, because all antibody fragment antigen binding (FAB) regions share a common structural motif of six hypervariable loops held in place by the closely packed constant and variable regions of the light and heavy chains.*[7]

What is recognition at the microscopic level? It is generally not the strong covalent bonding that makes molecules so persistent, but is rather a congeries of weak interactions between molecules that may include bonding types that chemists call hydrogen bonding, van der Waals or dispersion forces, electrostatic interactions (concatenations of regions of opposite charge attracting, or like charge repelling), and hydrophobic interactions (concatenations of like regions attracting, as oil with oil, water with water). These bonding types are the subject of much dispute, for they are not as distinct as scientists would like them to be.[8] In any case, the interactions between molecules are weak and manifold. Recognition occurs as binding, but it is essentially more dynamic than static. At body temperature, recognition is the outcome of many thermodynamically reversible interactions: the antibody can pick up a molecule, assess it, and then perhaps let it go. In the dance of holding on and letting go, some things are held on to more dearly.

Whatever happens has sufficient cause, in the geometry of the molecule, and in the physics of the microscopic attractions and repulsions between atoms or regions of a molecule. The paper remarks,

> *Four of these loops ... generally take up a hairpin conformation and the remaining two form more extended loops. X-ray analyses of protein-antibody complexes show that strong binding is achieved by the formation of a large and open interfacial surface (>600 Å) composed primarily of residues that are capable of mutual hydrophobic, electrostatic, and hydrogen bonding interactions. The*

*majority of antibody complementary determining regions (CDRs) contact the
antigen with four to six of the hypervariable loops.*[9]

The foregoing passage is a theory about the structure and function of antibodies,
but it is asserted with confidence and in precise detail. Standing in the background,
linking the world of the laboratory where small (but still tangible) samples of antibodies and proteins are purified, analyzed, combined, and measured, and the world
of molecules, are theories, instrumentation, and languages. There is no shortage of
theories here; indeed, we are faced with an overlapping, interpenetrating network
of theories backed up by instrumentation. These include the quantum mechanics of
the atom, and a multitude of quantum mechanically defined spectroscopies, chemistry's highly refined means for destructively or nondestructively plucking the strings
of molecules and letting the "sounds" tell us about their features.[10] There are equally
ingenious techniques for separating and purifying molecules, that we will loosely
term chromatographies. They proceed at a larger scale, and when traced are also the
outcome of a sequence of holding on and letting go, like antibody recognition.

Further, statistical mechanics and thermodynamics serve to relate the microscopic to the macroscopic. These theories are probabilistic, but they have no
exceptions because of the immensity of the number of molecules—10^{23} in a sip of
water—and the rapidity of molecular motion at ambient temperatures. Thus the
average speed of molecules "scales up" to temperature, their puny interactions with
light waves into color, the resistance of their crystals to being squeezed to hardness,
their multitudinous and frequent collisions into a reaction that is over in a second
or a millennium.[11]

These theories are silent partners in the experiments described in the paper,
taken for granted and embodied, one might say, in the instruments. But a further
dimension of the linkage between the two worlds is the languages employed by the
chemists, and that is what we now propose to examine at length.

II. Solving the Problem

The construction of "a calixarene with four peptide loops" serves two functions in
this paper. It serves as a simplified substitute for an antibody, though we doubt that
the intent of the authors is the design of potential therapeutic agents. More important, the calixarene serves to test the theory of antibody function sketched above:
is this really the way that antibodies work? The authors note that earlier attempts
to mimic antibodies have been unsuccessful, and propose the alternative strategy
which is the heart of the paper:

> ...the search for antibody mimics has not yet yielded compact and robust
> frameworks that reproduce the essential features of the CDRs. Our strategy is

to use a macrocyclic scaffold to which multiple peptide loops in stable hairpin-turn conformations can be attached.[12]

Stage 1a: The Core Scaffold

The experiment has two stages. The first is to build the antibody mimic, by adding peptide loops to the scaffolding of a calix[4]arene—a cone-shaped concatenation of four benzene rings, strengthened and locked into one orientation by the addition of small length chains of carbon and hydrogen (an alkylation), with COOH groups on top to serve as "handles" for subsequent reaction. (The benzene ring of six carbons is a molecule with a venerable history, whose structure has proved especially problematic for the languages of chemistry, as we point out later in the essay.) The authors write,

> *In this paper we report the synthesis of an antibody mimic based on calix[4] arene linked to four constrained peptide loops … Calix[4]arene was chosen as the core scaffold, as it is readily available and can be locked into the semirigid cone conformation by alkylation of the phenol groups. This results in a projection of the para-substituents onto the same side of the ring to form a potential binding domain.*[13]

Figure 16-1 is given to illustrate this description, as well as the following "recipe."

> *The required tetracarboxylic acid 1 was prepared by alkylation of calix[4] arene (n-butyl bromide, NaH) followed by formylation (Cl_2CHOCH_3, $TiCl_4$) and oxidation ($NaClO_2$, H_2NSO_3H).*[14]

The iconic representation offered is of a microscopic molecule, but the language is all about macroscopic matter, and it is symbolic. The symbolic language of chemistry is the language of formulas employed in the laboratory recipe. It lends itself to the chemist's bridging of the macroscopic and the microscopic because it is thoroughly equivocal, at once a precise description of the ingredients of the experiment (for example, n-butyl bromide is a colorless liquid, with a boiling point of 101.6° C, and is immiscible with water), and a description of the composition of the relevant molecules. For example, n-butyl bromide is construed by the chemist as $CH_3CH_2CH_2CH_2Br$; it has the formula C_4H_9Br, a determinate mass relationship among the three atomic constituents, a preferred geometry, certain barriers to rotation around the carbon-carbon bonds it contains, certain angles at the carbons, and so forth.[15]

The laboratory recipe is thus both the description of a process carried out by a scientist, and the description of a molecule under construction: a molecule generic

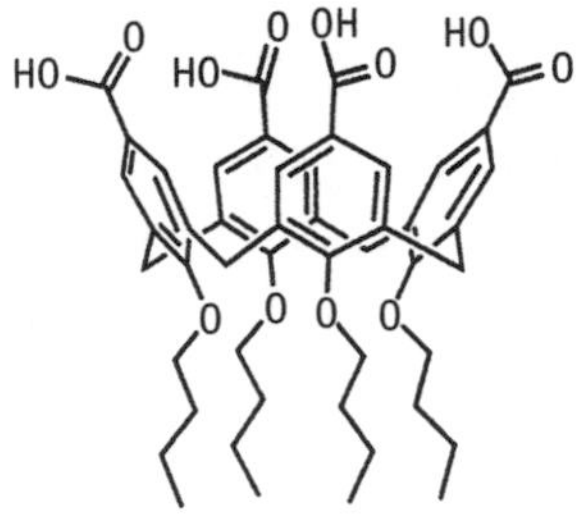

Figure 16-1. Structures 1-4 from Y. Hamuro, M. C. Calama, H. S. Park, and A. D. Hamilton, A Calixarene with Four Peptide Loops: An Antibody Mimic for Recognition of Protein Surfaces. *Angew. Chem. Int. Ed. Engl.* 36 (1997): 2680. © Wiley-VCH Verlag GMbH & Co. KGaA. Reproduced with permission.

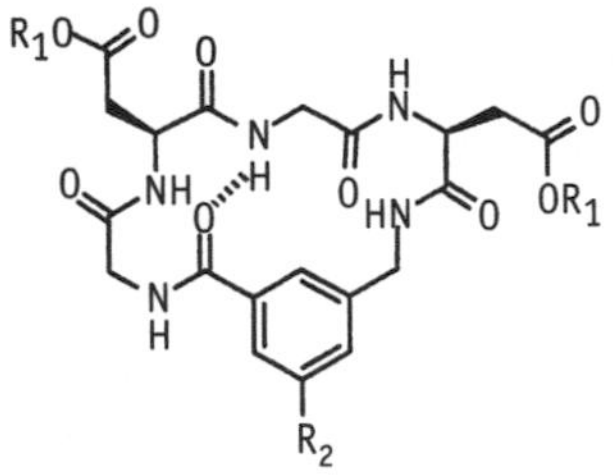

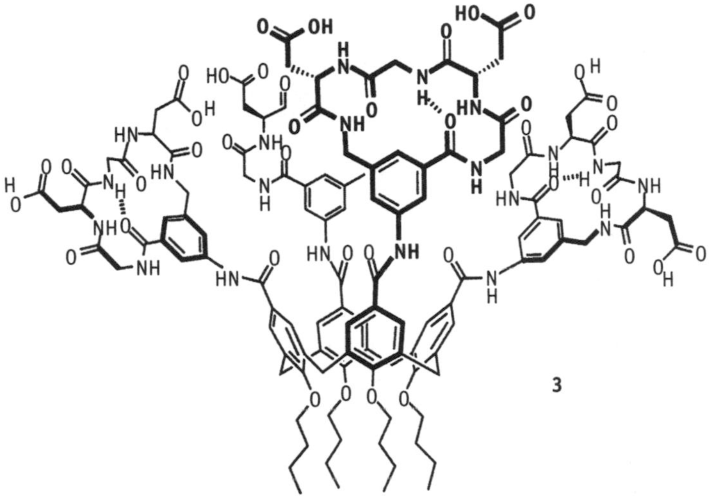

in its significance, since the description is intended to apply to all similar molecules, but particular in its unity and reality. There are parallels in other fields of knowledge. Thus in mathematics, the algebraic formula of a function applies equally to an infinite set of number pairs and to a geometric curve; its controlled and precise equivocity is the instrument that allows resources of number theory and of geometry to be combined in the service of problem-solving.[16] Likewise here the algebra of chemistry allows the wisdom of experience gained in the laboratory to be combined with the (classical and quantum) theory of the molecule, knowledge of its fine structure, energetics, and spectra.

But the symbolic language of chemistry is not complete, for there are many aspects of the chemical substance/molecule that it leaves unexpressed. (a) We cannot deduce from it how the molecule will react with the enormous variety of other molecules with which it may come in contact. (b) We cannot even deduce from it the internal statics, kinematics and dynamics of the molecule in space. Molecules identical in composition can differ from each other because they differ in constitution, the manner and sequence of bonding of atoms (tautomers), in spatial configuration (optical or geometrical isomers), and in conformation (conformers).[18] An adequate description of the molecule must invoke the background of an explanatory theory, but to do so it must also employ iconic languages. Thus, the very definition of the calixarene core scaffold given above involves a diagram. (It was also necessary for the authors to identify C_4H_9Br as n-butyl bromide, a nomenclature implying a specific connectivity of atoms.)

This diagram of calixarene is worth careful inspection, as well as careful comparison with its counterparts in the more complex molecules (for which it serves as core scaffold) furnished to us, the readers, by means of computer-generated images. First of all, it leaves out most of the component hydrogens and carbons in the molecule; they are understood, a kind of tacit knowledge shared even by undergraduate chemistry majors. The hexagons are benzene rings, and the chemist knows that the valence of (the number of bonds formed by) carbon is typically four and so automatically supplies the missing hydrogens. But this omission points to an important feature of iconic languages: they must always leave something out, since they are only pictures, not the thing itself, and since the furnishing of too much information is actually an impoverishment. In a poor diagram, one cannot see the forest for the trees. Not only must some things remain tacit in diagrams, but the wisdom of experience that lets the scientist know how much to put in and how much to leave out, wisdom gleaned by years of translating experimental results into diagrams for various kinds of audience, is itself often tacit. It can be articulated now and then, but cannot be translated into a complete set of fixed rules.

Secondly, the diagram uses certain conventions for representing configurations in 3-dimensional space on the 2-dimensional page, like breaking the outlines of molecules which are supposed to be behind other molecules whose delineation is unbroken. (In other diagrams, wedges are used to represent projection outwards from the plane of the page, and heavy lines are used to represent molecules which stand in front of other molecules depicted by ordinary lines.) Sometimes, though not in the context of a journal article, chemists show three-dimensional representation, such as "ball-and-stick" models. But then in addition, one may want to see more precise angles and interatomic distances in correct proportion and so resort to the images produced by x-ray crystallography. Or one may want some indication of the motion of the molecules, since all atoms vibrate and rotate. Arrows and other iconographies of dynamic motion are used in such diagrams. The cloudy, false-color, yet informative photographs of scanning tunneling microscopy come in here, as well as assorted computer images of the distribution of electrons in the molecule.[19]

Finally, the convention of a hexagon with a perimeter composed of three single lines alternating with three double lines to represent a benzene ring deserves a chapter in itself. This molecule has played a central role in the development of organic chemistry. No single classical valence structure was consistent with the stability of the molecule. Kekulé solved the problem by postulating the coexistence of two valence structures in one molecule. In time, practitioners of quantum mechanics took up the benzene problem, and to this day it has served them as an equally fecund source of inspiration and disagreement. The electrons in benzene are delocalized, that much people agree on; but the description of its electronic structure continues to be a problem for the languages of chemistry.[20]

An Interlude on Symbolic and Iconic Languages

Philosophers of science working in the logical positivist tradition have had little to say about iconic languages. Symbolic languages typically lend themselves to logical regimentation, but pictures tend to be multiform and hard to codify; thus if they proved to be indispensable to human knowledge, the logical positivist would be quite vexed. As any student of chemistry will tell you, conventions for producing "well formed icons" of molecules exist and must be learned or else your audience will misread them. But no single iconic language is the correct one or enjoys anything as precisely determined as a "wff" in logic. The symbolic language of chemistry is, to be sure, a precisely defined international nomenclature that specifies in impressive detail a written sequence of symbols so as to allow the unique specification of a molecule. But, significantly, the iconic representations of a molecule are governed only by widely accepted conventions, and a good bit of latitude is allowed in practice, especially a propos what may be omitted from such representations. Symbolic languages lend themselves to codification in a way that iconic languages don't.[21]

Symbolic languages, precisely because they are symbolic, lend themselves best to displaying relational structure. Like algebra, they are tolerant or relativistic in their ontological import: it doesn't so much matter what they pertain to, as long as their objects stand in the appropriate relations to each other. But iconic languages point, more or less directly, to objects; they are not ontologically neutral but on the contrary ontologically insistent. They display the unity of objects, a unity which might metaphysically be called the unity of existence. But there is no way to give an exhaustive summary of the ways of portraying the unity of existence; it is too infinitely rich and thought has too many ways of engaging it. We should not therefore jump to the conclusion that knowledge via an iconic language is impossible or incoherent: iconic languages despite being multiform employ intelligible conventions, they are constrained by the object itself, and they are made orderly by their association with symbolic language. An inference cannot be constructed from icons alone, but icons may play an essential role in inference.[22]

How does the iconic form of the chemical structure expressed as a diagram that displays atom connectivities and suggests the three-dimensionality of the molecule, bridge the two worlds of the chemist? The most obvious answer is that it makes the invisible visible, and does so, within limits, reliably. But there is a deeper answer. It seems at first as if the chemical structure diagram refers only to the level of the microscopic, since after all it depicts a molecule. But in conjunction with symbolic formulae, the diagram takes on an inherent ambiguity that gives it an important bridging function. In its display of unified existence, it stands for a single particular molecule. Yet we understand molecules of the same composition and structure to be equivalent to each other, internally indistinguishable. (In this, the objects of physics and chemistry are like the objects of mathematics.)

Thus, the icon (hexagonal benzene ring) also stands for all possible benzene rings, or for all the benzene rings (moles or millimoles of them!) in the experiment, depending on the way in which it is associated with the symbolic formula for benzene. The logical positivist in search of univocality might call this obfuscating ambiguity, a degeneracy in what ought to be a precise scientific language that carries with it undesirable ontological baggage. And yet, the iconic language is powerfully efficient and fertile in the hands of the chemist.

Now we can understand better why the kind of world-bridging involved in posing a problem/construction in chemistry requires both symbolic and iconic languages for its formulation. On the one hand, the symbolic language of chemistry captures the composition of molecules, but not their structure (constitution, configuration, and conformation), aspects which are dealt with better, though fragmentarily, by the many iconic idioms available to chemists. Moreover, the symbolic language of chemistry fails to convey the ontological import, the realism, intended by practitioners in the field. Hamilton et al. are not reporting on a social construction or a mere computation, but a useful reality: the diagram confidently posits its existence.[23] On the other hand, icons are too manifold and singular to be the sole vehicle of scientific discourse. Their use along with symbolic language embeds them in demonstrations, and gives to their particularity a representative and well-defined generality, sometimes even a universality.

We return to our reading of the Hamilton et al. paper.

Stage 1b: The Peptide Loops

Hamilton et al. chose cyclic hexapeptides to mimic the "arms" of the antibody because they can be modified so as to link up easily with the core scaffold, and because they form hairpin loops. "The peptide loop was based on a cyclic hexapeptide in which two residues were replaced by a 3-aminomethylbenzoyl (3amb) dipeptide analogue containing a 5-amino substituent for facile linkage to the scaffold."[24]

The recipe for constructing the peptide loops is then given; the way in which it couples the macroscopic and the microscopic is striking, for it describes a laboratory procedure and then announces that the outcome of the procedure is a molecule, pictured in diagram **2**.

> *The 5-nitro substituted dipeptide analogue was formed by selective reduction (BH₃) of methyl 3-amidocarbonyl-5-nitrobenzoate, followed by deesterification (LiOH in THF) and reaction sequentially with Fmoc-Asp-(tBu)-OH and H-Gly-Asp(tBu)-Gly-OH (dicyclohexyl carbodiimide (DCC), N-hydroxysuccinimide) to yield Fmoc-Asp(tBu)-5NO₂3amb-Gly-Asp(tBu)-GlyOH. Cyclization with 4-dimethylaminopyridine (DMAP) and 2-1H-benzotriazole-1-yl-1,1,3,3,-tetramethyluronium tetrafluoroborate (TBTU) was achieved in 70% yield, followed by reduction (H₂, Pd/C) to give the amino-substituted peptide loop 2.*[25]

The chemist working in the lab has constructed a molecule, at least a dizzying 20 orders of magnitude "below" or "inwards." To be sure, what was made was a visible, tangible material—likely less than a gram of it—but the interest of what was made lies in the geometry and reactivity of the molecule, not the properties of the macroscopic substance. So it is not by accident that the leap to the level of the molecule is accompanied by iconic language. Such language also accompanies the final step in the assembly of the antibody mimic.

Stage 1c: The Antibody Mimic

Four of the peptide loops are attached to the core scaffold; the laboratory procedure begins and ends with a pictured molecule. But this time the resultant new molecule is pictured twice in complementary iconic idioms.

> *Amine 2 was coupled to the tetraacid chloride derivative of 1 ((COCl)₂, DMF) and deprotected with trifluoroacetic acid (TFA) to give the tetraloop structure 3. The molecular structure of this host (Figure 16.1) resembles that of the antigen binding region of an antibody but is based on four loops rather than six.*[26]

To someone who understands chemical semiotics, the iconic conventions in diagram **3** (the tetraloop molecule called structure **3** in the quote above) do allow a mental reconstruction of the molecule. But the shape of the molecule is so important that the authors decide to give it again, in another view, in Figure 16.2. The figure is even printed in color in the original!

Why should the reader be offered another iconic representation? In part, it is part of a rhetorical strategy to persuade the audience of the cogency of a research

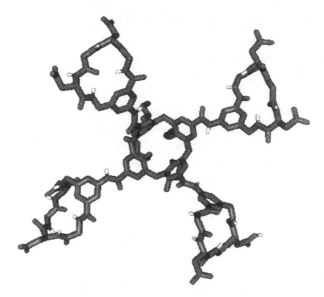

Figure 16-2 Figure 1 of Y. Hamuro, M. C. Calama, H. S. Park, and A. D. Hamilton,
A Calixarene with Four Peptide Loops: An Antibody Mimic for Recognition of Protein
Surfaces. *Angew. Chem. Int. Ed. Engl.* 36 (1997): 2681. © Wiley-VCH Verlag GMbH & Co.
KGaA. Reproduced with permission. See explanation in text.

program that involves mimicry. The computer-generated image of Figure 1 of
Hamilton et al. (our Figure 16.2) is actually the result of a theoretical calculation in
which the various molecular components are allowed to wiggle around any bonds
that allow rotation, and to reach a geometry that is presumably the most stable. In
that image, the general shape of the molecule (in particular the loopiness of the
loops) is beautifully exhibited, emphasizing its resemblance to an antibody. Note
that the experimentalist trusts the ability of a theoretical computer program to yield
the shape of a molecule sufficiently to insert it—in color—in a paper; that would
not have been the case twenty years ago.

Diagram 3 and Figure 1 of the Hamilton et al. paper are meant to be seen in tan-
dem; they complement each other. Both representations are iconic, though perhaps
Figure 1 is more so. Diagram 3 has a symbolic dimension due to the labels, and thus
serves to link Figure 1 to the symbolic discourse of the prose argument. Together
with the reproducible laboratory procedure—given in more detail at the end of the
article—Hamilton et al. give a convincing picture of this new addition to the furniture of the universe. There it stands: Ecce.

Stage 2

Once the antibody mimic has been assembled, it can be tested to see whether it in
fact behaves like an antibody, a test which, if successful, in turn provides evidence

supporting the theory of the action of antibodies invoked by Hamilton et al. Note the usefully—as opposed to viciously—circular reasoning here:[27] the antibody mimic correctly mimics an antibody if it behaves like an antibody; but how an antibody behaves is still a postulate, which stipulates what counts as the correctness of the antibody mimic's mimicry. To see if the antibody mimic, the base scaffold of calixarene with four peptide loops, will bind with and impair the function of a protein (the essence of what an antibody does), Hamilton et al. chose the protein cytochrome, an important molecule that plays a critical role in energy production and electron transport in every cell and has thus been thoroughly investigated. Moreover, it has a positively charged surface region that would likely bond well with the negatively charged peptide loops.

> We chose cytochrome c as the initial protein target, since it is structurally well-characterized and contains a positively charged surface made up of several lysine and arginine residues. In this study the negatively charged GlyAspGlyAsp sequence was used in the loops of 3 to complement the charge distribution on the protein.[28]

Note that the antibody mimic is referred to by means of the diagram 3. In a sense this is because the diagram is a shorthand, but its perspicuity is not trivial or accidental: as a picture that can be taken in at a glance, it offers schematically the whole configuration of the molecule in space. Its visual unity stands for, and does not misrepresent, the unity of the molecule's existence.

Does the antibody mimic in fact bind with the cytochrome? Their affinity is tested by an experiment which is neither analytical or synthetical, but rather a matter of careful physical measurement, an aspect of chemical practice central to the science since the time of Lavoisier. The "affinity chromatography" involves a column filled with some inert cellulose-like particles and cytochrome c linked to those particles.[29] The concentration of NaCl, simple salt, controls the degree of binding of various other molecules to the cytochrome c that is in that column. If the binding is substantially through ionic forces (as one thinks it is for the antibody mimic) then only a substantial concentration of ionic salt solution will disrupt that binding. At the top of the column one first adds a control molecule. It is eluted easily, with no salt. But the antibody mimic 3 turns out to be bound much more tightly—it takes a lot of salt to flush it out.

A second kind of chromatography, "gel permeation chromatography," gives more graphic evidence for the binding of cytochrome c to 3. In this ingenious chromatography the column is packed with another cellulose-like and porous fiber, called Sephadex G-50. The "G-50" is not just a trade name, it indicates that molecules of a certain size will be trapped in the column material, but molecules both larger and smaller will flow through the column quickly.

The results of this experiment are shown in Figure 2 of Hamilton et al., our Figure 16-3, replete with labeled axes. The vertical axis measures the absorption of light at a certain wavelength; this is related to the concentration of a species, the bound cytochrome c–3 complex. The horizontal axis is a "fraction number" that is related to the length of time that a given molecule (or compound? the equivocity here pervades chemical discourse) resides on the column. The pores in the Sephadex retard cytochrome c; it stays on the column longer (has a higher fraction number). The molecular complex of the mimic and cytochrome c comes out in a different peak, at lower fraction number. This means it is too large to be caught in the pores of the Sephadex, which in turn constitutes evidence for some sort of binding between the cytochrome c and the antibody mimic, creating a larger molecular entity.

So there is binding; but does it impair the function of the cytochrome c? Evidence for that is provided by reacting the cytochrome c with ascorbate (vitamin C), with which it normally reacts quite efficiently; here, on the contrary, it doesn't.

> *We have investigated the effect of complexation with **3** on the interaction of FeIII-cyt c with reducing agents. In phosphate buffer Fe^{III}-cyt c ($1.57 \times 10^{-5}M$) is rapidly reduced by excess ascorbate ($2.0 \times 10^{-3}M$) with a pseudo-first-order rate constant 0.1090 ± 0.001 (Figure 4). In the presence of **3** ($1.9 \times 10^{-5}M$) the rate of cyt c reduction is diminished tenfold ($k_{obs} = 0.010 \pm 0.001\ s^{-1}$), consistent with the calixarene derivative's binding to the protein surface and inhibiting approach of ascorbate to the heme edge (Figure 3).[30]*

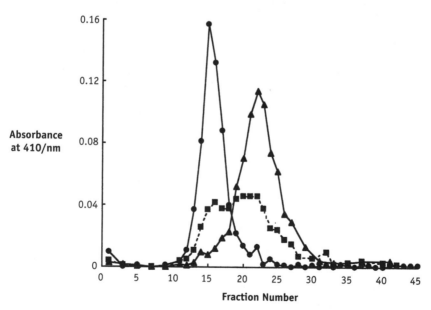

Figure 16-3   Figure 2 from Y. Hamuro, M. C. Calama, H. S. Park, and A. D. Hamilton, A Calixarene with Four Peptide Loops: An Antibody Mimic for Recognition of Protein Surfaces. *Angew. Chem. Int. Ed. Engl.* 36 (1997): 2682. © Wiley-VCH Verlag GMbH & Co. KGaA. Reproduced with permission.

Figure 4 of Hamilton et al. (our Figure 16-4) is another measurement, with the concentrations measured on the vertical axis, the time on the horizontal; it displays the outcome of an experiment on the kinetics of ascorbate reduction by cytochrome c, which supports the claim that the antibody mimic does impair the function of the protein, in this case its ability to react with ascorbate.

More interesting is another iconic representation, Figure 3 of Hamilton et al., our Figure 16-5, which is a picture of the antibody mimic binding with cytochrome c. Since the authors admit, "The exact site on the surface of the cytochrome that binds with **3** has not yet been established," this image is a conjecture; and it is the outcome of the same computer program that generated Figure 1 of Hamilton et al. It "docks" "a calculated structure for **3**" at the most likely site on the cytochrome c, where the four peptide loops "cover a large area of the protein surface." Figure 3 is a remarkable overlay of several types of iconic representation. The antibody mimic (at top) is shown pretty much as it was in Figure 1, but from the side. The atoms of cytochrome c are legion, and so are mostly not shown; instead, the essential helical loops of the protein are schematically indicated. But in the contact region, the atoms are again shown in great detail, not by ball-and-stick or rod representations but by tenuous spheres indicating roughly the atomic sizes or electron densities. The reader can make sense of these superimposed iconic idioms only by reference to a cognitive framework of words and symbols.

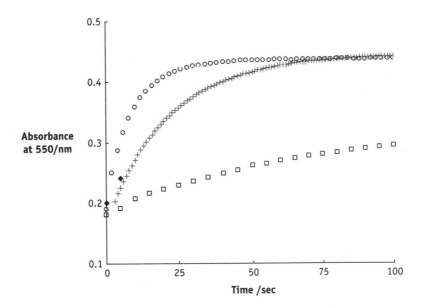

Figure 16-4 Figure 4 from Y. Hamuro, M. C. Calama, H. S. Park, and A. D. Hamilton, A Calixarene with Four Peptide Loops: An Antibody Mimic for Recognition of Protein Surfaces. *Angew. Chem. Int. Ed. Engl.* 36 (1997): 2682. © Wiley-VCH Verlag GMbH & Co. KGaA. Reproduced with permission. See explanation in text.

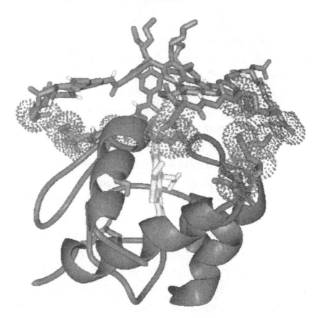

Figure 16-5 Figure 3 from Y. Hamuro, M. C. Calama, H. S. Park, and A. D. Hamilton, A Calixarene with Four Peptide Loops: An Antibody Mimic for Recognition of Protein Surfaces. Angew. *Chem. Int. Ed. Engl.* 36 (1997): 2682. © Wiley-VCH Verlag GMbH & Co. KGaA. Reproduced with permission. See explanation in text.

Iconic representations in chemical discourse must be related to a symbolic discourse; our access to the microscopic objects of chemistry, even our ability to picture them, is always mediated by that discourse rather than by our "natural" organs of perception. So the objects of chemistry may seem a bit ghostly, even to the practitioners for whom their existence is especially robust.[31] But conversely, symbolic discourse in chemistry cannot dispense with iconic discourse as its complement, nor can it escape its own iconic dimension. The side-by-side distinction, iteration, and concatenation of letters in chemical formulae echo the spatial array of atoms in a molecule. Otherwise put, the iconic combination of symbols often articulates otherness, which is exhibited by the things of chemistry as their spatial externality and proximity. The iconic array in Hamilton et al.'s Figure 3 also translates into spatial relations among symbols the temporal spread of stages of a chemical event, where otherness is priority or posteriority. Just as icons evoke existence, the unity of existence, so they evoke otherness as side-by-sideness, as externality. Identity and difference, *pace* the logicians, cannot be fully represented without the use of iconic as well as symbolic languages.

The icon in Hamilton et al.'s Figure 3 stands for a molecular complex that may or may not exist. It is a possibility, a guide to future research. For the authors of the paper, it is something they very much hope does exist, a wish that can perhaps be read in the bright, imaginary colors of the image. Chemical icons work their magic of asserting and displaying the unity of existence only when the symbolic discursive context and the experimental background allow them to do so.

Whatever remains still to be worked out, the authors of the paper declare a positive result, and its generalization to a broader research program.

> *The new type of synthetic host 3 thus mimics antibody combining sites in having several peptide loops arrayed around a central binding region. The large surface area in the molecule allows strong binding to a complementary surface on cytochrome c and disrupts, in a similar way to cytochrome c peroxidase, the approach of reducing agents to the active site of the protein. We are currently preparing libraries of antibody mimics from different peptide loop sequences and screening their binding to a range of protein targets.[32]*

III. Conclusions

Angewandte Chemie, where Roald Hoffmann found the article closely read in this paper, is no longer especially concerned with applied chemistry; indeed, it is arguably the world's leading "pure" chemistry journal. The December 15, 1997 issue of the journal in which the Hamilton article appears contains one review, two comments or highlights, several book reviews, and 38 "communications," articles one to three pages in length that in principle present novel and important chemistry.

Without question, the Hamuro, Calama, Park, and Hamilton article is a beautiful piece of work, deserving of the company it keeps in the pages of *Angewandte Chemie*; it caught Hoffmann's attention even though the subject is not one of his fields of specialization. But is this work typical of chemistry, and sufficiently so that any close reading of it might elicit generalities valid for the field? After all, it is not clear what counts as "typical" in a science whose topics range from cytochrome c, to reactions occurring in femtoseconds, to inorganic superconductors. And perhaps work that strives to redefine the boundaries of a science cannot fully represent what Kuhn called "normal science."

We believe nonetheless that the Hamilton et al. article exhibits many of the important features of most work in modern chemistry, especially in the way that it moves between levels of reality. On one line the authors of the article talk of a molecular structure, and on the next of a reaction; a certain linguistic item (symbol or icon) may stand for either or both. Theory and experiment, expressed in beautifully intertwined symbolic and iconic languages, relate the world of visible, tangible substances and that of the molecule. Is this sloppiness, an ambiguity that hard science must ultimately abolish? We think not.

The intuitive may be analyzed, the tacit may be articulated, but never completely and all at once: certain indeterminacies and logical gaps always remain, even as scientists achieve a consensual understanding of complex reality. Indeed, the indeterminacy and "gappiness" of knowledge may serve a useful purpose, in that it allows the double vision where creative endeavor often takes place. If, as we have claimed in this paper, chemists habitually think at both the level of macroscopic substances and their transformations in the laboratory, and the level of the statics and dynamics of microscopic molecules, the very equivocity of the field—the way it brings physics and mathematics into the service of chemistry—may be a source of its productivity. The logical gap between the two levels of description is never closed (by some kind of reduction), but rather is constantly and successfully negotiated by a set of theories embodied in instruments and expressed in symbolic and iconic languages.[33]

Precisely because these languages are abstract and incomplete (in the sense of being noncategorical, not capturing all there is to say and know about the entities they describe) they are *productively* ambiguous, and can be understood in reference to both the macroscopic and microscopic. This bridging function—carried out in different but complementary ways by symbolic and iconic idioms—is the special interest of this paper. We have tried to explain how it allows chemists to articulate and to solve problems, a task which often takes the form of imagining and then trying to put together a certain kind of molecule. Thus our account emphasizes what happens at the frontiers of knowledge rather than retrospective codification, and the investigation and creation of objects rather than the testing of theories.

Acknowledgments

A version of this paper was delivered under the title "Comment les langages symboliques et iconiques servent de passerelle entre les deux mondes du chimiste: une étude de cas de chimie bioorganique contemporaine," by the authors on April 19, 1998 at the Maison Rhône-Alpes des Sciences de l'Homme, Université Stendhal, Grenoble, as part of a year-long seminar on the Languages of Science. The authors would like to thank the organizers of that seminar, especially Professor Françoise Létoublon, for the invitation and their efforts at realizing a complex and truly interdisciplinary project. They are also grateful to Prof. Andrew Hamilton for supplying to them the original illustrations for his paper.

The nontrivial translation of our paper into French was masterfully accomplished by Carole Allamand. We thank her.

Emily Grosholz would like to thank the American Council of Learned Societies and the Pennsylvania State University for their support of her research as a Visiting Fellow at Clare Hall and a Visiting Scholar in the Department of History and Philosophy of Science at the University of Cambridge. She would also like to thank Professors François de Gandt (University of Lille) and Audrey Glauert (Clare Hall) for their useful suggestions on the essay, as well as Professor Jeremy K. M. Sanders and Dr. Darren Hamilton of the Department of Chemistry at the University of Cambridge for their enlightening discussion of the practice of chemistry.

Roald Hoffmann is grateful to Cornell University for a grant in support of his research, scientific and otherwise, and to Bruce Ganem for a helpful explanation of chromatography.

Notes

1. R. Carnap, *The Logical Syntax of Science* (London: Routledge and Kegan Paul, 1937).
2. (a) G. W. Leibniz, "Discourse on Metaphysics" (originally published 1686), in *Leibniz: Die philosophische Schriften*, vol. 4, ed. C. I. Gerhardt (Hildesheim: G. Olms, 1965), p. 427; G. W. Leibniz, "A New System of Nature" (originally published 1695), in *Leibniz: Die philosophische Schriften*, vol. 4, ed. C. I. Gerhardt (Hildesheim: G. Olms, 1965), p. 477; G. W. Leibniz, "Principles of Nature and Grace, Based on Reason" (originally published 1714), in *Leibniz: Die philosophische Schriften*, vol. 6, ed. C. I. Gerhardt (Hildesheim: G. Olms, 1965), p. 598; (b) E. Husserl, *Logische Untersuchungen* (Halle: Niemeyer, 1922); (c) T. S. Kuhn, *The Structure of Scientific Revolutions* (Chicago: Chicago University Press, 1970); and (d) M. Polanyi, *Knowing and Being* (Chicago: University of Chicago Press, 1960), and M. Polanyi, *The Tacit Dimension* (New York: Doubleday, 1966), and draw upon the earlier writings of both authors,
3. (a) R. Hoffmann, *The Same and Not the Same* (New York: Columbia University Press, 1995), and R. Hoffmann and P. Laszlo, Representation in Chemistry. *Angew. Chem. Int. Ed. Engl.* 30 (1991): 1; Chapter 14 in this book; and (b) E. Grosholz, *Cartesian Method and the Problem of Reduction* (London: Clarendon, 1991), and E. Grosholz and E. Yakira, Leibniz's Science of the Rational. *Sonderheft* 26 (Studia Leibnitiana, 1998).
4. See, for example, S. Shapin, Discipline and Bounding: The History and Sociology of Science As Seen Through the Externalism/Internalism Debate. *History of Science* 30 (1992): 333; S. Fuller, Can Science Studies Be Spoken in a Civil Tongue? *Social Studies of Science* 24 (1994): 143; A. Pickering, ed., *Science As Practice and Culture* (Chicago: University of Chicago Press, 1992); for a balanced analysis of the problem, see J. Labinger, Sciences As Culture: A View from the Petri Dish. *Social Studies of Science* 25 (1995): 285.
5. Y. Hamuro, M. C. Calama, H. S. Park, and A. D. Hamilton, A Calixarene with Four Peptide Loops: An Antibody Mimic for Recognition of Protein Surfaces. *Angew. Chem. Int. Ed. Engl.* 36 (1997): 2680. This work will be referred to as Hamilton et al. throughout this chapter. In our quotations we omit literature endnote numbers cited in this work.
6. Y. Hamuro, M. C. Calama, H. S. Park, and A. D. Hamilton, A Calixarene with Four Peptide Loops: An Antibody Mimic for Recognition of Protein Surfaces. *Angew. Chem. Int. Ed. Engl.* 36 (1997): 2680.
7. Y. Hamuro, M. C. Calama, H. S. Park, and A. D. Hamilton, A Calixarene with Four Peptide Loops: An Antibody Mimic for Recognition of Protein Surfaces. *Angew. Chem. Int. Ed. Engl.* 36 (1997): 2680.
8. For an introduction to chemistry and molecular interactions, see M. D. Joesten, D. O. Johnston, J. T. Netterville, and J. L. Woods, *World of Chemistry* (Philadelphia: Saunders, 1991).
9. Y. Hamuro, M. C. Calama, H. S. Park, and A. D. Hamilton, A Calixarene with Four Peptide Loops: An Antibody Mimic for Recognition of Protein Surfaces. *Angew. Chem. Int. Ed. Engl.* 36 (1997): 2680.
10. R. Hoffmann and V. Torrence, *Chemistry Imagined* (Washington, D.C.: Smithsonian Institution Press, 1993), p. 144.
11. M. D. Joesten, D. O. Johnston, J. T. Netterville, and J. L. Woods, *World of Chemistry* (Philadelphia: Saunders, 1991); P. W. Atkins, *The Second Law* (New York: Scientific American, 1984), P. W.

Atkins, *Molecules* (New York: Scientific American, 1987), and P. W. Atkins, *Atoms, Electrons, and Change* (New York: Scientific American, 1991); and Hoffmann, 1995.

12. Y. Hamuro, M. C. Calama, H. S. Park, and A. D. Hamilton, A Calixarene with Four Peptide Loops: An Antibody Mimic for Recognition of Protein Surfaces. *Angew. Chem. Int. Ed. Engl.* 36 (1997): 2681.

13. Y. Hamuro, M. C. Calama, H. S. Park, and A. D. Hamilton, A Calixarene with Four Peptide Loops: An Antibody Mimic for Recognition of Protein Surfaces. *Angew. Chem. Int. Ed. Engl.* 36 (1997): 2681.

14. Y. Hamuro, M. C. Calama, H. S. Park, and A. D. Hamilton, A Calixarene with Four Peptide Loops: An Antibody Mimic for Recognition of Protein Surfaces. *Angew. Chem. Int. Ed. Engl.* 36 (1997): 2681.

15. P. W. Atkins, *The Second Law* (New York: Scientific American, 1984); M. D. Joesten, D. O. Johnston, J. T. Netterville, and J. L. Woods, *World of Chemistry* (Philadelphia: Saunders, 1991); and R. Hoffmann, *The Same and Not the Same* (New York: Columbia University Press, 1995).

16. E. Grosholz, *Cartesian Method and the Problem of Reduction* (London: Clarendon, 1991), Chs. 1 and 2.

17. A reductionist might argue that given great computing power and perfected quantum mechanical calculations, one could start from a chemical formula and predict observations accurately. But in practice, the number of isomers for a given formula grows very rapidly with molecular complexity, so the goal is not realistic for a molecule the size of the calixarene. Moreover, complete computability may not be equivalent to understanding. Much of what a chemist means by understanding is couched in terms of fuzzy chemical concepts—the result of horizontal and quasi-circular reasoning—for which a precise equivalent in physics cannot be found (R. Hoffmann, Nearly Circular Reasoning. *American Scientist* 76 [1988]:182, Chapter 5 in this book, and R. Hoffmann, *The Same and Not the Same* [New York: Columbia University Press, 1995]; E. R. Scerri, Has Chemistry Been at Least Approximately Reduced to Quantum Mechanics? *PSA* 1 [1994]: 160; and E. Grosholz, Reduction in the Formal Sciences. *Proceedings of Conference on Physical Interpretations of Relative Theory IV* [London: British Society for the Philosophy of Science, 1994], p. 28.)

18. M. D. Joesten, D. O. Johnston, J. T. Netterville, and J. L. Woods, *World of Chemistry* (Philadelphia: Saunders, 1991); P. Zeidler and D. Sobczynska, The Idea of Realism in the New Experimentalism and the Problem of the Existence of Theoretical Entities in Chemistry. *Foundations of Science* 4 (1995–1996): 517; and R. Hoffmann, *The Same and Not the Same* (New York: Columbia University Press, 1995).

19. For an excellent account of the language of chemistry, and its parallels to linguistics, see P. Laszlo, *La Parole des Choses* (Paris: Hermann, 1993) as well as R. Hoffmann and P. Laszlo, Representation in Chemistry. *Angew. Chem. Int. Ed. Engl.* 30 (1991): 1, Chapter 14 in this book, and S. J. Weininger, Contemplating the Finger: Visuality and the Semiotics of Chemistry. *Hyle* 4 (1998): 3.

20. S. G. Brush, Dynamics of Theory Change in Chemistry: The Benzene Problem. *Studies in the History and Philosophy of Science* 30 (1999): 21; *Studies in the History and Philosophy of Science*, 30: 263.

21. L. Kvasz, History of Geometry and the Development of Its Formal Language. *Synthèse* 116 (1998): 141 helped the authors think about the distinction and the interactions between symbolic and iconic languages in mathematics and chemistry, but they disagree with Kvasz about the extent to which iconic languages may be codified.

22. G. G. Granger has an interesting discussion of the languages of chemistry in his book (*Formal Thought and the Sciences of Man* [Dordrecht: Reidel, 1983], Ch. III), where he focuses on the distinction between natural languages and formal languages. He makes the important observation that scientific language will always be partly vernacular and partly formal. Rejecting the claim that science might someday be carried out in a pure formalism, he writes, "The linguistic process of science seems to me essentially ambiguous: for if science is not at any moment of its history a completely formalized discourse, it is not to be confused with ordinary discourse

either. Insofar as it is thought in action, it can only be represented as an attempt to formalize, commented on by the interpreter in a non-formal language. Total formalization never appears as anything more than at the horizon of scientific thought, and we can say that the collaboration of the two languages is a transcendental feature of science, that is, a feature dependent on the very conditions of the apprehension of an object." However, Granger does not go on to consider the further linguistic aspect of chemistry, that is, its iconic aspect.

23. Y. Hamuro, M. C. Calama, H. S. Park, and A. D. Hamilton, A Calixarene with Four Peptide Loops: An Antibody Mimic for Recognition of Protein Surfaces. *Angew. Chem. Int. Ed. Engl.* 36 (1997): 2681.

24. Y. Hamuro, M. C. Calama, H. S. Park, and A. D. Hamilton, A Calixarene with Four Peptide Loops: An Antibody Mimic for Recognition of Protein Surfaces. *Angew. Chem. Int. Ed. Engl.* 36 (1997): 2681.

25. Y. Hamuro, M. C. Calama, H. S. Park, and A. D. Hamilton, A Calixarene with Four Peptide Loops: An Antibody Mimic for Recognition of Protein Surfaces. *Angew. Chem. Int. Ed. Engl.* 36 (1997): 2681.

26. Y. Hamuro, M. C. Calama, H. S. Park, and A. D. Hamilton, A Calixarene with Four Peptide Loops: An Antibody Mimic for Recognition of Protein Surfaces. *Angew. Chem. Int. Ed. Engl.* 36 (1997): 2681.

27. R. Hoffmann, Nearly Circular Reasoning. *American Scientist* 76 (1988): 182.

28. Y. Hamuro, M. C. Calama, H. S. Park, and A. D. Hamilton, A Calixarene with Four Peptide Loops: An Antibody Mimic for Recognition of Protein Surfaces. *Angew. Chem. Int. Ed. Engl.* 36 (1997): 2681.

29. For a description of chromatography, see P. Laszlo, in *Tresor. Dictionnaire des sciences*, ed. M. Serres and N. Farouki (Paris: Flammarion, 1997).

30. Y. Hamuro, M. C. Calama, H. S. Park, and A. D. Hamilton, A Calixarene with Four Peptide Loops: An Antibody Mimic for Recognition of Protein Surfaces. *Angew. Chem. Int. Ed. Engl.* 36 (1997): 2682.

31. See also P. Laszlo, Chemical Analysis As Dematerialization. *Hyle* 4 (1998): 29.

32. Y. Hamuro, M. C. Calama, H. S. Park, and A. D. Hamilton, A Calixarene with Four Peptide Loops: An Antibody Mimic for Recognition of Protein Surfaces. *Angew. Chem. Int. Ed. Engl.* 36 (1997): 2682.

33. P. Laszlo, Chemical Analysis As Dematerialization (*Hyle* 4 [1998]: 29) has cogently argued that in their practice of analysis, modern-day chemists "dematerialize" the substances they handle, so that the transactions of the contemporary laboratory mostly involve mental representations. He goes on to argue that our age of masterly synthesis doesn't achieve the rematerialization one might desire. While we think Laszlo verges perilously close to denying realism, his argument nevertheless is an intriguing one, and covers some of the same representational ground that we do.

17

How Nice to Be an Outsider

ROALD HOFFMANN

1

Every one of my scholarly/literary activities is outside literary studies as such. Yet to a varying degree all that I do is the subject of the amoeboid activities of the field. I also have, in principle, no vested interest in the flow of students into your departments [this was a lecture to an audience in comparative literature], nor do I have to worry about jobs for them, nor the level of remuneration of your sluggers and sometime pinch hitters. It seems to me that given this practical disinterest (reading Burke and Kant) I am ideally situated to make aesthetic judgments if not prognoses of the future of literary studies. Which is the reason, I suppose, that I was asked to do so.

But first let me count the ways in which I am marginal. First of all, I am a chemist, of the theoretical subspecies. I have done some good science, even shaped the way that chemists think of the motion of electrons in molecules, and how the electrons determine the shape and reactions of those persistent groupings of atoms we've learned to see without seeing. My and my collaborators' work is divulged, some of my colleagues would say preached, in over 450 scientific articles (our stock in trade, rather than books). Such "texts" have become the subject of a burgeoning field of literary studies of science. But no one would bother with my texts; they are individually unimportant (though what they collectively teach is of value; I think of my articles as chapters in a serialized text, but please don't tell the editors of the journals in which I publish). And perhaps when I write science I am too self-conscious of the central problem of representation for me to play the role of an innocent native (or his artifacts) awaiting the sage pseudo-anthropo/sociological investigation of the way I construct knowledge. Also the cognitive, intrascientific background needed to assess my papers is moderately formidable; there is a reason why chemists spend five years in graduate school... So, so far, I've escaped attention as an object of literary studies. I keep my fingers crossed.

Second, I have a modest career as a poet. It's much easier to make a living as a chemist (would I dearly like you to convince me that it is otherwise, by buying my books!), so the poetry is perforce a part-time vocation. Not much need be said about poetry as the subject of literary studies—past, present, and future. Fortunately for me, poetry of middling quality is not usually the object of literary studies. Unless it is written by figures of whom others have written...

Third, I write of chemistry, or I would prefer to say of the intersection of chemistry and culture—for various audiences. I do so for the ephemeral general public, for scientists who are not chemists, and, closest to my heart, just for you, my friends in the arts and humanities. I write for Lionel Trilling, who said:

> *Physical science in our day lies beyond the intellectual grasp of most men... This exclusion of most of us from the mode of thought which is habitually said to be the characteristic achievement of the modern age is bound to be experienced as a wound to our intellectual self-esteem. About this humiliation we all agree to be silent; but can we doubt that it has its consequences, that it introduces into the life of the mind a significant element of dubiety and alienation, which must be taken into account in any estimate that is made of the present fortunes of mind.*[1]

Some of my writing is "popular," some of it pretends to be scholarly. I also have a range of collaborations with artists. The outcomes are curiously positioned in-between art, literature, and science—an example is *Chemistry Imagined*, a kind of modern emblem book of chemistry that I've created with artist Vivian Torrence.

Expository writing about science has been a less popular subject of literary studies, I think. Curiously enough, the cognitive thornhedges around contemporary mainstream science have led literary scholars who seek to penetrate the barriers to rely much on just such expository writing. But critics have reflected little, I think, on the representational and narrative stances taken by the creators of the "better" popularized science genre. Some students of scientific texts and of scientists have become prisoners of the accessible metaphor. We have a curious situation that while humanists (and I) have been pushing scientists to accept the value of the metaphor within science, as a wellspring of creativity and an inevitable sidekick of just plain human thinking—while we have been desperately trying to do that, some people are applying insufficient caution to the knowledge received (in perforce metaphorical language) from science. Thomas Pynchon, a Cornell graduate who knows a lot of chemistry, writes: "The act of metaphor then was a thrust at truth and a lie, depending where you were: inside safe, or outside lost."[2]

Let me be specifically provocative: the interest, to put it mildly, expressed by humanists in chaos, fractals, and "order out of chaos" is totally out of proportion to the significance of these ideas in contemporary science. Now that is guaranteed to get me into trouble with some of my scientist friends, but I stand by it. Do ask some of your friends in mainstream science (who are not members of the Santa Fe Institute).

Triply an outsider, I have, however, great sympathy and empathy for your field. It's not only that I read, and sit in on your courses like any student. I also had the great fortune of listening to Mark Van Doren, Andrew Chiappe, and Donald Keene at Columbia in the fifties—sometimes I think I haven't grown up, I'm still in their classes... I do have as much fun as anyone reading of your internecine tribal conflicts, but that's just voyeurism. More seriously, theories of representation, semiotics and narrative are not only of intense interest to me, but they are personal ways to think about my science, within the science, and the presentation of that science to the outside. You might (only in weak moments, I know) think that your work is of little use—I tell you that I use it, in the most respectful way, i.e., without acknowledgment. So maybe I should take back what I said above about being disinterested. I'm very interested.

How nice to be an outsider, as I said. But as just a producer of science, literature, and genre, but not of literary studies, what can I then tell you—certainly nothing about the profession, but even about the purview of the enterprise? Let me essay two ideas about direction, and call these, loosely (1) a focus on audiences, and (2) the return of the subject.

2

I think that writing (and scientific research, and teaching) are best seen as existential human acts within an overlapping spectrum of audiences. In the beginning there is glimpse of understanding, in the case of poetry just a phrase, or even a word that seems right or is reaching out to be connected. In that genesis understanding forms in some inner dialogue between parts of me, me and an imagined audience of one, me and a blurred, ever-shifting audience of teachers (yes, those Columbia professors), gurus, the dead or absent father, in the lonely emotionally polyglot dialogue with the voices of skepticism and self-doubt, and of joy at understanding that are all me, all of me.

And if the voices fail the writer, he or she makes the existentialist mark on paper, the initial inscription. Without fail, that sets the ghostly audiences into motion; they rise, are driven, make themselves out of the desperate necessity to tell a story to someone. It's not all so dark—if I think I understand, as I sometimes do, I rejoice too. In any case, I want to tell it to others, "go tell it on the mountain." Not to the drawer, but to live people.

The partial understanding I have then emerges, most often in speech, in the setting of the scientific family, our research group. Then I write a paper, a technical (not too technical) paper. Now my idea is reified; I build arguments for it. But I'm still teaching—I don't see my audience, but I care for it, I care that it understands.

The motion from the chaos of my mind, to an idea voiced at a seminar, the paper submitted, criticized, published, to lecturing in a specialized course or an introductory one—that motion is all teaching. With empathy, I engage a spectrum of interested (and not so interested) audiences.

Now what does this have to do with literary studies? I think it would also be productive in analyzing meaning in literature to think of divergent and convergent, overlapping audiences. There is one in the writer's mind. It is hardly static, for the work develops in time. And so much of the peopling of the writer's conceived audience is subconscious—the psychic work of early childhood, turbulent flows of emotion whose origins are lost or suppressed.

The text emerges, somehow (it's a miracle it does), and audiences for it immediately spring up. First, there are friends and editors, who can be influential (think of Pound on Eliot). Then, if the work is fortunate enough to be published, there comes the audience of book reviewers and literary critics, sharpening knives or stuffing into pigeonholes, shaping meaning for their own purposes, and for consumption by others. If the writer is luckier still, there also emerges the singular and collective audience of thousands of readers. Each is indeed a multitude, for each receives the text within the well-populated mental landscape of his or her own psychic forces, literary predilections and sheer prejudice, not to speak of the state of their body when they read the work.

Audiences may be small or large, real or imaginary. If there be a perfect match (impossible) between the intended audience of the writer and the real audience of one reader, then actually nothing interesting happens. My soul sister doesn't need my poem. In my field, this would be routine, paradigmatic chemistry. But from the almost inevitable mismatch of audiences—intended, perceived, incoherently shaped—the kind of meaning springs forth that engenders change (Rilke's "Archaic Torso of Apollo"-kind-of-change) and compels a segment of the audience past the high of catharsis (which suffices) to the creation of the new.

I think it's intriguing to think about how overlapping audiences come to be, and what authors, critics, and society do to shape audiences.

3

Let me preface some comments on the return of the subject (who never left) by some observations on aesthetic motivation within science.

If one can make any generalization about the human mind, it is that it craves simple answers. This is no less true in science than it is in politics. The ideology of the simple reigns in science, whereas every real fact argues to the contrary. So we have the romantic dreams of theoreticians (e.g., Dirac) preferring simple and/or beautiful equations.

The intricacy of any biological or chemical process elucidated in detail points clearly in the opposite direction. Think of hemoglobin, a jumbled molecule of ten thousand atoms, whose beauty patently resides in its complexity, a complexity absolutely essential for its function.

There is a remarkable resistance on the part of scientists to coming to peace with the complexity of this beautiful and terrible world. Chemists are quite schizophrenic in this regard. On one hand they crave the elegantly simple, loving molecules that have the form of Platonic solids. And if they don't find it, they sculpt the facts and hypotheses into a simple shape, often using Ockham's Razor (another story, told elsewhere) for the purpose.[3]

On the other hand, I think chemists are more accepting of complexity. After all, we are squarely in the middle, dealing not with the smallest, nor with the largest. The middle is complex, it's where human beings are. Chemists like a molecule shaped like a dodecahedron. And they like the fact that a pheromone emerges as a perfumer's blend.

But now we must face up to a psychological problem. Simplicity, symmetry, order ride a straight ray into our souls. Perhaps (this is far out) we have evolved a psychobiological predilection for the qualities of the world that rationalize our existence as locally contraentropic creatures. But what if the world is determined to be—by us, by scientific us—to be complex, unsymmetrical, and chaotic? How do we find satisfaction, and I do mean psychological satisfaction, in such a world?

I think the answer is simple, at least for a chemist. We construct with ease an aesthetic of the complicated, we adumbrate reasons and causes. We do so by structuring a narrative to make up for the lack of Platonic simplicity. And then we delight in the telling of the story.

I would suggest that narrative becomes the substitute for soaring simplicity in the operative aesthetic structures of chemists, and—I think it's the same even for the most hard-core reductionist physicist. Continuing the story is the motive force for experimentation and weaving of theories.

Let me make the transition to literary studies by the following little story:

There appeared a nice, perceptive article about the language of science by a German critic. I liked what he said, and I wrote to tell him so. In the course of my letter I innocently made some remarks about narrative, much as above. For some reason that ticked him off, for he proceeded to tell me, in no uncertain terms, as they say, the following: Narrative is fine. But ever since Gustave Flaubert took a story of a certain woman, a woman the elements of whose story were no different from the stories of a thousand other fallen women, and when Flaubert transformed that story into a great novel—ever since then it was clear that the subject matter of a narrative was fundamentally unimportant.

I'm not sure what set that minor tirade off, but the critic's peeved reaction made me think about the role of the subject within the narrative tradition. In science the subject is pretty well-defined. The phenomena under study—a new turquoise-colored molecule, or the AIDS virus—are surely not quite as real and unyielding of construction as scientists speak and think of them. But they are much further removed still, I think, from the evanescent and ephemeral guises assigned them by the "social construction of science" gang. The truth lurks somewhere in-between; I am reminded how after I and Shira Leibowitz wrote what we thought was a clever article about how the military metaphors of drug design are damaging to our psyche and culture, and even to the logic of drug design[4]—how after we wrote this we got a letter from a friend who was struggling with cancer (and who soon passed away), who gently said that it was difficult for her to think of a "love not war" alternative metaphor when she was dying of the disease.

Some trends in literary studies have aimed to remove the purported subject of the text. First it's the author and his intentional fallacy who are banished. Then the subject; it is easy to subvert it. Even the reader, that ever-flexible vagrant can be done away with. But, strangely enough, these fictionalizable characters in the drama of modern literary theory—author, subject, reader, even critics—just keep reappearing, persistent ghosts that they are.

I suspect that we will see the resurgence of the subject, be it an emotion or a human being. It, she may not be as "real" as the subjects of the limited world of the scientist, who deals with the important subset of problems in this world that have solutions (rather than inexhaustible alternative resolutions; I value those too). But the subjects of literature—yes, Madame Bovary, my favorite stones (which are Archie Ammons' stones)—will be central to what I see in the future of literary studies. This is what M. H. Abrams has called a humanistic literary criticism, "one that deals with a work of literature as composed by a human being, for human beings, and about human beings and matters of human concern."[5]

Notes

1. L. Trilling, *Mind in the Modern World* (New York: Viking Press, 1972), p. 14. I owe this quotation to Fritz Stern.
2. T. Pynchon, *The Crying of Lot 49* (Philadelphia: Lippincott, 1966), p. 129.
3. See Chapter 4, Unstable, in this book.
4. R. Hoffmann and S. Leibowitz, Molecular Mimicry, Rachel and Leah, the Israeli Male, and the Inescapable Metaphor in Science. *Michigan Quarterly Review* 30(3) (1991): 382.
5. M. H. Abrams, What Is a Humanistic Criticism, in D. Eddins, ed., *The Emperor Redressed: Critiquing Critical Theory* (Tuscaloosa, AL: University of Alabama Press, 1995).

18

The Metaphor, Unchained

ROALD HOFFMANN

Scientists write, first of all for other scientists. It's not publish or perish, but rather that an open system of communication, a commitment (shading to an addiction) to telling others what you have done, is essential to the functioning of science.

The primary medium of communication in the profession is the peer-reviewed article. This, our stock in trade, has a ritual format with strong historical roots. Once more diverse, the language of published articles is now 85 percent English, or an approximation thereto. Declining mastery of language aside, it's probably okay for most papers to be written in a bare style, for the vast majority of more than 500,000 articles published in chemistry and related fields last year is highly specialized (and routine) science. I do wonder about the collective effect of so much stylistically undistinguished writing. Is more harm done by selling lesser science through good style (I'm not talking about hype), or by poor writing pulling down sound science?[1]

A second intersection of science and writing reaches out to nonspecialists. Here we have science journalism and the popularization of science. The best examples shape a genre onto itself. Some are authored by writers, by journalists or historians, and are just superb, as in K. C. Cole's tours through higher dimensions. But let me focus on practicing scientists who write in this mode. I would claim that when scientists themselves write for a general audience, their research is likely to improve. Why? Because writing sets free the oft-suppressed metaphor.

Genre

Paragons among the kind of general-audience books I have in mind are those of Oliver Sacks, Carl Sagan, George Klein, and Jacques Monod, all of whom are (or were) both distinguished scientists and gifted authors. In their volumes, stories of science are told in a strong narrative vein. In some, a philosophical framework is explicit; in others it remains for us to find. Such books have recently won Pulitzer

Prizes, National Book Awards and their worldwide equivalents. This recognition is something new in letters, and well-deserved.

Another facet of the genre is made up of articles written by scientists who lay out their research in popular terms. These authors write for many reasons. Some may be driven by the stick of outreach requirements from governmental granting agencies. But more often a carrot is at work—an invitation that cannot be but flattering, a lecture series that naturally suggests a published *precis*. Ultimately it doesn't matter what combination of pressures and incentives leads to writing an article for *American Scientist* or a similar publication. A process that initially appears painful grows quickly into the desire to do it again.

Metaphor

Short of research papers, the audience of the scientist-writer is not in one's own trade. So the author cannot use too much jargon; the gatekeepers will make sure of that. One must simplify or say it in another way. Metaphors, similes, analogies—all the ways human beings have devised to explain that A is sort of like B—come to the surface. If I want to explain the uncertainty implicit in measuring simultaneously the position and velocity of a moving electron using photons, I resort to a thought experiment that measures the same observables for a baseball, with, say, tennis balls thrown at it. As I think about how to explain the vibrational-translational energy transfer necessary for the greenhouse gas carbon dioxide, CO_2, molecules of which have absorbed infrared radiation, to heat the rest of the atmosphere (predominantly nitrogen, argon and oxygen), I envision the bending and unbending CO_2 molecule as a gym rat exercising, once in a while kicking an O_2 dumbbell that comes near.

These thought mappings (let's loosely call them metaphors) also pulse deep in the heart of science. By this I mean they exist in the daily practice of doing research—in the way scientists generate hypotheses, theories and experiments. But...people don't much admit to it. My observation is that scientists sanitize their papers to remove as many explicit admissions as possible of the fecund, generative utility of such metaphors. Why? Because metaphors are (mistakenly) thought to impress no one—they are not mathematicizable; they are less "rational" (see Figure 18-1).

Along comes science writing. Now the scientist needs to explain something to the partially literate masses. All of a sudden, the metaphor, previously suppressed, is set free. Its use is intuitive; in fact, it's desperately needed.

But there's more to letting loose the beast than merely lifting the lock. Infused with the red blood of real ideas, metaphor, simile and analogy become explicit. They are reified, and importantly so in the mind of the scientist-writer. He or she may have used the thought map to design an experiment, or try out an analysis. Yet few allow themselves to pursue it, fully. It may be their loss: A naked metaphor clearly

Figure 18-1 Science or art, it's not easy to build a new way of seeing. In Mark Tansey's metaphor for getting the new off the ground, Georges Braque and Pablo Picasso (or is it the Wright Brothers?) launch the cubist airplane. (*Picasso & Braque*, 1992 © Mark Tansey, courtesy of the Gagosian Gallery, New York.)

shows the analogy's limitations, its capacity for misinterpretation *and* its productive extensions. It aids its creator as well as its audience.[2]

Two-way Teaching

Science writing is inherently pedagogical. And the scientist-writer will be able to both express and understand the specialized science he or she does more clearly as a consequence of the act of writing. Let me explain.

Our minds are full of inchoate ideas, inklings and partial explanations. Once verbalized, at a research-group meeting, for instance, or in the process of writing a paper, the ideas become real. Being human, we then marshal support, adduce arguments. The scientific paper explains. It has to teach—and to teach one must use those slippery words, eternally straying, lacking fidelity to the idea. But it is only with words that the removed reader may be reached. I see no dichotomy between teaching and research, only a continually varying set of audiences.

Good science writing has the audience firmly in mind—it teaches you (and a good editor can help so much) to teach others. This is not the mindless teaching of techniques or arid tables of dates and names: That requires neither acuity nor imagination. Rather, the act of skillful writing schools its author in ways of explaining structure and significance, of explaining ideas. Which is just what you need to do good science.

Narrative

I can hear in my mind one reaction to what I have said: "Are not observations, objective facts and reproducible data the foundations of science? Does it not suffice to report these, without embellishment?" Well, no. Science cannot exist without narrative. And making the effort to write of science for the general public sensitizes the practicing scientist to the importance of telling stories.

I recently reviewed a paper which tried to embody Sgt. Joe Friday's laconic ideal ("All we want are the facts, ma'am"). It consisted of 25 tables taking up some 35 pages and a handful of written pages. The text, such as it was, effectively said, "this and that are true; just look at Table 16." The failure of such a paper is transparent. The facts are mute; people need words, spoken or written, to make sense of data.

There is an interesting dance here, in that data (observations, equations, structural formulas, spectra) are useless without the narrative, theoretical framework to make a story out of them. So one is open to the criticism that the narrative prejudices the content, or, in other words, is "theory-laden." But—and this is the dance—the exact language used, be it English, Japanese or Arabic, should not matter. The stories that are told aspire to the universal, or, to use Gunther Stent's idea, to the infinitely paraphrasable. The valuable stories (I would call them "myths," using the most respectful meaning of the word) are essences. And this is the lovely paradox: These essential stories are, in a way, stripped of the supposed subjectivity of language— subjectivity that is absolutely necessary to tell the story in the first place (and even more necessary for it to be believed).

Like metaphor, storytelling is not mathematical. Yet it also is essential to good science, for two reasons. First, when simplicity (always the first aesthetic criterion) fails, human beings prefer to organize their hard-won knowledge of reality in the form of a story. We find a pattern, which means we find a story. Second, the classical workings of the scientific method demand the formulation of not one but several alternative hypotheses. What is a hypothesis, if not a story? Better learn to weave not one, but many.

People love stories. The best science writing, such as the remarkable case studies in *The Man Who Mistook His Wife for a Hat* by Oliver Sacks, teaches us narrative. That skill, to tell a story, is most unlikely to be part of a technical education. Yet it is not lost on scientists.

Better Science Through Writing

I am convinced that I have become a better theoretical chemist, a better explainer of the common and strange things molecules do, because I had to teach undergraduate courses. And also because I chose to write about science for people who do not share my academic background. Metaphor, teaching, storytelling were set loose within me because I was addressing a general audience of students and readers. There was no formula for it—I wanted to catch and hold their interest, no more. This approach proved to be at once more natural and more effective than one comprised solely of facts, however rational their presentation.

They have no substance, these mental fetters that constrain metaphor and teaching and narrative in the communication of science. Break them. And when they are gone, still a scientist, you will understand better, see things more clearly, know what we cannot see.

Acknowledgments

An earlier version of this article was published in Correspondences, a newsletter of the Duke University Writing Program.

Notes

1. For a good introduction to the way science is written, see D. Locke, *Science As Writing* (New Haven: Yale University Press, 1992).
2. A readable account of the role of metaphor in science is T. L. Brown, *Making Truth: Metaphor in Science* (Champaign, IL: University of Illinois Press, 2003).

Part 3

ART AND SCIENCE

19

Art in Science?

ROALD HOFFMANN

Here (Figure 19-1) are two manuscript pages from articles I've written. And there (Figure 19-2) are the ways they appeared in print, in the *Journal of the Chemical Society: Dalton Transactions* and *Inorganic Chemistry,* two magazines you are unlikely to have read recently.[1]

The context of these images is the following: I'm a theoretical chemist. What you see is the initial draft and final printed version of fragments from two of the >500 articles I've written. Articles are the stock-in-trade of the professional scientist. By and large we do not write books; our achievements, such as they may be, are judged by these scholarly articles. In general they're written in English (well, really in a jargon that has some vague relationship to English), printed in journals with limited circulation (these, among the world's best chemistry journals, have circulations near five thousand each), glanced at only by other chemists, and read carefully by a few hundred people. On the basis of these articles my work is evaluated and I make a living.

That explains circumstantially Figure 19-2, the final printed pages. What about the manuscripts, Figure 19-1? Clearly these are collages. There are samples of writing in two hands on them; one is my own, the other that of the graduate student (David Hoffman) or postdoctoral fellow (Kazuyuki Tatsumi) who has worked with me on this research.[2] In science there is much, much collaboration. My papers typically have two or three coauthors. I pose the question, my coworkers and I discuss an approach to a solution, they do most of the tough work, we talk further, a presentation of intermediate results is made, they're off to test various unreasonable suggestions I make, they write a draft, and I revise it into a final paper. In what you see in Figures 19-1 and 19-2, each a page of the manuscript of the final paper, I've pasted in photocopies of a piece of my collaborator's draft that I decided to keep.

The actual drawings that the scientific journals print are reproduced from India ink originals on tracing paper. These are masterfully done (see drawings in Figure 19-2) by Jane Jorgensen and Elisabeth Fields, two illustrators who worked with me for many years. They trace the ink drawings from carefully designed pencil sketches made by me or my coworkers.

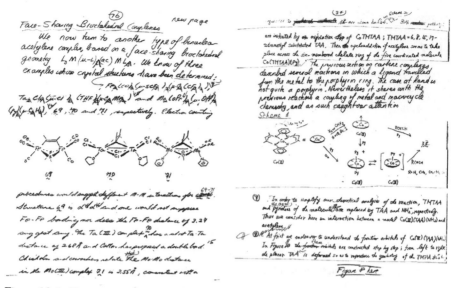

Figure 19-1 Two manuscript pages from articles of Roald Hoffmann.

Are these art, these collage-like manuscript pages and the final product? They look like science, and I've *told* you they are science. But what I would like to claim is that there is much more art in these assemblages of symbols than the scientist would admit or the artist allow.

Let us focus first on the most obvious visual feature of my printed scientific articles, and this is the preponderance of little drawings of molecules. These are "chemical structures." In a visual code they represent molecules. The representation is three-dimensional. And it is realistic, at least on the face of it. But is it?

The shape or structure of molecules is critical. Every chemical, physical, and biological property depends on the three-dimensional arrangement of atoms in space. For example, if water (H_2O) were "linear" (see A in Figure 19-3) and not "bent" (see B), as water really is, it would not be a liquid at ambient temperatures at the surface of the earth, and life as we know it would not exist. Another example: the mirror image of a molecule that is the essence of oil of wintergreen smells like spearmint.[3]

What chemists can learn, with the help of machines costing many thousands of dollars and a man- or woman-week of work, is the identity of atoms in a molecule, their three-dimensional structure, and the way they are connected to each other (H_2O and not H_3O; H-O-H and not H-H-O). The three-dimensional structure is presented usually as a ball-and-stick model, a typical example of which is shown in C in Fig 19-3. This molecule happens to be a phthalocyanine, representative of an important class of pigments that modern chemistry has added to the palette.

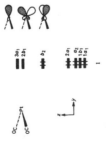

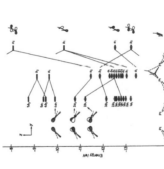
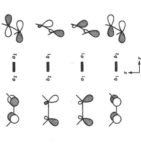

Figure 19-2 The way those pages appeared in the published articles.

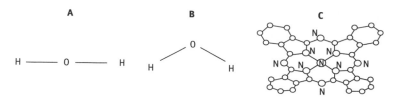

Figure 19-3 Water, A, the way it is not, and B, the way it is. C is a drawing of a model of a phthalocyanine.

It turns out, however, that this representation is ambiguous. The atoms don't sit still; they vibrate around certain preferred sites. And when we look at them, we don't see the nuclei, whose positions C represents, but the electron clouds around them. Chemists know this—the ambiguity of the model—but they will not admit to it unless forced to by an argumentative, perverse insider.[4]

But let us pass by that and continue. The model of the molecule is three-dimensional. The media available for disseminating its absolutely crucial structure are two-dimensional—a sheet of paper, a screen. And what's worse, the people who need to communicate this information are neither talented nor trained to produce effective two-dimensional representations; chemists are not sent to drawing classes. So what do they do? They improvise a primitive visual code, combining some elements of mechanical drawing with a code (a wedged line means *in front*; a dashed line, *in back*). And, with the aid of models, they indoctrinate into that code novices in the second year of college chemistry. It's quite miraculous that, from these primitive representations floating in some undefined space, chemists can reconstruct in their minds three-dimensional networks of some complexity. Here is testimony to the strength of symbolic codes and the inherent, irrepressible ability to see structures as three-dimensional.

I have written elsewhere of the symbol code of these structures and the peculiar relationship of the code to primitive art and the genres of caricature and cartoons.[4] What my colleagues have evolved is a method of representation that selects for emphasis some aspects of the model. Then they put a selected feature visually up front. If at another time they want to represent a different aspect of the molecule, no problem; just bring that part of it up front. It is no coincidence that photographs have found little use in chemistry journals (or anatomy books). Not that I want to argue that a photograph is a realistic representation. A photograph has *too much* detail and, at the same time, *not enough*—not enough of the essence of the molecule that one chemist is desperately trying to communicate to another.

In the drawings before you, Figures 19-1 and 19-2 (right) Kaz Tatsumi and I were faced with the problem of representing a molecule, a "macrocycle" with a cobalt atom at its center (think of those great Chinese blue glazes!). We first made the choice of showing all the atoms in the ring (but not the hydrogens at its periphery). Then we decided this was an inadequate (or overly detailed) representation and opted for a schematic ring with four lines, like spokes, to the metal in the middle. You see direct evidence here of choices being made, of representations altered for expressive purposes.

My claim is that these chemical structures are art—not great art, but art nevertheless. Even if their creators are unaware that they are producing art, even if they would deny the act, the "conceit" of being artists (which is what many scientists would call it, revealing thereby an interesting ambiguity toward art), what they are doing is the following: From a certain reality, that of a molecular model (which, like all realities, turns out to be on close examination a representation of a representation of...), the creators of these drawings try as hard as they can to abstract the essence. Then they attempt to communicate that essence to others, using a certain visual vocabulary. There is a concentration in what they do, an intensity that makes the object marked for communication come to life. There is also a distancing from the object (it's rendered from outside; it is remote) and a drawing in. Significant formal considerations—the relationship of the parts of a molecule to its whole— are essential.

An argument can be made that what is missing is (a) the chance, therefore unique, aspect of artistic creation, and (b) the affective realm, the play of the emotions, in this process of communication. To expand on the first point, which I think has some merit: while an artist's oeuvre reveals similarities, each work is different, a varied creation. The aleatory aspect, capitalized upon, is central. Scientific representations aspire, on the other hand, if not to anonymity, then to perfect paraphrase.[5] All those chemists who wind up drawing slightly different structures want other chemists to see the same molecule. And they do.

I will not argue too strongly with that. However, it has been my personal experience that, despite the announced or perceived intent of perfect paraphrasability, the creative moment in chemistry derives from a perception (often spatial) of a molecule in just one way and not another. We see that in the work of great synthetic chemists, master makers of molecules. The model turned in the hand in just one way, a redrawing of a structure with a certain unrealistic distortion, allowed them— and only them—to see it in a certain manner, to take it apart in the process of finding a startling way to put it together.

As for the emotional realm—well, I would agree that it is suppressed in the prescribed discourse of scientists. But first of all, to those privy to the code, that little free-floating picture can have tremendous emotional impact: something novel, something beautiful, a challenge to make, envy of the man or woman who made it.

Second, we have learned from literature and Freud what the consequences of suppression are. Here is a creative activity of human beings—science. Deep down it is driven by the same complex mix of psychic motives that drive any creation. The id will out. But the people who are doing this creative activity claim to be just reporting the facts and nothing but the facts. At best they may be fooling themselves; the very same impersonal, neutered language in which they choose to express themselves becomes charged with rhetorical impulses, claims to power, all the things they (we) foolishly thought we could suppress.[6]

Perhaps my argument here is overextended, for the emotional effect of the chemical representation is less obvious in the structures than it is in the *language* of the chemical article. And the printed pages shown in Figure 19-2 do not appear to be a spontaneous creation. It could be that the manuscript pages (Figure 19-1) fit the art model better. Their collage-like aspect certainly testifies to planning, to construction. But these word-image constructs have the feel of art, the pencil stroke made here and not there, the words (and pictures) crossed out. My sketch of a molecule is just that, a sketch; but as David Hockney could draw a "better beach chair" had he wanted to, I could have drawn a more realistic representation of my molecule. The information in that sketch suffices, at least to me. There is more expressive power in that little drawing than in my final finely-drawn product; it bears crude witness to my struggle to understand *and* explain, to conceptualize *and* articulate.[7]

There's no chance that any scientific journal would publish that initial sketch. Or even a "better" one, drawn by a more effective chemical artist. Perhaps that is the impoverishing aspect of this particular mode of human symbol transfer.

Notes

1. This article was written in 1990 for a journal of the Department of Art at Cornell University. The age of computer graphics has dated the introduction to this paper; rare is the graduate student who would know what tracing paper and India ink are.
2. Figure 19-1 is an early version of a figure in D. M. Hoffman, R. Hoffmann, and C. R. Fisel, Perpendicular and Parallel Acetylene Complexes. *J. Am. Chem. Soc.* 104 (1982): 3858. Figure 19-2 is an early version of a figure in K. Tatsumi and R. Hoffmann, Metalloporphyrins with Unusual Geometries. II. Slipped and Skewed Bimetallic Structures, Carbene and Oxo Complexes, Insertions into Metal-Porphyrin Bonds. *Inorganic Chemistry* 20 (1981): 3771.
3. For a good introduction to molecules and their shapes, see P. W. Atkins, *Molecules* (New York: Scientific American Library, 1987).
4. R. Hoffmann and P. Laszlo, Representation in Chemistry. *Diogène* 147 (1989): 24. Chapter 14 in this book.
5. G. Stent, Prematurity and Uniqueness in Scientific Discovery. *Scientific American* (December 1972): 84.
6. Some of these points are explored in R. Hoffmann, Under the Surface of the Chemical Article. *Angew. Chem. Int. Ed. Engl.* 27 (1988): 1593. Chapter 13 in this book.
7. Here I am grateful to Alexis Smith for a clarifying discussion.

20

Science and Crafts

ROALD HOFFMANN

I came to Penland to write.[1] The crafts were dear to me; first textiles, especially bobbin lace, which my wife made and collected, and taught me to look at. Then the Japanese ceramics to which Kenichi Fukui and Fred Baekeland introduced me. Followed by the protochemistry of dyeing with indigo from snail and plant sources, to me still the ideal bridge between science and culture. The tribute is to be seen around my house—my children's inheritance consumed as much by crafts as "high" art.

So it was easy to accept an invitation to come to Penland and write. Who knew what would come—I wanted to write poems, perhaps an essay. For the poems I've needed nature—not so much to write about as to shake me loose from the everyday worries of the (exciting) daily life I had in Ithaca. Nature was a path to concentration; I expected to find a different nature in the foothills of the Blue Ridge Mountains.

I would watch the crafts process. Maybe someone would even let me try something. Or ask me to tell them of the chemistry of their craft. I, in turn, would craft my poems out of the green hills.

But this is not what happened; here's what happened: I walk into Billie Ruth Sudduth's basketry class, and there's the whole group dyeing their canes, steaming pots of synthetic dye. I ask someone what they are doing, and she says, "Well, I'm getting ready for the upsetting," and then seeing the puzzled look on my face, patiently explains this old, wonderfully direct basketry term for bending the canes forming the base of a basket over themselves, so that they stand up (see Figure 20-1).

I walk uphill to the iron shop, clearly more of a macho place, watch an intense young man, lawyer become sculptor as it turns out, hammer out a hand on a swage block. Ben tells me that it's possible to burn away the carbon in the steel, and the iron would "burn" too, oxidize, in too hot a flame.

In a studio downhill, a young student carefully carves out the wax sprues (yes, I have to be told what these are—and I thought science was full of jargon!) that will eventually help him form silver leaves. His instructor, thinking I might be the useful

Figure 20-1 Upsetting.

sort of chemist, asks me about gases emitted when the "investment" hardens. Alas, I'm a theoretician, as impractical as they come within this profession.

There is no time to write, poetry suffers. The only nature I encounter are the ubiquitous fireflies as I walk back to my cottage late at night. I think of their wonderful luciferase chemistry, their rhythms and deceits—one carnivorous firefly species imitating another's flashing rhythm to…lure a male to death.

My heart is open. I am in thrall to these intent older and younger people, and the transformations they perform. I am not even discouraged that my own attempts at pots, under the tutelage of a great teacher, Paula Winokur, or my try at blowing glass, or forging iron, fall short. I need practice.

And I reflect on the kinship I have, as a scientist (and writer too, true) with the creators of crafts. The magic of Penland opens people to each other and their hands' work. But there are deeper ties.

Natural/Unnatural

In the context of environmentalist and ecological disputes, scientists and technologists are often branded as the makers of the unnatural. Aware of the shades of meaning, progressively negative, that accompany the words crafted, man- or woman-made, artificial, synthetic, unnatural, I use unnatural as a provocative extreme. Because that's how people see it. Sure, you could say everything people do is natural, because they/we, the makers, are. And we certainly deconstruct the natural/unnatural distinction every moment of our creative, transformative lives. Taking the natural, changing it. Making naturalistic shapes out of the most synthetic of materials. But I think it makes sense to distinguish the actions of human beings from those of the rest of nature, if we are to have a sensible debate on the environment.

My personal way to overcome this facile categorization into natural/unnatural is to ask people to think what's natural about a John Donne poem, Orson Welles at his most evil in "The Third Man," a Bach cantata, a desegregation law. These are acts of human creation, they enliven (and may hurt; a romantic affectation of artists is that all art is inherently positive). We have been put on earth to create. And...as human beings, the act of creation of molecules or poems must be coupled with an ethical assessment—will this hurt, will this heal?

At Penland, everyone, absolutely everyone, was into making the artifactual, into transforming, changing, modifying nature. From the wonderful sculpture of a ballerina's tutu made from birch bark, to the *shibori*-dyed cloth, to the shaved stakes worked into a basket, people were taking one thing (natural or synthetic) and transforming it into another. Yes, they did care about natural or synthetic dyes—maybe they'd use, for class purposes, synthetic dyes to color cloth or basket cane. At home they'd think again—some would stick with natural dyes; most I suspect would not.

The natural/synthetic story is fascinating in detail, and in the way it has provided an ethical dimension for crafts. People who make things for use (or beauty; could we live without it?) will go with any technical advance. No commercial fisherman will return to cotton nets after using nylon. Within three years of the coming of aniline dyes on the marker in Germany, they were used in Persian villages. To the detriment of the rugs, initially—some of the first dyes were corrosive to the wool. Navajo weavers unraveled Spanish red cloth, bayeta, to reweave it into their chief blankets.

Bright color has a way into the soul. And since synthetic dyes often are more intense in hue than natural ones, they have always tempted the craftsperson. I like the idea that the question of what dye to use now has an ethical dimension for the craftsperson. And, to complicate matters, that we have an inversion of the old class/color correlation—no longer are the rich more colorful.

In the company of craft people, I did not need my prefatory plea to recognize that we are all in the business of transformation. And we could move on—perhaps to the associated ethics, perhaps to the aesthetics.

Chemistry

Not only were the crafts people at Penland transforming the natural, but they were just plain doing a lot of *chemistry*. I've mentioned the dyes for textiles and baskets. In the magic of clay fired to a ceramic there was chemistry, also in the colors of the glazes. Higher up the hill was the land of perfervid metallurgy—people were not winning metals from their ores, but they were pouring bronze, burning out plastic or wax, annealing, tempering, etching, grinding, welding. I loved the sounds of the work. There were chemicals in the print shop, and the developer and hypo in photography.

There was concern about the health effects and safety of these processes. As there should be. The concern was amplified by the fact that people sometimes didn't know what they were working with; the ingredients were not specified (as little descriptive as "red earth" or "stabilizer"). Some people had hazardous material data sheets, others not. All were torn by the tension of wanting to use materials that expanded the range of what they could do, and not being sure of the biological effects of the new materials.

Chemistry is the art, craft, business, and now science of substances and their transformation. It was fun to see so many cryptochemists! Good, practical chemistry was being done left and right. In the usual way people have of thinking they are insufficient, and not having gone through a chemical apprenticeship, crafters were hesitant. And so a little afraid of me, a professional chemist. Little did they know that I, a theoretical chemist, was much more of a klutz in practical chemistry than they were!

The craftspeople I met also weren't quite aware of how much they, in fact, were like professional chemists. What I mean here is that both scientists and crafters alternate doing things carefully (measuring out that bevel, controlling the kiln temperature), and tinkering, trying things, trying them again to get a process to work.

Chemists also pursue matter in all its rich variety on the microscopic scale—they think of chemistry as being the transformation of persistent groupings of atoms we call molecules. What happens downscale, to the molecules and their constituent atoms, determines, as one moves upscale, what form macroscopic substances take on and how they transform. Their colors, crystalline shapes, their biological effects, their chemical reactions—all these have a molecular basis.

The creative people I met were moving on the macroscopic plane. Would it help them to know more, of the small structures inside?

How Much Does One Need to Know?

Not much. A lot. Just enough to create. I am speaking of the knowledge of the materials we work with, both as they "rest," and as we transform them. Is it important to know that steel is an alloy of iron and carbon, and that the carbon is there in several ways—part a solid solution in the interstices of metallic iron, part in discrete Fe_3C and Fe_5C_2 compounds. Should one care that the chemical structure of indigo is the one shown in Figure 20-2 below, and that to have it bind better to wool and linen one has to "reduce" the molecule to a colorless form, which, once absorbed into the biopolymers, is oxidized back to a molecule colored "...like unto the sea and the sea is like unto the sky and the sky is like unto the sapphire, and the sapphire is like unto the Throne of Glory..." as Rabbi Meir said of the wondrous dye?

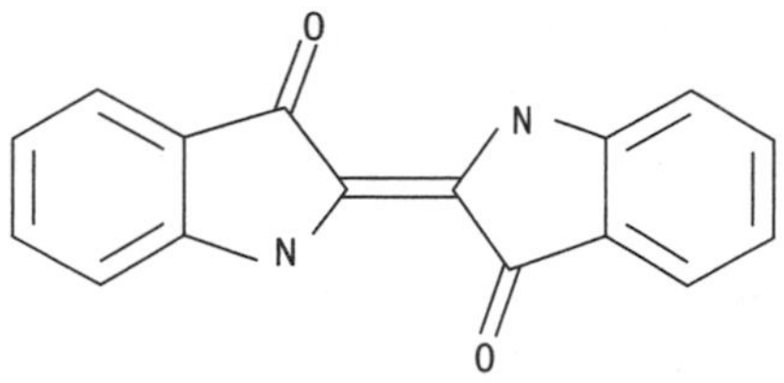

Figure 20-2 The indigo molecule.

The art is wonderful, the overall change possessed of sufficient mystery to make the spirit soar when the blue of the dye reappears as the oxygen of the air hits the wool. Need we care about what happens on the molecular level?

Curiously, the question "Should I know?" or rather "Should I learn deeper?" is there in science as well. The special context there is reductionism—a worldview (unrealistic and unworkable in my opinion) that science bought into early on. By reductionism I mean the description of a hierarchy of sciences, and a definition of understanding in terms of a reduction from one science to another. So behavior is to be understood in terms of biology, biology is to be understood in terms of chemistry, chemistry in terms of physics. Given the premium on understanding, in this hierarchy there is no question that there is more value in going deeper. And deeper is defined in terms of reduction.

Actually, this kind of reasoning is often used as a rhetorical avoidance strategy for those unwilling to broach the real world—how matter might behave upscale.

In that real world, now of the practicing chemist, things operate very much as they do for the craftsperson. Chemists analyze, to be sure. But much of their activity is creative, the uniquely chemical matter of synthesis. A few hundred thousand new compounds are made each year. Why? For all kinds of reasons. For example, to make an anti-tumor agent, isolated from the bark of a yew tree in the lab, so one wouldn't kill yew trees by stripping their bark. Thus for a specific use. Also to "sell" these molecules. And for fun. There's no utility in a molecule shaped like an icosahedron, made all out of boron and hydrogen. When carbon wants to have four bonds going out toward the vertices of a tetrahedron, what would it take to induce those bonds to align themselves in the four directions of a square? The syntheses of chemists here are driven by beauty. And, incidentally, not only the facile beauty of Platonic polyhedra, but that beauty much harder to learn to love, in crafts or science, that of rococo intricacy.

The mix of utility and beauty as motivation pervades science, as it does the crafts. And, in both fields, utility and beauty may be uneasy partners.

In making a molecule, the synthetic chemist often uses processes that he or she does not fully understand (to a reductionist's satisfaction). So there's a catalyst, a metal powder primed by another chemical, and that catalyst adds two hydrogens to a molecule just there, and not here. And while we don't know exactly how it does that, it does it so efficiently and reproducibly. The practical chemist often says, "I'll take that, let someone else find out how it works." My métier is actually calculating how it *does* work, on the atomic level.

Here's a practical argument for trying to understand, at every level, in crafts or science: The synthesis, or carving a pattern of grooves into iron, is going great. But one day the catalyst fails to do its expected magic. And the next batch of steel just fails to give those temper colors. What does one do? Throws away that catalyst, that block of steel. Tries another. And then, that fails too. There must be a reason, which will not be revealed by prayer or anger—there is an argument here for trying to comprehend, at least piecewise, enough to fix something when it goes wrong. As it will.

But I think the primary argument for understanding is ultimately psychological and aesthetic, rather than practical. Paulus Berensohn writes:

> The molecules of clay are flat and thin. When they are wet they become sticky with plasticity and hold together as in a chain. A connecting chain. I like picturing that connection in my head.
>
> I am making my connection with clay. Clay turns me on and in. It seems clearer and clearer that I was drawn to clay by its plasticity. For it is plasticity that I seek in my life. To be able to move into new and deeper forms as well as make them. Making the connection and being plastic.[2]

Knowing that in the fibers that will form paper is cellulose, gleaning from the arrangement of atoms in that molecule its kinship to rayon and to sugar or starch—that knowledge may be of little direct use to the craftsperson. But I think it makes all of us feel better—for understanding pleasures the mind.

Knowledge not only satisfies, but it also bolsters the mind when things don't work—when the flux pulls away from the metal, or the paper cracks. The intuition to try something else comes from knowledge subconsciously assimilated. One can go on, there are reserves of intuition to take a new tack. Knowledge also counters alienation—that of art from science, that of us from our materials and tools. These molecules are similar, they are different. They share some things, differ elsewhere. We see the world as connected. As making at least a little sense. And go on to make the next thing, as we are driven to do.

With or Against

You must go with the material, of course. The clay needs to be dried before it is fired, the glass and metal annealed. I watched Greg Fidler at Penland shape a bulb he had blown on a rounded wood form, flatten it a little, extend the thick neck separating it from another, connected bulb, soften the neck, and wait patiently for the moment when he would swing the blowpipe with the glass attached in a near circle. The neck stretched and the round bulb bent over as the steady swing was completed, in that moment nestling gently into the space the other, flattened yet slightly curved shape made for it. There was suspense in that swing. And Greg knew when he could do it (see Figure 20-3).

In his New York studio I watched a master sculptor, Daniel Brush, recreate Etruscan granulation, a way of attaching thousands of tiny gold spheres to a flat or curved gold sheet. The spheres loosely glued in place, he sprinkled the surface of the spheres with a copper salt, and heated it. As the object approached the softening temperature of gold, Daniel had all of about one second in which the alloyed copper metal formed on the sphere surfaces, melted, a tad below the gold, ran down the outer surface of the sphere and formed a perfect weld at the juncture with the flat. Had he stopped heating a second earlier, there would be no bond. A second later the spheres would just melt.

"Going with" comes from observation. While working. Which builds into competence. And is ultimately transformed into intuition, body and mind intertwined.

But the will must be there to make out of the natural both the useful and the transcendent. I think of Antonio Gaudi's dragon in the gate to the Finca Güell in Barcelona. What iron ever wanted to be such a fierce segment of our imagination?

Figure 20-3 Mirrored Bowing Series, blown glass, 2003, by Greg Fidler. Used by permission of the artist.

I think of Egin Quirim and Cosmos Damian Asam, concocting a stucco angel, on a fat cloud, just soaring out of a wall toward us in their Bavarian rococo church.

In science, as in craft, the master just knows what filter will effect the separation, intuits the flux to be used to make a solid state reaction run. And the apprentice learns. Intuition begins in trial and error, respecting the richness of matter and its changes. Homage is paid to chance, serendipity can be courted when invention stagnates.

But ultimately one tries to make matter do what it had not done before. This, incidentally, is what distinguishes chemistry from other sciences, and puts it close both to engineering and the arts. Maybe that's where crafts are too! We chemists make new molecules, a few hundred thousand of them every year. With the intention to do no harm, if not heal. And yet some of them, like the chlorofluorocarbons that damage the ozone layer, hurt. But then creation has always been a risky business, and I'm not just talking about procreation. In the Popol Vuh, the book of the Quiche Maya of present day highland Guatemala, are told stories of several creations that went astray.

I saw a paper recently in which a German chemist reported the making of a line of six carbon atoms, bound to a line of three osmiums, each bearing several carbon monoxides. An unnatural assembly itself, it could be traced back to the naturally occurring element osmium, carbon from petroleum sources, and natural/synthetic (and poisonous) carbon monoxide. I saw that one line of atoms was tensed and curved, more than the other. Hard to know who to blame—carbon or osmium? One of my graduate students, Pradeep Gutta—how possessive we grow of our apprentices—said "Hey, how about making a big circle out of it?" He had thought of the tension of an arc, of allowing it to play out that tension by completing a circle. Off he went to the computer, to build a model of the electron motions in that circular ribbon of osmium, carbon, and CO.

Pradeep is most certainly going with what the molecule "wants" to do. And he is transforming it (on paper in our case, for we are theoreticians; but we have such great faith in our experimental colleagues—it will be done!) into something new and beautiful. And who knows, maybe useful.

A master smith

said: comply, but
contend—make

hard soft, hard
again, beat blade

and girder into
rabbit's ear and

morel. Love, oh
love for steel too,

is built sweet out
of strict desire,

for the you, that
is not you. You.[3]

Like a Horse and Carriage

Is there an analogue to the art-craft discussion in science? For one aspect of craft there is an easy scientific counterpart—it is the experience of experimental work, using tools. Take DNA. One had to be able to separate biomolecules, and build X-ray diffractometers, before the structure of DNA could be deduced theoretically by two young helixeers 50 years ago.

But what is art in science? Is it theory, viewed in the general sense as the building of frameworks of understanding? It can't just be theory. I think the analogue of art is the imaginative faculty, which makes scientists creative. It does not come to the fore in deductive thinking, or applications of that easy idol of science, Ockham's Razor. Art is in the formulation of far-out hypotheses, in seeing connections between the seemingly unconnected, in designing instruments and experiments.

Science, a European invention, is a system for gaining reliable knowledge by the interaction of curious yet fallible human beings, who are obliged to tell others what they have done. Science also mandates a continuous dipping back and forth between reality (gauged by our senses and instruments) and flights of theoretical fancy.

So the system of science enforces links between the art of hypotheses and the craft of the instruments. You just can't have one without the other—a theory not tested will not be accepted, and reports of experience without trying to understand it (without a theoretical framework) are unreadable. It sure looks like you can't have the art of science without its craftsmanship.

Another interesting intersection is around the idea of utility, a subject not without dispute in the crafts. When is a teapot not a teapot? Does it matter whether a shape sells? The crafts were always of commercial value, they were a profession. And there were middlemen, even way back then. Half the students at Penland when I was there were making a living (sorry, trying to do so) from their crafts.

Utility at first sight poses a problem. For many thoughtful theories of art conclude that the concentration, intensity, and unity in an art object can only be appreciated if one is *dis*interested in its value, whether monetary or utilitarian. I hear snickering on how this applies to the objects consecrated in our temples of high art. That aside, and admitting the corruptive power of money and familial (or political) relationship, I think it is too harsh to deny to the beauty-crafting personal bond between object and human that daily use creates. Of a teapot, or my Harris tweed jacket.

There is an interesting utility/knowledge for knowledge's sake (substitute "art" for "knowledge") tension underlying all of modern science. We need support—once it was called patronage. Yet we resent it when the government, the modern patron, tries to direct our work with a dollared carrot—come work on "star wars" and you'll get support! At the same time we forget a little about the meliorative aspect, the desire to help people, to leave the world a little bit of a better place than it was before. In general applied scientists don't get much prestige in academia. And yet one out of two assistant professors in molecular biology and materials science is running after that new start up company.

I look at a mask on my wall by Alaskan native Evans Apatiki—carved whalebone, polar bear hair around it. It evokes its animal construction, its ritual use. And the mask is as expressive and constructed as an Ernst Barlach sculpture. Or as the synthesis of vitamin B_{12} by R. B. Woodward and Albert Eschenmoser with 99 friends—each step necessary, executed with improvisational aplomb, a thing of beauty.

Hands and Minds Combined

From time to time, we in chemistry are put under pressure to teach an introductory course without a laboratory. Couldn't you just talk about the logic of chemistry; wouldn't some molecular models and good stories suffice? At one such meeting, there rose to our defense a print maker. He said, "There's a difference between talking about lithographs and making one," and sat down.

There is no question that the crafts are about hands and the senses, especially vision and touch. And sounds too—at Penland I loved the unexpected roar of the iron furnace, scissors snipping through paper, even the buzz saw (at a distance). With a guiding mind, and yes, with tools and chemicals, a photograph is developed, printed, pasted into a book.

And science is about tools, and handwork too. Though the heroic figures of physics are by and large theoreticians (Fermi and Rutherford are the exceptions), the practice even of this quite mathematical science is largely experimental. The tools may be fancier, all those laser spectroscopes. But on the "optical bench" are carefully mounted mirrors, machined vacuum chambers, and yes, even now, blown glass containers. All designed and made artifacts.

For the crafts and for science, this—that both thinking and doing are engaged and cooperating—is our finest link. The world is disintegrated—separating mind and body. We cater to the mind through a novel or a Bach Cello Suite on a CD. And to the body through the long sanding of the walls before painting, or those Nautilus machines. Craft and science, both, integrate mind and body.

Could one imagine making a bracelet, linked silver triangles with an inlaid braid, without planning it out, making a mold for the triangles (all different), hammering in the decoration? The synthesis of a molecule shaped like a necklace—yes, there are such—begins with a plan. Which has to be changed a few times as one moves along, for things do go wrong. But the molecule is also a macroscopic substance, a solid, crystalline, each crystal the blue of aquamarines in a real necklace. And being something real and substantive, this necklace-shaped molecule must be made. It happens, in a wondrous ballet of all the glass vessels you can conjure up, the sequences of heating, stirring, of bubblings, filterings, stinky solutions and mother liquors. It's a long day's night to make it, bracelet or molecule.

At the end, there's craftsmanship, the proud, cunning work of human hands and mind, joined in the service of creation.

Notes

1. The Penland School of Crafts in Penland, NC is a very well-known, 90-year-old center for education in the crafts. I was there for two weeks in 2002.
2. P. Berensohn, *Finding One's Way with Clay* (New York: Simon and Schuster, 1972).
3. Part of a poem by R. Hoffmann, "With or Against," unpublished.

21

Molecular Beauty

ROALD HOFFMANN

My wife and I were on our way to Columbus, Ohio. After I settled on the airplane, I took out a manuscript I was working on—typical for the peripatetic obsessive chemist. Eva glanced over and asked, "What are you working on?" I said: "Oh, on this beautiful molecule." "What is it that makes some molecules look beautiful to you?" she asked. I told her, at some length, with pictures. And her question prompted this essay.

What follows is an empirical inquiry into what one subculture of scientists, chemists, call beauty. Without thinking much about it, there are molecules that an individual chemist, or the community as a whole, consider to be the objects of aesthetic admiration. Let's explore what such molecules are, and why they are said to be beautiful.

In the written discourse of scientists, in their prime and ritual form of communication, the periodical article, they've by and large eschewed emotional descriptors. Even ones as innocent as those indicating pleasure. So it is not easy to find overt written assertions such as "Look at this beautiful molecule X made." One has to scan the journals for the work of the occasional courageous stylist, listen to the oral discourse of lectures, seminars, the give-and-take of a research group meeting, or look at the peripheral written record of letters of tenure evaluation, eulogies or award nominations. There, where the rhetorical setting seems to demand it, the scientist relaxes. And praises the beautiful molecule.

By virtue of not being comfortable in the official literature—in the journal article, the textbook or monograph—aesthetic judgments in chemistry, largely oral, acquire the character of folk literature. To the extent that the modern-day subculture of chemists has not rationally explored the definition of beauty, these informal, subjective evaluations of aesthetic value may be inconsistent, even contradictory. They are subfield (organic chemistry, physical chemistry) dependent, much like the dialects, rituals or costumes of tribal groups. In fact the enterprise of excavating what beauty means in chemistry seems to me to have much of the nature of an anthropological investigation.

But this is not going to be your typical seemingly detached critical analysis revealing with surgical irony the naive concepts of beauty held by a supposedly sophisticated group of people. The honesty and intensity of the aesthetic response of chemists, when they allow themselves to express it, must be taken positively, as a clue to an unformulated good, as spiritual evidence, as a signpost to record, to empathize, to make connections with other aesthetic experiences.

Aesthetic judgments made by chemists about chemistry are perhaps more cognitively informed than aesthetic judgments in the arts (more on this below). Which ensures that those judgments are jargon-laden. But I'm certain that people outside of chemistry can partake of what makes a chemist's soul jump with pleasure at the sight of a certain molecule.[1] It's worth trying to see the motive force for all that intense, disinterested contemplation.

The Shape of Molecules

Let's begin with the obvious, which was not accessible to us until the twentieth century, namely, *structure*. Molecules have a shape. They are not static at all, but always vibrating. Yet the average positions of the atoms define the shape of a molecule. Geometry can be simple, or it can be exquisitely intricate. Structure **1** is a molecule with a simple shape, dodecahedrane. This $C_{20}H_{20}$ polyhedron (the polyhedron shows the carbons; at each vertex there is also a hydrogen radiating out) was first made in 1982 by Leo Paquette and his co-workers.[2] It was a major synthetic achievement, many years in the making. The Platonic solid of dodecahedrane is simply beautiful and beautifully simple. Molecule **2** has been dubbed manxane by its makers, William Parker and his co-workers.[3] Its shape resembles the coat of arms of the Isle of Man. And molecule **3** is superphane, synthesized by Virgil Boekelheide's group.[4] All are simple, symmetrical, and devilishly hard to make.

1.   **2.**   **3.**

Let's try a structure whose beauty is a touch harder to appreciate. Arndt Simon, Tony Cheetham, and their co-workers have made some inorganic compounds of the formula $NaNb_3O_6$, $NaNb_3O_5F$, and $Ca_{0.75}Nb_3O_6$.[5,6] These are not discrete molecules but extended structures, in which sodium, niobium, and oxygen atoms run on in a small crystal, almost indefinitely. Below is one view of this truly super molecule, **4**.

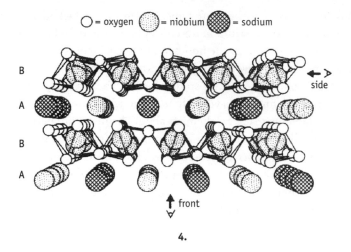

4.

Some conventions: the white balls are oxygens (O), the stippled ones niobium (Nb), the crosshatched ones sodium (Na). The perspective shown chops out a chunk from the infinite solid, leaving it up to us to extend it, in our mind, in three dimensions. That takes practice.

Deconstruction aids construction. So let's take apart this structure to reveal its incredible beauty.

In drawing 4 we clearly see layers or slabs. One layer, marked A, is shown in structure 5. It contains only niobium and sodium atoms. The other layer, B (structure 6), is made up of niobium and oxygen atoms arranged in a seemingly complex kinked latticework. Let's take on this B layer first.

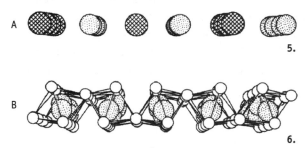

The building block of the slab is an octahedron of oxygens around a niobium. One such idealized unit is shown in drawing 7, in two views. In 7a, lines (bonds) are drawn from the niobium to the nearest oxygen. In 7b these lines are omitted, and instead the oxygens are connected up to form an octahedron. Which picture is right? Which is the true one? Sorry—both are. Or, better said—neither is. Three-dimensional molecular models, or their two-dimensional portrayals, which is what we have before us, are abstractions of reality. There is no unique, privileged model of a molecule. Instead, there is an infinite variety of representations, each constructed to capture some aspect of the essence of the molecule. In 7a the essence is deemed to lie in the chemical bonds, a pretty good choice. These are Nb-O; there are no O-O bonds. Yet portrayal **7b** draws lines between the oxygens. This representation seeks after another essence, the polyhedral shapes hiding in the structure. Graphically, forcefully, **7b** communicates to us that there are octahedra in this structure.

You may wonder where these octahedra are in the complex structure of $NaNb_3O_6$. Well, let's take the octahedra of drawing **7** and rotate them in space, to the viewpoint shown in drawing **8**. If you compare **8** with the middle piece of layer B (shown in structure **6**), you will see a certain resemblance.

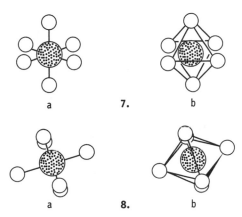

a 7. b

a 8. b

Now consider the structure of the layer. First, a large, semi-infinite number of such octahedra are linked into a one-dimensional array by sharing opposite edges. Three views of such an edge-sharing octahedral chain are presented in drawing **9**.

top   front   side

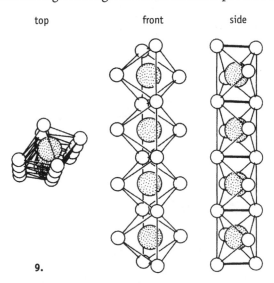

9.

One of the three views is from the same vantage point as in **4** or **6**. Let's call that "top." The two other views are roughly from the "front" and "side," the viewpoints so marked in the original drawing, **4**. The shared edges are emphasized by darker lines in the side view.

If you compare the top view of one of these infinite chains in **9** with the view in **6**, you will see a difference—the niobiums are receding from you in a neat straight line in **9**, but are "staggered" in pairs in **6**. Indeed drawing **9** is an idealization. One of the stacks in the real structure is carved out in drawing **10**, shown in the same top view as in **6**, but also from the front and the side. The motion of the niobiums off the centers of the oxygen octahedra, and an associated asymmetry of the oxygens, are clearly visible.

top   front   side

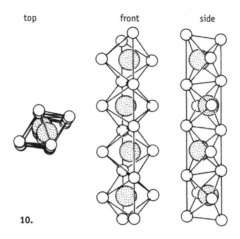

10.

Would you like to know why there occurs this departure from ideality? So did we. A piece of the answer is to be found in a paper that Maria José Calhorda and I have written.[7] For the moment let's accept this symmetrical asymmetry as one of those complexities that makes life interesting.

Next the one-dimensional chains of octahedra combine to generate the full B layers by sharing two opposite vertices with identical chains. They could have done so in a nice "straight" way (see **11**, a top view of a line of such vertex-sharing octahedra). But they don't; they "kink" (drawing **12**) in a less straightforward but still symmetrical way. One gets the feeling that nature is insubordinate...What really is going on, though, is that we, in the weakness of our minds, fix on the first, most symmetrical suggestion of how things might be.

top

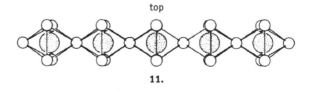

11.

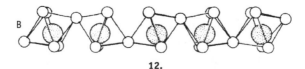

12.

We now have layer B, this fantastic slab (repeated over and over in the crystal) of infinite, one-dimensional, edge-sharing octahedral chains, in turn stitched up to a two-dimensional slab by sharing vertices. What about layer A?

Drawing 4 shows that layer A is made up of needle-like lines of sodium and niobium. We might think these atoms are equally spaced, but this molecule has another surprise in store for us, as the front view of layer A indicates (13). Whereas the sodiums *are* approximately equally spaced, the niobiums clearly are not. They pair along the vertical direction (this pairing is obscured from the "top" vantage point of 4 or 5), so that there are distinct short (2.6A) and long (3.9A) Nb • • • Nb separations. The short one is very short, substantially shorter than in pure niobium metal.

front view of layer A

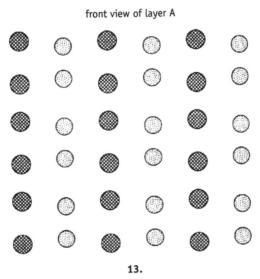

13.

Why do the niobiums pair? In the study we've done of the way electrons move in these compounds, we find that the pairing is driven by a desire-cum-necessity to form Nb-Nb bonds along the needle.[7] There are even Nb-Nb bonds, not shown here, between the niobium atoms of stacks A and B.

Other links stitch up the layers. For instance, the niobiums in layer A are not floating in empty space. They are at a bonding distance from the oxygens of two bordering B slabs. In fact, as drawing 14 shows, each pair of niobiums in a line in A nestles comfortably into an array of eight oxygens from layer B. The layers are connected up—this is not a one- or two-dimensional structure, but a true three-dimensional array in which substructures of lower dimensionality are embedded.

Now we've toured the structure. The beauty of this aesthetic object resides in its structure, which is at once symmetrical and unsymmetrical. The beauty is in the incredible interplay of dimensionality. Think of it: two-dimensional slabs are assembled from infinite one-dimensional chains of edge-sharing octahedra of oxygens around niobium, which in turn share vertices. These two-dimensional slabs interlink to the full three-dimensional structure by bonding with one-dimensional

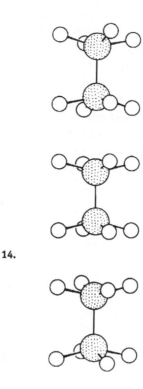

14.

needles of niobium and sodium. And then, in a final twist of the molecular scenario, these one-dimensional needles pair up niobiums, declining to space equally. The $NaNb_3O_6$ structure self-assembles, in small black crystals, an aesthetic testimonial to the natural forces that shape the molecule, and to the beauty of the human mind and hands that unnaturally brought this structure into being.

Kant

It's hard to escape the feeling that a chemist appreciative of the dimensional ins and outs of $NaNb_3O_6$ is doing just what Kant described in the following familiar words:

> *He who feels pleasure in the mere reflection upon the form of an object . . . justly claims the agreement of all men, because the ground of this pleasure is found in the universal, although subjective, condition of reflective judgments, viz. the purposive harmony of an object (whether a product of nature or of art) with the mutual relations of the cognitive faculties (the imagination and the understanding).*[8]

But let's go on, to look at another source of molecular beauty.

Frogs About to Be Kissed

Could one say much by way of approbation for molecule **15**? Not at first sight. What are those dangling $(CH)_{12}$-Cl chains at left? Or the unsymmetrical $(CH_2)_{25}$ loop at right, or the NH_2? The molecule is, if not ugly (there are no ugly molecules, says this most prejudiced chemist), at least plain. It's not an essential component of life, it's not produced in gigakilogram lots. In fact its purpose in life is not clear.

15.

The last sentence contains a clue to what makes this molecule, a frog that is a prince, beautiful. Chemistry is molecules, and it is chemical change, the transformations of molecules. Beauty or elegance may reside, static, in the very structure, as we saw for the molecule $NaNb_3O_6$. Or it may be found in the process of moving from where one was to where one wants to be. Historicity and intent have incredible transforming power; this molecule is beautiful because it is a way point. Or as they say in the trade, an intermediate. So: *quo vadis*? To a catenane, structure **16**, two interlocking rings of carbon atoms, not chemically combined but held together like the links of a chain.

16.

Why should people try to make a catenane? For no particular reason. For the best reason, because none was made before. How to make it? Here's one strategy, which I will term a statistical one. A typical chemical reaction is a cyclization, schematically, written out in structure **17**. If we run the cyclization of a long chain, some fraction of the time—purely by chance—the chains will be entwined, or a chain will be threaded through an already formed ring, in such a way that a catenane will be formed. Remember how small molecules are, how many (10^{20}) there are in a typical pot. In all this multitude, statistics have a chance to work. E. Wasserman actually realized this in 1960, synthesizing a catenane for the first time.[9]

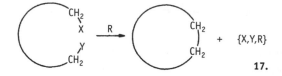

17.

The statistical procedure works, albeit inefficiently. There are other ways to craft the catenane topology. One beautifully conceived synthesis is due to Schill and Lüttringhaus.[10] Their logical scheme is summarized in chart **18**. The starting point is a molecule with lots of specifically disported functionality. In chemistry, a functional group is a set of bonded atoms whose properties are more or less invariant from one molecule to another. The most important of these properties is chemical reactivity, the "function" of the group. To put it another way, in the context of doing chemistry on a molecule, functional groups are the *handles* on a molecule. The transformation of functional groups, and particularly the predictability of their reactions, are a crucial element in the conceptual design of syntheses in organic chemistry. Common functional groups might be R-OH (alcohols), R-COOH (organic acids), R-COH (aldehydes), R-X (X = F, Cl, Br, I, or the halides), where R is anything. The substituents X, Y, and Z in structure **18** are functional groups.

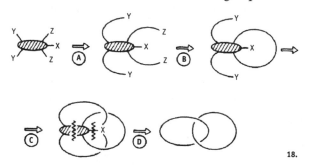

18.

Step A in **18** elaborates the Y and Z functional group handles into long chains. Step B is a closing up, or "cyclization," of one set of chains. Step C is a different kind of cyclization, another linking up, now of the other set of chains to the remaining core functionality X. Step D, perhaps several steps, is a fragmentation, in which the core from which the reaction was initiated, on which all was built, is now mercilessly torn apart, revealing the catenane.

Several points about this process are important: (i) the synthesis is architecture, a building, and (ii) as such it requires work. It's easy to write down the logical sequence of steps as I have. But each step may be several chemical reactions, and each reaction a more or less elaborate set of physical processes. These take time and money; (iii) the architectonic nature of the process almost dictates that the middle of the construction have molecules that are more complex than those at the beginning and the end.[11] Note (iv) the essential topological context of chemistry. It's evident not only in the curious topology of the goal, a catenane, but in the very process of linkage that pervades this magnificent way of building.

To return to the reality of the specific, the moment scheme **18** is laid out it is clear what molecule **15** is. It's the crucial point in the middle, after step B, before step C. It's poised to cyclize, the chlorines at the end of the $(CH_2)_{12}$ chains set to react with the NH_2 group. The synthesis by Schill and Lüttringhaus begins with molecule **19** and ends with catenane **20**, in which a ring of 28 carbons interlocks with another ring with 25 carbons and one nitrogen. But while getting to structure **20** is sweet, it should be clear that what is important is "getting to," the process. That process is reasonably linear (although **18** makes us think it's more linear than it is). One might suppose that any step in a linear chain of transformations (a→b→c→d→e) could claim primacy of significance. Indeed the steps in a synthesis may differ vastly in their difficulty, and therefore in the ingenuity invested to accomplish them. The unpublished lore of chemistry abounds with tales of a fantastically conceived, elegant synthesis in which the very last step, thought to be trivial, fails.

19.   20.

Nevertheless, I will advocate a special claim for a molecule somewhere in the middle of the scheme, the molecule most complicated relative to the starting materials and the goal; the molecule most disguised, yet the one bearing in it, obvious to its conceiver but to few others, the surprise, the essence of what is to come.[12] It's the col of complexity, and the only way from this beautiful molecule is on, on.

A Position on Utility

The last section, if it correctly describes a prevalent feature of beauty in the mind of the chemist (and I assure you it does), departs substantially from a Kantian perspective.[13] There is *Zweckmässigkeit* in abundance in molecule **15**, but it is powered by *Zweck*, the catenane.

Detachment has been central in analytic theories of aesthetics. Some frameworks have introduced a stronger quality, disinterest. To Kant an object that is of utility (for instance the catenane precursor, not to speak of an antibiotic or sulfuric acid, made in a mere 250 billion pounds worldwide this year), whose valuation is not sensually immediate but requires cognitive action, cannot qualify as being beautiful. As several commentators have pointed out, this is a rather impoverishing restriction on our aesthetic judgments.[14] It seems clear to me that knowledge—of origins, causes, relations and utility—enhances pleasure. Perhaps that cognitive enhancement is greater in scientific perusal, but I would claim that it applies as well to a poem by Ezra Pound. But let's go on.

As Rich as Need Be

Look at molecule **21**. It seems there's nothing beautiful in its involuted curves, no apparent order in its tight complexity. It looks like a clump of pasta congealed from primordial soup or a tapeworm quadrille. The molecule's shape and function are enigmatic (until we know what it is!). It is not beautifully simple.

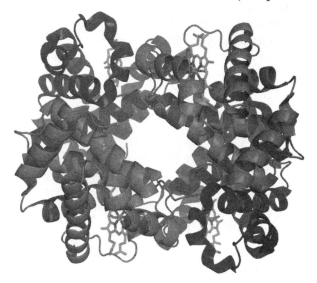

Complexity poses problems in any aesthetic, that of the visual arts and music as well as chemistry. There are times when the *Zeitgeist* seems to crave fussy detail—Victorian times, the rococo. Such periods alternate with ones in which the plain is valued. Deep down, the complex and the simple coexist and reinforce each other. Thus the classic purity of a Greek temple was set off by sculptural friezes, the pediments, and the statues inside. The clean lines and functional simplicity of Bauhaus or Scandinavian furniture owe much to the clever complexity of the materials and the way they are joined. Elliott Carter's musical compositions may seem difficult to listen to, but their separate instrumental parts follow a clear line.

In science, simplicity and complexity always coexist. The world of real phenomena is intricate, the underlying principles simpler, if not as simple as our naive minds imagine them to be. But perhaps chemistry, the central science, is different, for in it complexity is central. I call it simply richness, the realm of the possible.

Chemistry is the science of molecules and their transformations. It is the science not so much of the hundred elements, but of the infinite variety of molecules that may be built from them. You want it simple—a molecule shaped like a tetrahedron or the cubic lattice of rock salt? We've got it for you. You want it complex—intricate enough to run efficiently a body with its ten thousand concurrent chemical reactions? We've got that too. Do you want it done differently—a male hormone here, a female hormone there; the blue of cornflowers or the red of a poppy? No problem, a mere change of a CH_3 group or a proton, respectively, will tune it. A few million generations of evolutionary tinkering, a few months in a glass-glittery lab, and it's done! Chemists (and nature) make molecules in all their splendiferous functional complexity.

Beautiful molecule **21** is hemoglobin, the oxygen transport protein. Like many proteins, it is assembled from several fitted chunks, or subunits. The subunits come in two pairs, called α and β. Incredibly, these actually change chemically twice in the course of fetal development, so as to optimize oxygen uptake. The way the four subunits of hemoglobin mesh, their interface, is requisite for the protein's task, which is to take oxygen from the lungs to the cells.[15]

One of the hemoglobin subunits is shown in structure **22**. It's a curled up polypeptide chain carrying a "heme" molecule nestled within the curves of the chain. All proteins, not just hemoglobin, contain such polypeptide chains (see structure **23** for a schematic formula), which are assembled in turn by condensation of the building block amino acids, shown in structure **24**. These come in about 20 varieties, distinguished by their "side chains" (R in structures **23** and **24**). A typical protein, the hemoglobin β-chain is made up of 146 amino acid links. Here is richness, reaching out to us! Think how many 146-link chain molecules there could be given the freedom to choose the side chains in 20 possible ways. The incredible range of chemical structure and function that we see in those tiny molecular factories, enzymes, or in other proteins, derives from that variety. The side chains are not adornment, they make for function.

The protein folds, the diversity of the side chains provides opportunity, the particular amino acid sequence enforces a specific geometry and function. Extended pieces of hemoglobin curl in helical sections, clearly visible in **22**. At other places the chain kinks, not at random, but preferentially at one amino acid, called proline. The globular tumble of helical sections, nothing simple, but functionally significant, emerges.

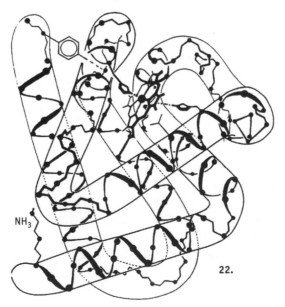

NH$_3$

22.

```
        R    O              R'   O             R"   O
        |    ||             |    ||            |    ||
  — N — C — C — N — C — C — N — C — C —
     H  H              H    H             H    H
```

23.

```
    H          R    O
     \         |    //
      N — C — C
     /         |    \
    H          H    OH
```

24.

Significant in what way? To *hold* the molecular piece that binds the oxygen, and to *change*, in a certain way, once the oxygen is bound. The O_2 winds its way into a just right pocket in the protein, and binds to the flat, disk-shaped heme molecule. Heme's structure is shown in **25**. The oxygen binds, end-on, to the iron at the center of the heme. As it does, the iron changes its position a little, the heme flexes, the surrounding protein moves. In a cascade of well-engineered molecular motions the oxygenation of one subunit is communicated to another, rendering that one more susceptible to taking up still another O_2. Here, in **26**, is an imagined scenario of the geometrical changes upon oxygenation.

25.

That bizarre sculpted folding has a purpose, in the structure and function of a molecule critical to life. All of a sudden we see it in its dazzling beauty. So much so that it cries out "I've been designed"; "For this task, I'm the best that can be." Or, if you're so inclined, it testifies to a Designer.

Beautiful? Certainly. The best, fashioned to a plan? Hardly. It only takes a moment to get us back to earth, a few bubbles of CO, the lethal, odorless product of incomplete combustion of fires and car exhausts. Carbon monoxide fits into the same wondrously designed protein pocket, and it binds to hemoglobin several hundred times better than oxygen.

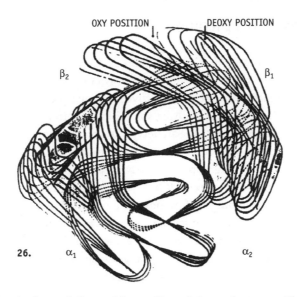

OXY POSITION   DEOXY POSITION

β_2 β_1

26. α_1 α_2

So much for the best of all possible worlds and the evolutionary Plan. As F. Jacob has written, "Nature is a tinkerer."[16] It has a wonderful mechanism for exploring chance variation, and, until we came along, much time on its hands. While it was banging hemoglobin into shape there wasn't much CO around. So it didn't "worry" about it. Actually the story, the story of molecular evolution, is more complicated, more wondrous still. It turns out that there is always a bit of CO around in the body, a natural product of cellular processes. Heme, free of its protein, binds CO much better than hemoglobin. So the protein around the heme apparently evolved to *discourage* CO bonding a little. Not enough to take care of massive doses of external CO, just enough to allow the protein to take up sufficient O_2 even in the presence of naturally produced CO.[17]

Iconicity

A word might be in place here about the preponderance of visual representations of molecules in this exposition. Could I be overemphasizing the *picture* of the molecule in analyzing the pleasure chemists take, at the expense of the *reality* of molecules and their transformations?

The relationship of the signifier to the signified is as complex in chemistry as in any human activity. The problem is discussed in substantial detail elsewhere.[18] The empirical evidence for the importance of the structural drawings that crowd this paper is to be found on any page of a modern chemical journal. Typically, 25% of the area of the page is taken up by such drawings (so this paper *looks* like a typical chemical paper). The structures that decorate chemist's articles are recognized by them as imperfect representations, as ideograms. But in the usual way that representations have of sneaking into our subconscious, these schematic diagrams merge with the real world and motivate the transformations that chemists effect in the laboratory.

Archetypes and Epitomes

Classical notions of beauty do have a hold on us. Central to Plato's and Aristotle's notions of reality was the ideal of a universal form or essence. Real objects are an approximation to that form. Art, in dismissive moments, was to the Greek philosophers mere imitation, or more positively, something akin to science, a search for the essential core. Concepts such as the archetype and the epitome figure in the Greek aesthetic. And they are to be found in chemistry today. The archetype is the ideal simple parent molecule of a group of derivatives, say methane, CH_4, and not any of the myriad substituted methanes CRR'R"R'''which make life interesting.

The epitome is something typical, possessing the features of a class to a high degree. It is this concentration of feeling which I want to focus on, for it is one of the determinants of beauty in chemistry. Molecule 27 (made by Clark and Schrock, structure determined by Churchill and Youngs) is such an emblem, but the background needs to be set for its compressed beauty to emerge.[19]

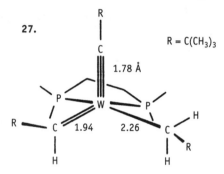

Molecules exist because there are bonds, the electronic glue that binds atoms together into molecular aggregates. In organic chemistry bonds come in several types single as in ethane (structure 28), double as in ethylene (structure 29), triple as in acetylene (structure 30). The plain English words tell the story: double is stronger than single, triple stronger still. The lengths of the bonds follow their strength, for chemical bonds act much like springs. Thus the atoms held together by a triple bond are more tightly bound, the bond between them shorter in length than a double bond; the latter one in turn is shorter than a single bond.

You can mix bonds, that is put several (classically up to four) bonds on a carbon, or two single ones and one double, or a single and a triple, or two double bonds (structures **31–34**). But because carbon has the capability of forming (in general, more on this later) only four bonds, you cannot have molecule **35**, a carbon atom with a single, double and triple bond to it.

31.   **32.**   **33.**

$$-\overset{\displaystyle |}{\underset{\displaystyle |}{C}}- \qquad \overset{\diagdown}{\underset{\diagup}{C}}= \qquad -C\equiv$$

$$=C= \qquad\qquad \overset{\diagup}{\underset{\diagdown}{\diagup\!\!\!\!\diagdown}}\!\!\!\equiv$$

34.   **35.**

Not so for metals. The "transition metals"—chromium, iron, manganese, cobalt, nickel, rhenium, tungsten, etc.—have the capacity to form up to 9 bonds. The chemistry of metal to other element bonds, especially the metal-carbon single bond, is nearly sixty years young, that of metal-carbon double and triple bonds younger still. This is the burgeoning realm of organometallic chemistry. The concentrated beauty of structure **27** lies in that it is a molecule in which one and the same tungsten atom forms a single WC bond, a double one and a triple one. And two bonds to phosphorus, for good measure. Incidentally **27**'s official name is [1,2bis(dimethylphosphino) ethane](neopentylidyne)(neopentylidene)(neopentyl)tungsten(VI)!

Such bonds are present, individually, in many molecules made in the last four decades. But in structure **27** they're all in one. The epitome, for that is what molecule **27** is, intensifies what it exemplifies by concentrating several disjoint examples into one. Its psychological impact is more than the sum of its parts; by such concentration it enhances our aesthetic response.

Novelty

Note the strong intrusion of the cognitive into judgments of the beauty of the molecules in the previous section. Their beauty is dependent, to use the Kantian term, they are very special members of a class. Still more dependent, in fact stretching outside of the limits of what is usually considered a viable aesthetic quality, is what characterizes the molecules of this section. It is *novelty*. I would claim that in the minds of chemists the new can leap the gap between "interesting" and "beautiful."

Science certainly subscribes in its very structure to the idea of innovation. It may be discovery—understanding how hemoglobin, discussed above, works. Or finding out how pre-Colombian Andean metalsmiths electroplated gold without electricity.[20] It may be creation—the synthesis of the catenane **16**, or the tungsten compound **27**.

We are addicted to new knowledge, and we value it. (Therefore it's interesting to reflect on the generally conservative tastes of scientists in art, or their astonishment that some other people don't view scientific or technological innovation as an absolute good.) At the same time most chemistry builds slowly. It is paradigmatic science, routine if not hack work, extending step by patient step what has been done before. Chemists appreciate this patient work, it allows them to read a new issue of a journal quickly. Yet it is inevitable that they grow just a bit bored by its steady drone, its familiar harmony.

Then, all of a sudden, from the plain of fumaroles, a geyser of fire reaches for the sky. It's impossible not to look at it, it is a hot intrusion on the landscape of the mind, as beautiful as it is new. A surprising, unexpected molecule.

Two examples come to mind, in counterpoint to the organometallic epitome. The long accepted inability of carbon to form more than four bonds is the fertile volcanic ground from which grow millions of natural and synthetic products, all the beauty of life and the democratizing utility of modern chemistry. But it is not holy ground, this four-coordination of carbon.

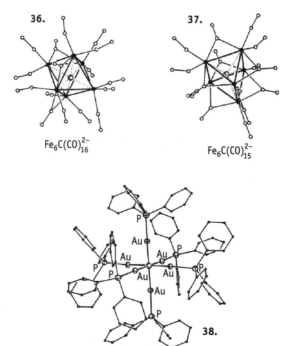

36.

$Fe_6C(CO)_{16}^{2-}$

37.

$Fe_6C(CO)_{15}^{2-}$

38.

Some time ago, inorganic chemists made molecules **36** and **37**, called metal carbonyl clusters.[21] In these the metal atoms form a lovely symmetrical polyhedron—an octahedron of irons in structure **36**, a trigonal prism of rhodiums in **37**. Around that polyhedron, seemingly loosely scattered on its periphery, are carbon monoxides, called carbonyls. And in the center, captive, encapsulated, resides a single carbon atom. It is connected to, equidistant from, all *six* metals around it. In the 1980s, H. Schmidbaur and his co-workers in Munich made a not unrelated, spectacularly simple molecule, **38**.[22] In it, central, is a lone carbon. Bonded radially to it are six ligands, $AuPR_3$ groups. There is a plus two charge on this molecule.

In these compounds (molecules **36–38**) carbon patently forms six bonds. This is the surprise, the shock, the full impact of which should (but hasn't yet) hit every maker of carbon compounds.[23] It makes these metal carbonyl clusters and the $C(AuPR_3)_6^{2+}$ molecule beautiful. They are new, interesting and lovely.

And they pose questions. How can carbon form six bonds? Are the old ideas wrong? Not entirely, for when we look at how the electrons move in these molecules we find that these are different kinds of bonds, perhaps weaker individually than normal carbon's four classical bonds. Theory expands to accommodate the new; the novel in time will become routine, only to be shaken by the unforeseen violator of the new set of rules.

Empirical Chemical Aesthetics and Formal Theories

I have hardly exhausted the capacity of molecular creation to please the human mind. Molecules can be beautiful because of the wondrous quantized motions they undergo, truly a music played out in tones, harmonics and overtones that our instruments, now measuring instruments, hear. They may be beautified by their miracles—who would deny it to penicillin or morphine? Or more lowly, they may be as beautiful as the ten billion pounds of phosphoric acid, H_3PO_4, manufactured every year. You're more likely to have heard of the rougher guys, the spectacular hydrochloric, nitric and sulfuric siblings. But this quiet one is responsible for a good part of the essential phosphorus in your DNA.

Much philosophic tradition would object to including the utilitarian perspective as an aesthetic criterion. I've mentioned above some objection to this limiting view. Here I would only add the empirical testimony of the practitioners—be they deluded, naive or not—chemists really do think that *use* is an element, not the most important one, but an element of chemical beauty.

But perhaps it's time to stop here and take another tack. Let's posit that we've discovered in this anthropological study of chemistry a reliable sampling of the qualities the experts/natives use as attributes of beauty. They, the chemists, we, for I am among them, have an aesthetic. Maybe we don't call any molecule ugly, but some molecules are more beautiful than others. Does our aesthetic, our way of assigning beauty, have something in common with the aesthetics characteristic of other parts of human experience, those of games, of business, of love, but especially of art?

I'm not going to answer this question here to anyone's satisfaction, but it's worth asking, isn't it? A fundamental problem, of course, is that aesthetics is not a closed chapter of philosophy. Rival theories abound, indeed the dialogue shifts with time, much as the subject of its discussion. Nevertheless, one could proceed by seeing how the concept of beauty in chemistry fits or doesn't fit into the existing (fashionable?) aesthetic frameworks erected by philosophers. This has been done to some extent in the previous sections, but let me approach the problem more directly here.

For instance Monroe Beardsley supposes that the aesthetic response (to a work of art, which is an artifact intended to elicit such a response) entails importantly a degree of detachment on the part of the viewer (listener, etc.), and the elements of intensity, unity and complexity in the object viewed.[24] His argument deserves deeper exposition than these fuzzy words, but it seems to me that the chemist's aesthetic response entails many of Beardsley's factors. As for detachment, a concentration that envelops, well—the only people I've seen more detached than chemists looking at molecules are computer hackers or Pachinko players. Intensity has been discussed in the previous section, in the context of molecule 27. It said a lot, economically. Unity is by and large absent in the exemplars selected by me. They stand alone. But, implicitly, these structures cannot be viewed as beautiful except in the context of knowledge of other molecules. And if they are totally new, they impose a stress on existing theories to assimilate their brash flaunting of not fitting in. New molecules incite theory, which is the unifying, framework-building way the chemist makes connections.

I hesitate on complexity, not because it is unimportant (remember hemoglobin and all your enzymes) but because I see so clearly the aesthetic strength of simplicity. The parent molecule, the symmetrical molecule, the reaction that goes under wide conditions, the simple mechanism, the underlying theory expressed by a single mathematical equation—these have beauty-conferring value.

However, there is a thread that runs through the tokens of chemical beauty that inclines me to another aesthetic philosophy, which is that of Nelson Goodman.[25] Goodman views science and art both as cognitive processes, differing perhaps only in their intensity or degree of elaboration or manipulation of symbols. And one is certainly struck by the cognitive element in all these appreciations of the chemist, in our reactions to molecules. *We feel* that these molecules are beautiful, that they express essences. We feel it emotionally, let no one doubt that. But the main predisposition that allows the emotion, here psychological satisfaction, to act, is one of knowing, of seeing relationships. I took apart $NaNb_3O_6$ into chains of octahedral and layers, and related it so to other materials. I saw the catenane synthesis planned, and so grew to love the molecule at its high pass. I *know* what hemoglobin does, therefore I care about it. And the molecules in the preceding section are clearly fascinating because they standout, or soar.

Perhaps we should not press too hard to fit the multifarious manifestations of chemical beauty into tight categories or theoretical frameworks. Even if we were to agree on a definition of beauty, what would it gain us? As M. H. Abrams has pointed out, saying that X is beautiful is almost the dullest thing one can say about X.[26] One needs to describe the object's attractiveness. I hope that this essay has done so, partway, for molecules.[27]

These products of our hands and minds, beautiful molecules, appeal directly to the mind. For a chemist, their line into the soul is direct, empowering, sometimes searing. They are natural, hemoglobin like a fern unfurling, like the cry of a duck on a winter lake. They are synthetic (or if you like artifactual, man or woman-made, unnatural)—the catenane, Schrock's tungsten epitome, like the Shaker tune " 'tis a gift to be simple," like Ogata Korin's screens.

Notes

1. For a general introduction to molecules see the beautiful book by P. W. Atkins, *Molecules* (New York: Scientific American Library, 1987).
2. L. A. Paquette, R. J. Temansky, D. W. Balogh and G. Kentgen, Total Synthesis of Dodecahedrane. *J. Am. Chem. Soc.* 105 (1983): 5446. Dodecahedrane was recently synthesized in a very different way: W.-D. Fessner, B. A. R. C. Murty, J. Wörth, D. Hunkler, H. Fritz, H. Prinzbach, W. D. Roth, P. von Ragué Schleyer, A. B. McEwen and W. F. Meier, Dodecahedrance aus [1.1.1.1] Pagodanen. *Angew. Chem.* 99 (1987): 484.
3. M. Doyle, W. Parker, P. A. Gunn, J. Martin and D. D. MacNicol, Synthesis and Conformational Mobility of Bi-cyclo (3,3,3)undecane(Manxane). *Tetrahedron Letters* 42 (1970): 3619.
4. Y. Sekine, M. Brown and V. Boekelheide, [2.2.2.2.2.2] (1,2,3,4,5,6) Cyclophane: Superphane. *J. Am. Chem. Soc.* 101 (1979): 3126. For an immensely enjoyable tour of the coined landscape of chemistry see A. Nickon and E. F. Silversmith, *Organic Chemistry: The Name Game* (New York: Pergamon Press, 1987). Structures **1-3** are drawn after this source.
5. J. Köhler and A. Simon, $NaNb_5O_5F$—eine Niob-Niob-Dreifachbindung mit 'side-on'-Koordination durch Nb-Atome. *Angew. Chem.* 98 (1986): 1011.
6. S. J. Hibble, A. K. Cheetham and D. F. Cox, $Ca_{0.75}Nb_3O_6$: A Novel Metal Oxide Containing Niobium-Niobium Bonds. Characterization and Structure Refinement from Synchrotron Powder X-Ray Data. *Inorganic Chemistry* 26 (1987): 2389.
7. M. J. Calhorda and R. Hoffmann, Dimensionality and Metal-Metal and Metal-Oxygen Bonding in the $NaNb_3O_6$ Structure. *J. Am. Chem. Soc.* 110 (1988): 8376.
8. I. Kant, *Critique of Judgment*, trans. by J. H. Bernard (New York: Hafner, 1951), p. 28.
9. E. Wasserman, The Preparation of Interlocking Rings: A Catenane. *J. Am. Chem. Soc.* 82 (1960): 4433. For an elegant Möbius-strip approach to catenanes see R. Wolovsky, Interlocked Ring Systems Obtained by the Metathesis Reaction of Cyclo-dodecene. Mass Spectral Evidence. *J. Am. Chem. Soc.* 92 (1970): 2132; D. A. Ben-Efraim, C. Batich, and E. Wasserman, Mass-Spectral Evidence for Catenanes Formed via a 'Mobius-Strip' Approach. *J. Am. Chem. Soc.* 92 (1970): 2133. See also the ingenious work of J.-P. Sauvage and J. Weiss, Synthesis of Dicopper (I) {3} Catenanes: Multiring Interlocked Coordinating Systems. *J. Am. Chem. Soc.* 107 (1985): 6108.
10. G. Schill and A. Luttringhaus, Gezielte Synthese von Catena-Verbindungen. *Angew. Chem.* 76 (1964): 567.
11. For some leading references on complexity in chemical synthesis, chemical similarity and chemical "distance," see S. H. Bertz, Convergence, Molecular Complexity, and Synthetic Analysis. *J. Am. Chem. Soc.* 104 (1982): 5801; M. Wochner, J. Brandt, A. von Scholley and I. Ugi, Chemical Similarity, Chemical Distance, and Its Exact Determination. *Chimia* 42 (1988): 217; J. B. Hendrickson, Systematic Synthesis Design. 6. Yield Analysts and Convergency. *J. Am. Chem. Soc.* 99 (1977): 5439.
12. For an inspiring, original presentation of the achievements of organic synthesis see N. Anand, J. S. Bindra and S. Ranganathan, *Art in Organic Synthesis,* 2nd ed. (New York: John Wiley and Sons, 1988).
13. For leading references see P. Guyer, *Kant and the Claims of Taste* (Cambridge, MA: Harvard University Press, 1979) and D. W. Crawford, *Kant's Aesthetic Theory* (Madison, WI: University of Wisconsin Press, 1974).

14. See Crawford, *Kant's Aesthetic Theory* (Madison, WI: University of Wisconsin Press, 1974), pp. 49–51 and N. Carroll, Beauty and the Genealogy of Art Theory. *Philosophical Forum* 22 (1991): 4.
15. For more information on hemoglobin see R. E. Dickerson and I. Geis, *Hemoglobin* (Menlo Park, CA: Benjamin/Cummings, 1983) and L. Stryer, *Biochemistry,* 3rd ed. (San Francisco: W.H. Freeman, 1988). I owe much to these books and to the original work of Max Perutz.
16. F. Jacob, Evolution and Tinkering. *Science* 196 (1977): 1161.
17. R. E. Dickerson and I. Geis, *Hemoglobin* (Menlo Park, CA: Benjamin/Cummings, 1983) and L. Stryer, *Biochemistry,* 3rd ed. (San Francisco: W.H. Freeman, 1988).
18. R. Hoffmann and P. Laszlo, Representation in Chemistry. *Diogène* 147 (1989): 24. Chapter 14 in this book.
19. D. N. Clark and R. R. Schrock, Multiple Metal-Carbon Bonds, 12. Tungsten and Molybdenum Neopentylidyne and Some Tungsten Neopentylidene Complexes. *J. Am. Chem. Soc.* 100 (1978): 6774; M. R. Churchill and W. J. Youngs, Crystal Structure and Molecular Geometry of $W(\equiv CCMe_3)$ $(=CHCMe_3)(CH_2CMe_3)(dmpe)$, a Mononuclear Tungsten (VI) Complex with Metal-Alkylidyne, Metal-Alkylidene, and Metal-Alkyl Linkages. *Inorganic Chemistry* 18 (1979): 2454.
20. H. Lechtman, Pre-Columbian Surface Metallurgy. *Scientific American* 250 (1984): 56.
21. B. F. G. Johnson and J. Lewis, Transition-Metal Molecular Clusters. *Advances in Inorganic Chemistry and Radiochemistry* 24 (1981): 225; J. S. Bradley, The Chemistry of Carbidocarbonyl Clusters. *Advances in Organometallic Chemistry* 22 (1983): 1.
22. F. Scherbaum, A. Grohmann, B. Huber, C. Krüger and H. Schmidbaur, "Aurophilie" als Konsequenz Relativistischer Effekte: Das Hexakis(triphenylphosphanaurio)methan Dikation $[(Ph_3PAu)_6C]^{2+}$. *Angew. Chem.* 100 (1988): 1602.
23. Some of these molecules were foreseen theoretically. See, for instance, P. von Ragué Schleyer, E.-U. Wurthwein, E. Kaufmann, T. Clark, and J. A. Pople, CLi_5, CLi_6, and the Related Effectively Hypervalent First-Row Molecules, $CLi_{5-n}H_n$ and $CLi_{6-n}H_n$. *J. Am. Chem. Soc.* 105 (1983): 5930; D. M. P. Mingos, Molecular Orbital Calculations on Cluster Compounds of Gold. *Journal of the Chemical Society (Dalton Transactions)* (1976): 1163.
24. M. C. Beardsley, *Aesthetics,* 2nd ed. (Indianapolis, IN: Hackett, 1981); M. C. Beardsley, *The Aesthetic Point of View,* ed. M. J. Wreen and D. M. Callen (Ithaca, NY: Cornell University Press, 1982).
25. N. Goodman, *Languages of Art,* 2nd ed. (Indianapolis, IN: Hackett, 1976).
26. M. H. Abrams, personal communication. See also G. Sircello, *A New Theory of Beauty* (Princeton: Princeton University Press, 1975), p. 118.
27. I'm grateful to Maria José Calhorda, Paul Houston, Barry Carpenter, Bruce Ganem, Noël Carroll, two reviewers, and Donald W. Crawford for their comments and help in the preparation of this work, and to Sandra Ackerman for her editorial assistance. This article derives from four "Marginalia" that I wrote for *American Scientist* 76 (1988): 389, 604; *American Scientist* 77 (1989): 177, 330. The ink drawings were beautifully done, as always, by Elisabeth Fields and Jane Jorgensen. Max Perutz gave permission to reproduce one drawing of hemoglobin. Another is taken from the spectacular, illuminating graphics of hemoglobin designed by Irving Geis.

Part 4

CHEMICAL EDUCATION

22

Teach to Search

ROALD HOFFMANN

George Pimentel was a wonderful man, whose heart and soul were in chemistry.[1] And just as much in research, in which he excelled, as in teaching. From his writing it is clear that he did not separate the two. Nor do I, which is why I am happy and proud to be associated with an award given in George Pimentel's name, and especially one in chemical education.

I will speak of two themes:

- The inseparability of teaching and research. And for that matter, of chemistry and the world.
- The necessity of chemists to teach broadly, to speak to the general public. And the tensions that arise in the process.

But before I launch into these subjects, let me say some words about how I feel about teaching and receiving this award. Whatever success I have had I owe to teaching. The logic or rhetoric of teaching underlies my research within chemistry and my writing outside of chemistry. As I began to think about this, I felt suddenly a little less guilty about receiving an award in chemical education.

Let me tell you why I felt—feel—guilty. What am I—viewed by the community of chemistry as a researcher whose work has received ample recognition—doing getting an award that should be given to those who have toiled so hard, dedicating their lives to chemical education? When there aren't too many of these awards around...

A second source of guilt for me is that I suspect that a significant component in the thinking of the Pimentel award committee was my role in making the Annenberg/CPB television course in chemistry, *The World of Chemistry*. I *was* a member of the team, indeed, and my soul and sweat went into the project. But the part I played—more than just being a pretty face, true—was in fact much smaller than the parts of several other people, who really deserve recognition. I will tell you about those people in time. I do feel guilty about receiving this award, but my guilt is assuaged just a little by pride in the fact that I have not only taught thousands, but

I have also taught others to teach. I have taught, subtly, the research community in chemistry that teaching strategies are productive in research. And I have contributed, I think, to the growing respect for teaching in the community of chemistry at large.

1

First let me address the issue of teaching and research. A damaging misconception about modern universities is that research dominates and diminishes teaching, and that the tension of balancing (unsymmetrically) the twain is unhealthy. Defenders of the universities argue that the two functions are complementary and that research or scholarship enhances the quality of teaching. I go further: I say research and teaching are, quite literally, inseparable. And they are symbiotic.

One root of the error, I believe, lies in thinking of learning in terms of place rather than audience. Places (classrooms, labs, library carrels) are, indeed, circumscribed, but the audiences of learning (undergraduates, graduate students, faculty, our minds) always shift, overlap, and enrich each other, like the colored glass bits of a kaleidoscope.

As I reflect on the possibility of a separation of research and teaching, I look at my research group. We meet twice a week—four graduate students, four postdoctoral associates, and I. One time we talk about the incredible, fertile literature of chemistry, while in the other session one of the people in the group reports on work in progress. We also ask why marzipan pigs are popular in Denmark, explain all those football and baseball metaphors in colloquial English to our foreign group members, and try to guess who is likely to be the author of those scurrilous referee's comments on our last paper. In these group meetings half the time I'm giving a monologue; the rest of the time the hardly shy remainder of this research family speaks. Is that research, is that teaching?

I travel to the University of British Columbia to lecture about my work—about making and breaking bonds in the solid state. Ninety percent of the audience consists of graduate students, with a sprinkling of undergraduates. I talk to *them*. Is that research, is that teaching? I think the answer in both cases is yes. It's research and it's teaching.

Teaching and research are inseparable. The struggle to do both well enriches our personal intellectual lives, and enhances our contributions to society.

I am certain that I have become a better researcher and a better theoretical chemist because I've had to teach undergraduates. When I began at Cornell, I thought I knew all about thermodynamics, all those beautiful partial differential equations that relate the derivative of A with respect to B to C. But thermodynamics is a subject of great richness, with practical common-sense roots (steam engines, the boring of cannon) and a mathematical structure of breathtaking sophistication. I had followed only the latter and hadn't really understood the full empirical beauty of "thermo" until…I had to explain the subject to students *without* the crutch of the mathematical apparatus. The more I taught beginning classes, the more important it became to me to explain. The rhetoric of pedagogy permeated my research. I think those in the community of chemistry who know my work will recognize what I mean.

I think there is nothing unique to me in all this. I believe that rather than treating research and teaching as disparate activities, it is more productive to cast the discussion in terms *of audiences* for creative work in science or the humanities.

In the beginning is research or discovery, a gleam of the truth, or of a connection, within an individual's mind. Actually I've experienced such moments, and so have others, most often not in isolation but in discourse with another person. Or when I sit down to write a paper, before me the draft or progress report by one of my students.

In fact, understanding already formed in the inner dialogue between parts of me, me and an imagined ideal audience of one, or of a multitude, in the lonely dialogue with the voices of skepticism and self-doubt that are all me, all of me.

> *Deep in*
> *it's a docile crowd*
> *most of the time, lazing*
> *around, waiting for the train*
> *of concentration to haul a few words*
> *onto paper. It listens, then it stirs, the one*
> *that speaks in many voices, to say:*
> *these are just words, falling limp*
> *into the untensed space they need sculpt, or:*
> *make me understand.*
> *They hate my compromises.*
> *Here and there they offer up a phrase.*
> *In their babble I hear the voices*
> *of my teachers rise from a page or cafe. Sometimes*
> *one speaks with an accent—I think*
> *it's my father, it's him, the world*
> *I have to please.*
> *For them I leave no word unturned.*
> *For it I sing, tone-deaf that I am,*
> *the song that frees itself within.*

In the next stage the audience expands to my research group. In the process of talking to them my understanding of the discovery deepens, takes a stronger hold on reality. Then I write a technical paper. Now my audience is out of my control. Writing is the message that abandons, as Jacques Derrida has called it. I can't grab that removed reader in Poznań or Puna and tell him no, you must read it this way and not that way. It has to be all there, in the words with which I struggle. It has to be there—the substance of what I found and the argument to convince him or her, the absent reader. I write for that audience from a position of substantive ignorance about them. I don't know their preparation, their level of sophistication, their willingness to work to reach enlightenment! It begins to sound an awful lot like teaching.

To me, the writing of a research paper is in no way an activity divorced from the process of discovery itself. I have inklings of ideas, half-baked stories, a hint that an observation is relevant. But almost never do I get to a satisfactory explanation until I have to, which is when I write a paper. Then things come together, or maybe I make them come together. F. L. Holmes has argued convincingly the same point, that scientific writing and scientific discovery are not disparate activities. In an analysis of draft manuscripts of a Lavoisier memoir on respiration he "…could watch important ideas emerging, growing, changing form or decaying during the evolution of a scientific paper."

An invited technical seminar introduces another audience. Sure, I want to impress my colleagues; claim precedence, power; please real or surrogate parents. Many things go on subliminally in the course of any talk. Yet most of all I want to impart real, significant, new knowledge. But the audience includes people of disparate backgrounds. The organic chemists may not know much about my present loves, which are surface and solid state chemistry. Depending on their background, different parts of the audience may attach different meanings to the plain English words at my disposal. There are many graduate students here. I want to teach all, convince all. Remarkably, incredibly, we can do it—speak to many audiences at the same time. That's what teaching is all about.

To me, the steps from a research seminar to teaching a graduate course, then an undergraduate one, are small moves in interacting with the continuous, overlapping spectrum of audiences. In the theater of the mind the audience is always shifting, never constant. There are different strategies (call them tricks, the stuff of experience), that one applies with audiences of young people that one might not try in a research group meeting. But the similarities of pedagogical strategy across the spectrum of teaching/research far exceed the differences.

The spiritual rewards for opening a person's mind, sharing new-found knowledge, are also quite similar. I've taught introductory chemistry many times, to thousands of students. There is the same unmitigated pleasure that hits me when I detect, on an examination or by the nonverbal signs students give in lectures, that someone has understood the magnificent and simple logic of the mole, so that he or she can tell me how much sulfur there is in a pound of sulfuric acid.

To return to my main point, I wish to argue that the desire to teach others, enhanced by being *obliged* to teach others, leads to greater creativity in research.

The rhetorical imperative operates to make a scientist or scholar examine widely the potential responses (objections?) of his or her audience. Having to teach enlarges one's encounters with real audiences and therefore sharpens the imagined audience one engages in the inner dialogue in the course of research.

I do not mean to imply that you need to be a researcher to teach well, nor that you absolutely must teach to do research well. A reviewer of this paper appropriately reminded me that the "vast majority of high school and college teachers who contribute mightily to innovation in our field... are not researchers." I recall the tremendous success of the graduates of City College in New York (which 60 years ago had little research activity) and the many small colleges that are the baccalaureate source of our best researchers. And there are many talented researchers working in industry and government who have little occasion to teach. I respect the multiplicity of professional styles.

As my friend R. Freis has pointed out, following St. Thomas Aquinas, teaching is truly a cooperative art. It *works together* with the nature of the student as learner, knower, apprentice, in order to bring that nature to its perfection. Teaching is clearly also a rhetorical act. But it is more than mere persuasion because of its empathetic, reflexive aspect being cooperative. How could a mind that faces up to the problem of teaching a novice something new and difficult possibly avoid using the same strategies in explaining to itself something still more new, more difficult? Which is what people call research.

2

I want to try to illustrate to you what I mean by the rhetoric of teaching influencing my research style. To do that I've picked a paper entitled "A 2,3-Connected Tellurium Net and the Cs_3Te_{22} Phase," written by Qiang Liu, Norman Goldberg, and myself (*Chemistry, a European Journal* 2 [1996]: 390). Our work grew out of a paper we saw by Sheldrick and Wachhold in a February 1995 issue of *Angewandte Chemie*,[2] who reported a new Cs_3Te_{22} compound. Now the chemistry of tellurium is very rich. For instance, in the Cs-Te system some nine binary cesium telluride phases ($CsTe_4$, $CsTe_5$, Cs_2Te, Cs_2Te_2, Cs_2Te_3, Cs_2Te_5, Cs_3Te_2, Cs_5Te_3 and Cs_5Te_4) had been reported earlier, and two more have been made since.

The beautiful structure of the Sheldrick and Wachhold compound (Figure 22-1) displays a number of unusual features. Discrete crown Te_8 rings can be easily identified in Figure 22-1. Though such eight-membered crown-shaped molecules are well known for sulfur and selenium, they had not been previously observed for tellurium. Also apparent are infinite two-dimensional sheets that are formed by Te atoms and that include one Cs atom per six telluriums. Each Cs atom in the $CsTe_6$ sheet is located in the center of a large square of 12 Te atoms. The structure may also be described as consisting of two different types of layers: $CsTe_6$ sheets separated by layers of $CsTe_8$ crowns: $[CsTe_8]_2[CsTe_6]$. If one assumes the Te_8 rings to be neutral

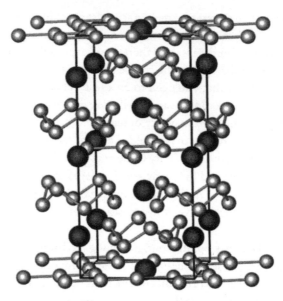

Figure 22-1 The structure of Cs_3Te_{22}.

molecular entities and assigns the valence electrons of cesium fully to the only atoms left, the tellurium sheets, the compound may be described as $[Cs^+]_3[Te_8]_2[Te_6^{3-}]$. The Te_6^{3-} net is definitely electron-rich.

The pattern of the $CsTe_6$ sheet is remarkable (Figure 22-2, looking down the *c*-axis onto the sheet; the darker and larger spheres are Cs, the light ones Te). This is a rare net; the C_4 axis is the principal symmetry element present (aside from two-fold rotation axes and the mirror plane containing the sheet itself).

So far, you see an intriguing structure. That was apparent to the authors and readers of the initial report; they saw the same beautiful structure that I show you. Next we, as theoreticians, did the stuff of our trade, a calculation of the electronic structure of the three-dimensional material. The outcome is shown in Figure 22-3. It is a so-called band structure, showing the energy levels of the molecule.

Now if the pedagogical imperative were not important for me and my group, I think I would have (in an alternative universe, I can't imagine doing so here) stopped pretty much with an analysis of the bonding, perhaps worried about stability, and reached the conclusion that the material might be a conductor. But in my real world of trying to *understand* this big molecule, of trying to see its connection to everything else in the molecular world, that band structure is just the beginning. I look at that incredible net with fourfold symmetry, and I see in it two kinds of Te atoms. One is linear, bonded to two other Te atoms. Call this Te2. The other, which we call Te3, is T-shaped, bonded to three Te atoms. It is important to note here that the Te2 and Te3 notation does not refer to a crystallographic numbering; it is our way of reminding ourselves of the coordination environment of each Te.

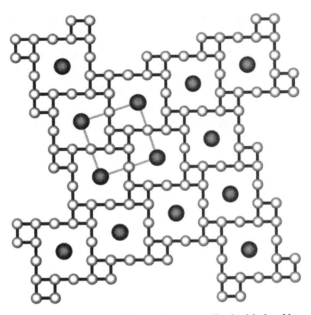

Figure 22-2 A top view of the $CsTe_6^{2-}$ net. One unit cell is highlighted by a square.

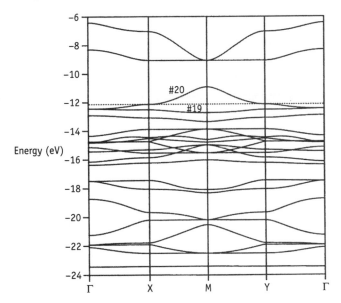

Figure 22-3 Band structure for the Te_6^{3-} sheet, along various directions in the Brillouin zone, showing the energy levels of the molecule.

I think about these; where else have I seen two- and three-coordinate tellurium or its analogues? Where else have I seen tellurium squares?

Well, for two-coordinated main-group EX_2 molecules, both bent (H_2O, H_2Se, H_2Te, and Te_3^{2-}) and linear configurations (XeF_2 and I_3^-) are possible. Why is Te2 linear in this sheet?

The T shape of Te3 reminds one of the BrF_3 molecule, and it does occur in a number of other extended tellurium structures.

As far as squares go, there aren't that many main group element squares around. E_4^{2+} species (E = S, Se, Te) are known, as is Bi_4^{2-}, and they are isoelectronic with electronically happy, $C_4H_4^{2-}$. To my knowledge there are no square hypervalent molecular groupings with halogens, noble gases, or metals.

As we wrote this paper, I felt it essential to construct our understanding of the extended structure through molecular models and bonding schemes drawn from model molecules. Which is what we did. We began by looking at a simplified model for Te2 by calculating a Walsh diagram (i.e., how the energy levels of this triatom varied with bending at tellurium) for H_2Te^{n-}. We found (not surprisingly) that the preferred geometrical configuration of H_2Te^{n-} depends strongly on its electron count. The molecule prefers a bent geometry when it is neutral, as expected. And the triatomic H_2Te^{2-} is linear, analogous to a hypervalent H_2Xe or F_2Xe. We also looked at a more realistic model for the atomic environment of Te2 in the solid, Te_3^{n-}.

Next I will actually quote a piece of our paper (omitting references, of which there were many), not because it is important, but because it helps me make two points:

> A connection needs to be made here to the classical and well-characterized linear triiodide I_3^-. This species is, of course, isoelectronic to Te_3^+, as is the related XeF_2. The bonding in I_3^- or XeF_2 is very well understood—we have in these molecules an electron-rich three-center bond. If one omits the s orbital on the central atom from the bonding, one expects the level pattern at left in [Figure 22-4], while if the s orbital is included we get the pattern at right. Note in either case that one and only one I-I-I antibonding orbital remains unfilled...[3]

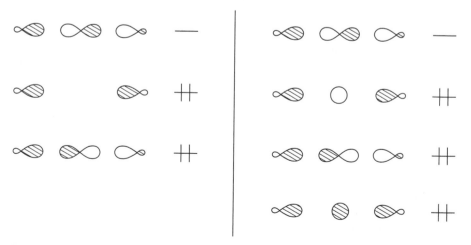

Figure 22-4 The orbitals involved in the three-center bonding in Te_3^+, I_3^-, and XeF_2.

Why do I quote this? First, the subject just happens to be related to work that George Pimentel did. He and Rundle first explained qualitatively the bonding in triiodide anion and related electron-rich compounds. Second, I see this section as an example of the teaching imperative influencing research. In this section and in the paper as a whole, I am intent on drawing the connection to electron-rich three-center bonding. And I will *not* assume that everyone has seen it. So I repeat an orbital level scheme—that's part of my teaching-in-research strategy—even if that level scheme has been in the literature before. I repeat it because that orbital scheme is part of the story; the story is incomplete without it; I am anxious to get into the mind of the poor graduate student assigned by the professor to talk about this paper at the next group meeting; I'm interested in teaching that graduate student, and— you know, it actually helps *me* understand this bonding if I explain it in detail, as if I were teaching...

You can be sure that such pedagogically driven paper-writing strategies are as disliked by some reviewers and some journal editors as they are appreciated by the young researchers who read these papers. Sometimes it has not been easy to get such teaching-research narratives published. But I persevere and sneak them in. As here.

Let me show you another piece of this paper which makes some reviewers see red:

> The T shape reminds one of the BrF_3 molecule, whose bonding is described qualitatively in Figure 22-5. Note the formal F⁻ nature of the "axial" fluorines. We see two lone pairs on the Br, a "normal" equatorial Br-F bond, and electron-rich three-center F-Br-F axial bonding. BrF_3 is clearly related to SF_4 and XeF_2. A tellurium analogue (Figure 22-5) would be Te_4^+.

> Let's look at the structure at hand in still another way. Each Te2, linear, is hypervalent and (if it were maximally hypervalent) could be assigned an electronic structure such as that shown in Figure 6, and a formal charge of -2. Each Te3 can be assigned a locally hypervalent structure (Figure 22-6) and a -1 formal charge. With these charges throughout the net we would have a charge per formula unit, $(Te3)_4(Te2)_2$, of -8. However, the actual charge

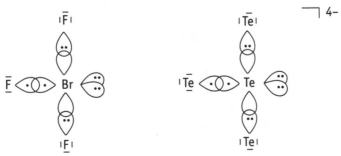

Figure 22-5 The hypervalent BrF_3 and Te_4^+.

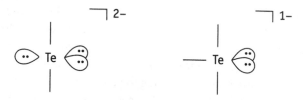

Figure 22-6 Maximal hypervalence in the Te$_6$ net would lead to a 2− charge on the linear telluriums (left) and a 1− charge on the T-shaped tellurium (right).

is only -3! In other words, our Te$_6^{3-}$ net is hypervalent (as the T-shaped Te3 and linear Te2 indicate), but it is not "maximally hypervalent" in the sense of as many electrons as these hypervalent geometries would allow. It is this intermediate reduction stage that makes the electronic structure of Te$_6^{3-}$ truly nonclassical and requires a delocalized bonding description.[3]

What annoys some reviewers about this section is that most simple of chemical intuitions, electron counting, here done in public.

There is much more in this paper: discussions of an alternative structure (see Figure 22-7) and possible fragmentations of the net, and the suggestion (on the basis of the computed electronic structure) of two unknown compounds, [CsTe$_6$]$^-$ [CsTe$_8$]$^+$ or CsTe$_7$ and [CsTe$_6$]$^{3-}$[CsTe$_8$]$_3^{3+}$ or Cs$_2$Te$_{15}$. Both should have structures similar to that of Cs$_3$Te$_{22}$, but composed of layers of CsTe$_6$ sheets and CsTe$_8$ units in 1:1 and 1:3 ratios, respectively. The CsTe$_7$ phase should be metallic. Happily, one of these compounds has been made, and it almost (not quite) has the simple structure we predicted.

Now I must rein in my enthusiasm for this wonderful molecule and return to the subject of this lecture. I do hope that by this example I have illustrated what I mean when I say that the rhetoric of teaching has influenced my research. And I hope to have enticed you to look at tellurium's weird and fascinating chemistry.

3

Now I turn to what I see as the reasons for talking to the public about science, and the difficulties of doing so. I do not mean to exclude personal and structural reasons. (By these I mean the sheer fun of the study of matter and its transformations, the fact that most of us are employed to teach chemistry—with attendant obligations, and that we do need to train professional chemists.) These things we understand well. They may be fun or a chore, but they are intrinsic to our profession. I want to address in a more general, reflective way, the reasons why we must teach in the broadest possible way.

First, there is public and political concern about money spent on science. The public ultimately supports our research through tax dollars. The informed citizen

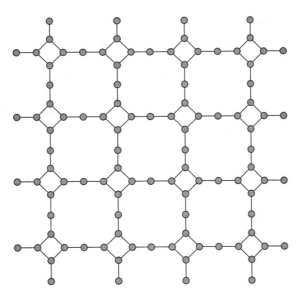

Figure 22-7 A possible distortion of the tellurium net to a hypothetical D_{4h} Te_6^{3-} sheet.

will let the talented expert carry out his or her basic research. He will accept even that technological benefit is not guaranteed. Such a citizen will accept even a measure of vagueness about what is done and will take the excitement of the scientist as a sign of creative activity. For a while. But at some point we have to tell people (not the least among them being our parents and spouses) what it is that lures us back to work nights and Sundays; why it's thrilling to open a new issue of the *Journal of the American Chemical Society,*

Second, chemophobia is rampant. There is a negative image of chemistry and chemists as the producers of the unnatural, the toxic. But at the same time, every survey shows that the public views scientists with high regard and trusts them. This is not irrationality, just nice human inconsistency. People are not machines. They can both value and be afraid of something at the same time. But certainly the more people know (of substances, of molecules) the less likely they are to be afraid of what we create.

Third, if we do not know the basic workings of the world around us, especially those components that human beings themselves have added to the world, then we become alienated. Alienation, due to lack of knowledge, is impoverishing. It makes us feel impotent. Not understanding the world, we may invent mysteries, new gods, much as people once did around lightning and eclipses, St. Elmo's fire, and volcanic sulfur emissions. I feel the growth of antirational movements (although to me religion is most certainly not one of these) in the form of cults and of interest in the occult and astrology. These are modern-day reactions to the mysteries with which science has surrounded human beings.

My fourth and last point of concern about chemical illiteracy and teaching chemistry concerns democracy. Ignorance of chemistry poses a barrier to the democratic process. I believe deeply that "ordinary people" must be empowered to make decisions—for example, on genetic engineering, waste disposal sites, dangerous and safe factories, and which addictive drugs should or should not be controlled. They can call on experts to explain the advantages and disadvantages, options, benefits, and risks. But experts do not have the mandate; the people and their representatives do. The people have also a responsibility. They need to learn enough chemistry to be able to resist the seductive words of, yes, chemical experts who can be assembled to support any nefarious position you please.

There are diverse audiences for what we want to say:
The children and young people in our schools,

- who will be citizens and not scientists (99%),
- who might be scientists; who may be motivated to become such by what we teach,
- the small unknown subset who will decide the future of the country and of science.

Not children, but

- the proverbial general public, those who watch soap operas religiously and whose labors move the country on,
- the politicians and authorities: and I don't necessarily have a bad view of them; I respect the compromises they must make and that I can avoid,
- the merchants and business men and women who power our economy,
- our friends in the arts and humanities and religion, the shapers of the spirit, and
- other scientists.

It seems that each audience needs a different approach. It seems impossible to speak to all. But please don't despair. Every teacher has the experience that, as difficult as it seems, it *is* possible to speak to several audiences at the same time.

What are the problems we face in teaching science? What are the attitudes people bring to our subject? Let me describe some:

- Science is complicated, too difficult for a "normal" person to understand. Science is for smart people.
- Science doesn't concern me. It's not going to determine if Syracuse does or does not win the NCAA championship, or whether I can buy my beer (or DVD).
- Science is boring; it is definitely not fun.
- Science is done by rich people—old, white, European men with beards.
- Science is for people who don't want to talk to other people but would rather play with computers or things.
- Science creates the unnatural, the dangerous.
- Science kills the lovely hawk circling the sky and dissects him. Science is unpoetic, inhumane.

These are caricatures and extremes; but I think you recognize them, don't you?

One can attempt to counter these attitudes one by one. But I would rather make some observations that point to an approach rather than to a specific strategy.

- These are *perceptions* about science (mostly wrong), not realistic assessments. That's okay; we live by ideas and things of the spirit as much as we do by matter. Please accept the strength (and even sincerity) of a perception even when it is at some level wrong.
- What we do not understand, we usually find uninteresting and sometimes are just afraid of. What we understand, we may find interesting and may (not always) be unafraid of. This applies to art as well as science—think of attitudes to Stravinsky's music of the beginning of the twentieth century. And this is why we must teach, in myriad ways.
- People like facts; but they really love a good story, well told. Telling stories is very old. Stories are human. They are about perceptions. Through them a shared understanding forms.

So I think that one way into people's hearts and minds is to tell stories of science. In school it may be done in the midst of a logical development of a subject. To a general audience it's a way to normalize, humanize, science and build a piece of understanding. I urge you to explore the power of a story well told in your teaching.

4

I return to the making of our video course, *The World of Chemistry* (see Figure 22-8). The course was conceived by Isidore Adler of the University of Maryland and Nava Ben-Zvi of the Hebrew University of Jerusalem. In the 1980s, they approached the Annenberg/CPB project, which eventually funded the major part of this course. Adler and Ben-Zvi, together with University of Maryland physical chemist Gilbert Castellan, Margot Schumm of Montgomery College, and Mary Elizabeth Key, then of St. Albans School in Washington, formed the "academic team" that conceived the content and supervised the production of the programs. Richard Thomas was the imaginative executive director and producer of the series, working together with the able technical staff at the Educational Film Center, Annandale, Virginia, an Emmy Award-winning production company. This Pimentel award is for them as well. Sadly, Izzy Adler and Gil Castellan are no longer with us.

The World of Chemistry is intended for a junior college, four-year college, or remote-learner audience. It can serve as a complete course in chemistry, but it can also be used as a supplement for courses at any level, secondary school or university, as a resource for young people or industrial workers, or just as entertaining viewing for the citizen-at-large.

Figure 22-8 Making "The World of Chemistry."

Each program contains the following components: (i) two chemists who appear in nearly all the segments: I myself, as the presenter or series host, and Don Showalter of the University of Wisconsin at Stevens Point. He has all the fun, since he gets to do the spectacular demonstrations. There are also (ii) one or two lively interviews in each segment, (iii) some computer animation (no blackboard!), and (iv) weaving it all together over fast-moving montages and footage illustrating the concepts taught, a narration.

Some of the programs teach very directly—so one of the 26 is on "The Mole," and two on "The Driving Forces" and "Molecules in Action," which explain on the macroscopic and microscopic levels why chemical reactions occur. Other programs describe important chemistries: chemistries of color, of metals, and of reactions on surfaces. There is a whole program devoted to "The Chemistry of the Environment."

In professional television, which is what these programs are, you get what you pay for to a certain extent. These programs were produced at a per-minute cost only half that of a typical U.S. *Nova* program, or only one-tenth that of *Cosmos*. So whereas my friend and colleague Carl Sagan stood in front of the ruins of the library of Alexandria in Egypt, I got sent as far from Washington, DC, as Baltimore, Maryland, to stand on a wintry day on a tank car of sulfuric acid! Actually you'll see more in these programs than one might have judged from their cost. They were made with the dedication, sweat, and mental energy, mostly unpaid, of a remarkable team of people.

One observation that I would make is that what started out as a weak point of the production turned out to be one of its great strengths. We did not have the money to shoot 10 to 15 minutes of each half-hour from scratch (i.e., illustrating through our own filming the concepts what we wanted to teach). So we made do with skillfully edited "stock footage." This is a euphemism for free or inexpensive film clips obtained from government or industrial sources. An industrial public relations consultant could often identify the source of a scene from the five seconds it is on the screen. Now this stock footage could have been a weakness. It isn't what we would have done in that ideal, unlimited-budget world of which every filmmaker dreams. But by being forced to use (fairly, without commercialization) scenes supplied by others, especially industrial sources, our programs acquired a "real world" feel. I can put it another way: in how many chemistry courses in the world does the instructor show students five minutes of steelmaking and three minutes of the Hall process? We do, and the visual impact is tremendous.

I want to make some observations on the tensions I see arising in the use of television to teach.

First, a philosophical question: can one teach via television? Or does a switch go on in our minds that this is entertainment?

Second, the process is incredibly expensive. So much of one's time is spent in fund-raising. But this is as typical in the arts as it is in science. My friends Ivan Legg and the late Paul Gassman worked very hard on raising the funds to a sequel to *The World of Chemistry*. We failed. The industrial-community, our own industry, was not supportive of our efforts.

Third, the television medium is inherently journalistic. As such it leads to excessive mythologizing of individuals and of the way things happened. This hurts.

Fourth, there is the continuous treachery of simplification, of making little compromises. In the end it doesn't leave the person making those compromises (me, and maybe I was too sensitive) feeling good about either himself or the process.

Finally, the power of images is incredible, so there was a temptation to be driven by images, which at times did violence to the real intellectual achievement of chemistry—the slow but ingenious marshaling of indirect evidence to build a framework of incontrovertible reality for molecules and their transformations. Chemistry is the most marvelous example I know of knowing without seeing.

But it was worth it! I am proud that we created the first video course in chemistry ever made. The credit for the programs goes to our team, whose members' names I've mentioned above. I wish they could have shared in this award.

It is time here to reassert my confidence in what we do. We teach chemistry: the art, craft, science, and business of substances (now known to be molecular) and their transformations. We introduce young people to the molecular science, awakening in their minds the ability to deal with the balance of simplicity and complexity that characterizes chemistry. We believe (I feel confident it is not only I who thinks this way) that chemistry instruction at every level must be done in the context of a liberal arts education, fighting compartmentalization all the way and connecting chemistry to economics, literature, history, society. To *culture* in the broadest sense. We believe that the student is best served by consistently being led to value discovery and true understanding. It's not easy, but there is nothing I love more than teaching. As you do. Thanks for honoring me with the Pimentel Award!

Acknowledgments

Significant parts of this lecture are taken from articles I have written, which have been published in *American Scientist, Chemical and Engineering News, Chemistry, A European Journal,* and the *Journal of College Science Teaching.* I am grateful to Norman Goldberg and Greg Landrum for their help in the construction of this paper and to Derek A. Davenport and another reviewer for their suggestions.

Notes

1. The ACS Award in Chemical Education that I received was named after the late George Pimentel, an excellent physical chemist and superb educator. This essay was my award address.
2. W. S. Sheldrick and M. Wachhold, Discrete Crown-Shaped Te_8 Rings in Cs_3Te_{22}. *Angew. Chem. Int. Ed. Engl.* 34 (1995): 450.
3. Q. Liu, N. Goldberg, and R. Hoffmann, A 2,3-Connected Tellurium Net and the Cs_3Te_{22} Phase. *Chem. Eur. J.* 2 (1996): 390.

23

Some Heretical Thoughts on What Our Students Are Telling Us

ROALD HOFFMANN AND BRIAN P. COPPOLA

There is a time, twice a year, when those of us who teach introductory courses sit down in a comfortable chair, pour ourselves a middling portion of single malt Scotch whisky, and begin to read the comments that students write about our teaching. For the overall ratings, numerical in nature, we can bear to wait—the computer will dutifully compile these single point undifferentiating indicators.

What we settle down to read are the "free-style comments," where the students are encouraged to write (anonymously, of course) what they think of the book, the exams, and, of course, of the lecturer. Many, not all, universities give students the opportunity to express themselves in this way. Some of us have learned to avoid asking silly questions with predictable responses, such as "What is the best part of the course?"

So we sit down, perhaps turning on some Chopin to complement the whisky, and face those student responses. Many are positive, as (with a trace of mild astonishment) "I didn't think I'd like chemistry, but Prof. Coppola made it fun!," "I actually enjoyed going to the lectures," or "I didn't get a very good grade, but I sure learned a lot." It's not always easy for a student (or us) to say a word of praise, to give thanks graciously harder still.

Positive feelings generally wash over us leaving small marks. Happiness is often diffuse. But pain is sharp—the small pain of a torn cuticle, the stronger incapacitating pain of a broken bone. Or, negating the validity of the familiar litany "sticks and stones..." the mental anguish of reading an evaluation such as "Prof. Hoffmann spends all his time on digressions, relating chemistry to politics, history, God knows what else. Who cares how hemoglobin or catalytic converters work? I want to know what's on the MCATs." Or "I got an A by memorizing equations and doing exam problems that were exactly like the problems that I had seen on the previous tests..." Or, "As far as I am concerned I did not need to go to class."

Now this hurts—ergo the whisky and music; it hurt last time too... Our reaction comes in part from this inability to weigh appropriately emotional praise and criticism. Differentiating among the negative comments, we can easily forgive the simple nastiness of resentment released under cover of anonymity. We are more wounded when the students condemn exactly what we are most proud of in the educational process: we finally got this course right! More than merely the course contents, that neutral list comprising the syllabus, we more importantly developed the spirit of our science (chemistry in culture, and chemistry as culture, as it should be at a liberal arts university) and the process of its construction (stressing understanding and discovery). We finally understood (and thought we succeeded in communicating) that multiculturalism embraces all parts of the university experience, and is as inclusive of intellectual constructs, such as chemistry, as it is of the traditional social ones. Then to get such comments really, really hurts.

We could counter, and lash out at the immature young people, at societal pressures and at all the things that make for their wrong attitude toward learning. Better we release our anger on them than on those dear to us... Or we could take another sip of the Lagavulin and reflect on what we can learn from the students' comments, from just those comments that wound most.

As teachers, we invest a great deal of our professional intellectual lives trying to see beneath the surface of what we encounter. What drives our curiosity is trying to understand core phenomena or motivations that give rise to what we see. That is, we try, even if we don't always succeed, to be attentive and insightful learners. For we believe that a high road to effective teaching is to be a good learner when analyzing a students' work or perspective.[1] This is as satisfying an intellectual challenge as authorship or laboratory research.

An effective analogy that one of us (B. P. C.) has created for demonstrating that anybody who takes on the "teaching" role must think (to learn) before despairing about ignorance, is given here:

You are teaching multiplication. To probe the students' mastery of the subject, you give an examination. To which one student provides the following answers:

$2 \times 2 = 4$
$1.1 \times 11 = 12.1$
$3.5 \times 1.4 = 4.9$
$-1 \times 0.5 = -0.5$
$-3 \times 0.75 = -2.25$
$2 \times 4 = 6$

What do you do? You can shake your head and say "How can a student who can multiply noninteger and even negative numbers make such a mistake?" Or you can decide to learn from what the student's response is telling you. And revise your educational strategy accordingly.

The student has done nothing wrong, except...to think that multiplication is addition.

Teachers and students meet in the classroom to fulfill the terms of a tacit covenant of instruction. There is more to it than being paid to teach—we sacrifice whatever else we could be doing during that hour when we teach, or even when we read their comments, to confront a simple question: "Am I being understood?" We learn from books and other media (oh, how imperfectly via these comment sheets!) at our convenience, but in classrooms teachers and the taught come together for just the kind of feedback that is unique to our conversational profession. All classroom pedagogy revolves around ways for the faculty to learn "Am I being understood?" Students want to know this too: "Are we being understood?"

So...we force ourselves to listen to students who have confronted the subject matter and ideas we have so painstakingly (and, we hope, eloquently) provided. But the students have not constructed the same understanding that we have...of the subject, its ambience, and its process. This may be sad, but it is true, as those comments of theirs so painfully reveal.

Of course we understand that our own appreciation continued to grow after the first—or tenth—time that we turned our thinking to the subject. And especially it grew when we finally needed to teach it. We do not expect novices to surpass us in their first round. But we must also not dismiss what we may learn from their unique perspective as less experienced learners. We listen unwillingly, for we are sure that we are right. But we try, because they are right, also. In collaborative communities, the distinction between who is the "teacher" and who is the "learner" becomes blurred, if not wholly imaginary to begin with. Here's what we think we hear:

The students are telling us that you don't have to understand everything in chemistry to learn and use the science.

Yes, we'd like them to understand, and we have designed our course so as to emphasize the process of understanding. But learning in chemistry is (a) a curious mixture of proof (real proof), and of belief (accepting on faith, trusting that someone else has proved, or that proof might be forthcoming if one advances in the subject). And that learning is (b) sequential, in an intriguing, intellectually inconsistent way—it proceeds by first understanding something, then memorizing something else, then using the mathematical expression of what was understood in a rote or algorithmic (yes, unthinking) way so as to solve a real problem. We develop a tacit tolerance for the fundamental inconsistencies that define the edges of our understanding. All this, mixed up with occasional necessary bouts of memorization and a nomenclature that has pretensions of being systematic.

As mature learners, we include as many strategies as we can in our arsenal for inquiry. Progress does not occur because we have excluded memorization, but rather because we recognize when memorization is precisely the most effective strategy to use. As much as we would like to enact a truly Socratic dialogue with undergraduates, the reality of teaching thousands of students has made this impossible. It may be that the only potentially authentic thinking in on-your-feet creative situations we place students into our examinations. Regardless of any rhetoric we provide in class, our examinations transmit the learning objectives that are targeted for comment by students.

Let's take an example: We derive the ideal gas equation, $PV = nRT$, by historical or experimental appeal to the individual Gas Laws (of Boyle, Charles, and Gay-Lussac). We and the students "understand" the formula (how limited that understanding is, how unreal the ideal, becomes clear in a physical chemistry course). We see the formula in our minds, its beauty in the chemistry and physics it so succinctly summarizes for us. We go on to use it in a myriad of problems, from balloons to equilibria, from determining molecular weights to thermodynamic cycles. And in using it we do not go back in each instance to the derivation. We use it as we need it, as a formula.

The reason we shouldn't get angry at students who say they got by "just memorizing the formula" is that they are just shading their response—very probably they understood a lot, but then chose to emphasize the formulaic use. We think that as much as we value *sophia* and understanding, that knowledge and learning also involve a component of suspending understanding, or at least pushing it into the background. We ask the reader to recall the problems of 40 years ago with "the new math" in primary education.

There's an even broader lesson, we think:

> **You don't have to understand everything in order to (a) operate as a normal successful human being in this world, or (b) even to do creative work of the highest degree.**

Once again, we have to begin by saying ever so clearly that we value real understanding, that knowledge is an absolute good. And the special contribution from formal education, schools and universities is centered, we believe, in their being the place where connections between general educational and professional training objectives are constructed and maintained. Elsewhere in life, other imperatives, often economic, dominate.

However, it is clear that technical training is of great pragmatic value even in the absence of the connections forged at a university. Practice and experience suggest that this is the way of the world: we usually learn to use technology's products quite separately from the underlying context, and we can make successful and productive, humane, contributions without even being aware of any appendant knowledge. Driving an ambulance to an accident site does not require a cognitive awareness of the thermodynamics of combustion; the thermodynamics operate just fine without us. We use calculators to help us do arithmetic, and we choose to need to understand how learning arithmetic allows us to make the necessary judgments about the outcomes of button-pushing, while at the same time we choose not to understand things about batteries, liquid crystal displays, the manufacture of silicon chips and the marketing of calculators. Performing a specific task on an assembly line can be done well when the laborer is completely unaware of the other tasks on the line or even the object being assembled. Sometimes that is the learner's choice, quite democratic and informed, also.

Let's jump to the creative act in our science. The synthesis of a new antitumor agent, the perfection of a new industrial process that avoids the use of a harmful solvent, may both involve a heterogeneous catalyst. The catalyst does something reproducible, taking, say, an olefin, and epoxidizing it specifically on one of the two olefin faces. We may have a vague idea how this works on the molecular level, but should we suspend use of the reaction until we really understand the catalyst mechanism? That would be just as silly as to ask Archie Ammons to tell us the metallurgy of the keys of the ancient typewriter that he uses before he writes a poem.

The pressure to understand everything betrays a simplistic reductionist worldview. As one of us has expounded (perhaps tiresomely) elsewhere, reductionist (or vertical) understanding is just one way of knowing the world. The other (call it horizontal) way is to understand the world, quasi-circularly if you insist, in terms of the concepts that have evolved in the field under consideration, concepts as complex and seemingly poorly defined as what one is trying to understand.[2] So a telephone that makes a call to an ambulance is accepted as a communication device, working or not working, able to place a call here but not there. It is paralyzing (if not useless) to start to think of the workings of the telephone in a reductionist manner when it is time to call an ambulance. Understanding at some level is definitely needed to fix the telephone, still more complete understanding to create a better telephone.

Jean-Marc Lévy-Leblond makes the important point[3] that we should not wring our hands in despair when we see the results of "ignorometry," all those surveys which show us how ignorant most people are of science, or of history, or of geography.[4,5] The very same "scientifically illiterate" people drive automobiles pretty well, use word processors, microwave ovens, and lawn mowers. Ignorant by one measure, they know quite a lot of the real world, learning just enough to function as normal, productive citizens. Lévy-Leblond remarks, "Should we not start by admitting and admiring these achievements, instead of denying and lamenting the failures?"[6] As scientists, we're not that superior when we interact with machines or tools of higher complexity.

Craig Nelson asks the provocative question: What is the shape of the earth?[7] Two plumb lines separated by any distance on the surface of our planet are not parallel, yet the flat earth assumption is manifest in architecture. What is the shape of the earth? Round? No. A sphere? Hardly. An ovoid? Only if you blur your eyes and don't watch over time. Nelson's point reminds us that our very best theories are only the latest version of Flat Earth, and only better by decimal places of agreement with what is observed, not by "truth" in an absolute sense.

The intrinsic beauty of a model is tied to its ability to function, to deliver useful information upon which we may act.[8,9] Good models inspire productive experimentation rather than retire lab coats. One important thing to remember about models is a tenet of General Semantics, attributed to Alfred Korzybski: "The Map is not the Territory,"[10] which was inspired, according to popular mythology by René Magritte's "This is not a pipe."[11,12] However heretical (and in one way incorrect) it seems, chemistry instruction would benefit from an explicit understanding that "H_2O" is not "water."[13,14]

Still another lesson from our students, one we don't want to hear:

Compartmentalization is an effective strategy for the workings of the world, and may be for learning.

We tell them of the Haber-Bosch process as an example of Le Châtelier's principle at work, and can't pass up (at least some of us can't) talking about catalysts in general, and relating Fritz Haber's tragic story, and how it took the talented engineer Carl Bosch to convert Haber's discovery into a real process, and, for good measure, telling them of nitrogenase as well. We tell the students about solubility constants and illustrate the subject by discussing commercial water softening and the composition of kidney stones. We feel good as we do this, for we have served the goal of a liberal arts university, and have connected up different parts of chemistry. This seems so essential in an age of specialization.

Some students like this. More yawn, and tell us we confuse them with the digressions: "Just tell us what we need to know on the next test..." They're wrong, of course. The unity of the world, not only chemistry, will catch up with them. The world only looks disintegrated because they learned, from us, that this met their educational needs. After all, we are the ones who chose to write test questions (or textbook chapters) about solubility constants without much mention of their wondrous applications. Our students are only eighteen, too focused in on a profession, and see us as a barrier between them and medical school, or as a useless burden on the way to being an engineer or running a farm.

But in a way they are right. First of all, the lesson of the animal cell or the Volvo assembly line is that specialization and compartmentalization work. There is a reason (efficiency, not divine design) for the nucleus storing the DNA, the potassium channel letting through just that ion and no other. Second, analysis works as a learning strategy, breaking a complex whole a synthesis of vitamin B_{12} by the body or by Woodward, Eschenmoser and 99 friends, the Haber-Bosch process mechanism into more comprehensible building blocks (almost an argument for reductionism!) Analysis is inherently compartmentalizing. Third, it is difficult, indeed sometimes confusing, to deal with the whole. Learning the pieces is a strategy for comprehending the whole. You can't see the forest without the trees, either!

The counterargument is clear. The real test of understanding is to use the pieces to build a whole, even more so to construct wholes different from the one we initially disassembled. If you learn only what is in the compartments, or one task on the assembly line, if you don't push your way through to assembly and integration, you...will be stuck in the pieces, on the assembly line.

Schools and universities need to be inclusive of the broadest menu of choices. Craft, knowledge, and cunning have been fragmented—too much so, we think. Universities must be the places where the answers to reintegration's questions can be found. Indeed, even assembly lines have gone reintegrative: in many manufacturing plants workers learn to perform many tasks and, in some cases, groups take collective responsibility for the whole product. Can we do less? Disciplinary separation that leads to cultural isolation threatens to remove reintegrative choices from the menu of formal education. We can choose to do this, *mea culpa...nostra culpa;* but let us first make sure that we realize there is a decision to be made.[15]

It is time here to reassert our confidence in what we do. We teach chemistry— the art, craft, science, and business of substances (now known to be molecular) and their transformations. We introduce young people to the molecular science, awakening in their minds the ability to deal with the balance of simplicity and complexity that characterizes chemistry. Both of us believe that chemistry instruction at every level must be done in the context of a liberal arts education, fighting compartmentalization all the way and connecting chemistry to economics, literature, history, society, to culture. And chemistry must be recognized as culture, in the broadest sense. We believe that the student is best served by consistently being led to value discovery and true understanding, rather than being restricted to memorization as the only way of knowing. And, yes, we take a paternalistic viewpoint that we—not the two of us, but the community—know a little more than the student of what is essential and valuable in the science taught.

It has grown dark. A second glass of that marvelous Scotch brew of water and grain (a little bit of chemistry, too), tasting of peat and iodine, the color of heather on the hills at a certain time of year, that second glass will get us in trouble. It's time to finish reading what our students say. And perhaps we don't need that second glass, after all. Perhaps the sting of the students' words comes from our own willingness to stop at the surface of their comments; or perhaps we stop at the surface because we fear what we imagine lies even deeper—we imagine they just don't like us. The experience that we have as teachers, and the effort we have put into our teaching, really do not mean that we know how best to wake up the qualities that reside dormant in each student's mind. Our view is that we should interpret our experience as a license to listen, and to learn ourselves. Whether the course turned out a little better or a little worse, and especially if we think we finally got it right, we still can learn something from our students.

Notes

1. B. P. Coppola and D. S. Daniels, Structuring the Liberal (Arts) Education in Chemistry. *Chemical Educator* 1 (1996): 1.
2. R. Hoffmann, *The Same and Not the Same* (New York: Columbia University Press, 1995).
3. J.-M. Levy-Leblond, About Misunderstandings. *Public Understand. Sci.* 1 (1992): 17.
4. See also R. Hoffmann, Ignorance, Ignorantly Judged. *New York Times* (September 14, 1989), p. A29.
5. See references quoted by Levy-Leblond, About Misunderstandings. *Public Understand. Sci.* 1 (1992): 17.
6. J.M. Levy-Leblong, About Misunderstandings. *Public Understand. Sci.* 1 (1992): 17.
7. C. E. Nelson, "The Status of Science in a Changing Society: Uncertainty, Diversity, and Teaching," lecture at the University of Michigan chapter of Sigma Xi, December 9, 1994.
8. N. Oreskes, K. Shrader-Frechette, and K. Belitz, Verification, Validation, and Confirmation of Numerical Models in the Earth Science. *Science* 263 (1994): 641.
9. N. Goodman, *Languages of Art* (Indianapolis, IN: Hackett, 1976).
10. A. Korzybski, *Science and Sanity* (Lancaster, PA: International Non-Aristotleian Library, 1933).
11. R. Magritte, *Magritte* (Oxford: Phaidon, 1979).
12. M. Foucault, *Ceci n'est pas un pipe: deux lettres et quatre dessins de René Magritte* (Montpellier, France: Fata Morgana, 1973).
13. S. Leibowitz and R. Hoffmann, Signs and Portents: No Parking in the Courtroom. *Diacritics* 21(1) (1991): 2.
14. R. Hoffmann and P. Laszlo, Representation in Chemistry. *Angew. Chem. Int. Ed. Engl.* 30 (1991): 1. Chapter 14 in this book.
15. B. P. Coppola and D. S. Daniels, Mea Culpa: Formal Education and the Disintegrated World. *Science and Education* 7 (1998): 31.

24

Specific Learning and Teaching Strategies That Work, and Why They Do So

ROALD HOFFMANN AND SAUNDRA Y. McGUIRE

The two of us have been teaching and helping others to teach chemistry at every level—from high school teachers to undergraduate and graduate students to university faculty—for over four decades. From that experience have come a number of teaching and learning tactics that we find effective in facilitating student learning. Initially improvised, these strategies are more than gimmicks, for they have proven themselves in practice. Here we share some of them.

Since we are inclined to be reflective as well as pragmatic, we've also sought out in recent advances in cognitive psychology, and in the scholarship of teaching and learning, insight into why these approaches work. We think through why they are of use in those most magical and mystical processes of learning and teaching any subject, not just chemistry. And we also spell out potential problems.

Caring deeply for student learning entails keeping an eye out for what works for others. Perforce, this means borrowing and adapting. Thus a potential injustice in our account is that credit may not be given to the real innovators. Frankly, we do not know where some of the strategies we suggest originated—in examples by others, or out of our own improvisations as we struggled to become better teachers. Many people have independently come to similar practices.

Some of what we write is addressed to teachers, some to students. This is deliberate. Cognizance of learning strategies benefits teachers, and awareness of teaching strategies can help learners understand the motives of teachers. Teaching and learning are a double flame.

Six Learning Strategies

1. TAKE NOTES BY HAND, REWORK THEM THE SAME NIGHT

Take notes by hand, even if the class notes are being provided by the instructor or a for-profit service. Even if they are web-cast. Preferably not later than the evening of the class day, rewrite your notes, by hand, amplifying their content.

Notice that this process involves two stages—taking the original notes and then rewriting/reworking them. There are various note-taking systems,[1] including the Cornell Note Taking System,[2] the mapping method, the outline method, etc., that students can learn. During the rewriting stage, it is important that you not just recopy your notes, but rather both condense and extend them where appropriate, paraphrasing them "in your own write." So that you make the meaning your own.

The question of whether taking notes on a laptop or by hand is more effective is a contentious one.[3] We think taking notes by hand works best, but our preference may be due to our age and educational experience. A real concern, however, is that much of the information in science courses is graphical and based on mathematical equations. Students find it difficult, if not impossible, to jump from words to chemical structures, graphs, and equations if they are taking notes on a computer. We see that students who take computer notes waste extraordinary amounts of time in the frustrating task of making sure formulas and structures are drawn correctly by digital methods.

Why this works: It is now well established that active engagement in the process is imperative for learning to occur.[4] When students take their own notes, they are engaged, in real time, and their minds focus on the task. For kinesthetic learners, the movement involved in taking notes facilitates learning.

The process of paraphrasing and rewriting the notes shortly after lecture helps to transfer the information from short-term to long-term memory.[5] If the rewriting is delayed longer than twenty-four hours, much of the information needed to flesh out the notes taken in class will have disappeared from memory. And…it is so much better that gaps in understanding surface in the engaged rewriting of notes, rather than the night before an exam.

Potential problems: Students may feel that they do not need to rewrite their notes if they understood the material in class. And it takes time to do so. However, the review that comes with the rewriting deepens learning and facilitates long-term retention of the information.[6]

2. MISSED A LECTURE? GET YOUR NOTES FROM A LIVE PERSON

If you must miss a class, rather than simply download the notes from a webpage, get the notes from a fellow student.

Aside from being a great way for men and women to meet (in every combination), this strategy is another way into group discussion and learning. It is important to develop relationships with other class members and to form study groups (see below) early in the course.

Why this works: During discussion of class notes, much learning takes place. A typical scenario: Student A (the one who missed lecture and is borrowing the notes) says "I don't understand this part of what you wrote," to student B, the note taker. Because B is a fellow student, A is comfortable asking her the question, while

A might be reluctant to ask it of the course instructor. B explains, and is, of course, *ipso facto* engaged in the most salutary of learning actions, teaching.

Potential problems: The note taker may not understand, or may propagate a misconception. Additionally, some people are just too shy to ask another human being.

3. OPTIMIZE LEARNING FROM HOMEWORK AND TEXT EXAMPLES

Most students do their homework in solitude (or as much of that as a residence hall room allows) by trying to follow text examples of similar problems. But often the text examples are not exploited for the learning opportunities they provide.

Here is a simple and effective strategy for approaching text examples and assigned homework:

- Do the obvious: study the text and lecture information relevant to the problems.
- Treat the examples in the text and in lecture notes as homework problems. Read the problem statement in the example, but do not look at the answer, cover it up. Now work the text example.
- Compare your approach to the text's, not just your answer. Is the example problem solved by a method identical or close to yours? If not, yet your answer is correct, don't be afraid to continue using your method. But try to understand the text's. There are often several ways to do a problem.
- Answers to homework, provided by the instructors, of course, should *not* only be numbers—answers should always include ways of working each problem. If they do not, the instructor and teaching assistant should be encouraged (that's putting it mildly) to provide complete solutions, even alternatives. A problem set solution (and examination solutions, the most carefully read information in the entire course) is a teaching opportunity.

Why this works: Students develop the essential skill tested by all exams—the ability to work a problem without using a model of how it should be worked. This approach to homework focuses on *methods* rather than final *answers*. Furthermore, alternative methods are explored so that students learn to be agile, flexible thinkers. This method also affirms a student's intellectual intuition.

Potential problems: Students may be tempted to peek at the examples, a solution manual or a website rather than spend the time to figure out how to work the problems themselves.

4. STUDY FIRST BY YOURSELF, THEN IN A GROUP, THEN BY YOURSELF

The idea here is simple: First, you should try to do a homework problem or prepare for an exam on your own. Then, the collective wisdom of a group is enlisted. Three to six fellow students who have each done their best to digest and absorb difficult material are powerful resources for each other. Finally, you must return to solving the problem set or facing the exam on your own. The sequence here is important.

Why this works: Not all instructors are comfortable with homework done in groups, but our experience is that groups are very effective, both for problem set solutions and studying for exams. Do-it-yourself is the primary principle of active learning. But groups can help resolve the occasional blind spot—usually, someone in the group will know how to do the problem. Social constructivist learning theorists have shown that meaningful learning results from small study groups with two crucial features: discussion and problem solving activities.[7,8,9] Finally, to cement their learning, and with existential courage, the students must face the material by themselves.

The potential problem: After a period of optimal group functioning, groups inevitably organize themselves along lines drawn by personality characteristics, in particular on the active/passive axis. A group's value quickly diminishes for a passive personality; such students will too easily fall into a pattern of merely listening. Groups may converge on a wrong approach; this is less likely to occur when students are encouraged to use their resources to double check all information. And it will not happen when all students feel equally empowered to contribute to the group discussion.

There is available information on group study[10] that will raise the likelihood that the group will be effective for all members. Several websites provide excellent tips on forming and running successful study groups.[11]

5. ENTER THE TESTER'S MIND

Make up practice quizzes and tests for each other.

Why this works: One of us tells his students that "the only way you will get into my mind about the exam is...to try to get into my mind. That means to do what I do, and make up an exam." Creating a practice exam involves not only the selection and organization of all the material (including choices about what is representative and what is important) but also discussion of the exam in a group setting. Its value as a learning tool cannot be overstated. Usually, one group member, on seeing another student's trial exam, cries out "The professor (we suspect more colorful characterizations would be used) would never ask that!" Others react, and a discussion emerges about what material is important and what is not. That's just what we want to encourage.

Another way to enter the tester's mind is by *teaching* the material, one student to another. When one of us asks instructors attending faculty development workshops when they began to develop a deep understanding of the conceptual structure of their discipline, most say that it did not happen until they began teaching. The other author got an A+ in a graduate school thermodynamics course, but never really, really understood thermodynamics until he had to teach it.

Potential problems: This takes considerable time and discipline on the part of students. And role-playing, such as is involved in taking the teacher's place, is not easy for some.

6. SET ATTAINABLE GOALS

Perhaps there is something to learn from standard psychological advice on how to help oneself out of (subclinical) depression. It's not that studying is depressive; but if you are spinning your wheels and studying does not lead to learning, the process can share some symptoms with depression—a perceived inability to act, for instance. For this reason, it is important to (a) tackle small, achievable tasks and (b) try to focus on other people, not yourself. When you teach others, you step outside of yourself, and interestingly that becomes for *you* a path to learning.

The relevant advice in constructing exercises (and tests) is to move slowly, from simple problems, to more complicated, integrative ones. Teaching assistants should work in the same way in their recitation sections.

Why this works: Success, self-achieved, builds confidence, and so is a very powerful motivator. When you attempt to reach a goal that is within your grasp, a wonderful cycle of initial success, more effort, and additional success is put into motion. And helping other students moves one past disappointment about not getting things right oneself. Usually, you can help someone else get going; the gratification is motivating for both parties in such an exchange.

Potential problem: What is easily attainable for one student may be very difficult for another. Conversely, an attainable goal for one student may be trivial for another, requiring no effort whatsoever. This may result in a student assuming that learning chemistry is effortless, and s/he will not develop the learning skills necessary for performing more cognitively demanding tasks.

Six Teaching Tactics

1. A CONTRACT, NOT A CURVE

We recommend instructors grade on a contract with the students, whereby grades are based on a combination of a major absolute performance component (examinations and quizzes) and a minor—as small as possible—"curved" part of the course (such as labs and other multi-section pieces).

The only reason for curving should be fairness—if several graders are involved, for instance. The grading criteria must be made known to the students at the beginning of the course. An instructor might say to the students: "An A of some kind is 85% mastery as judged by various components, a B is 75%, a C is 60%, and 50% is passing. I will not raise the borderlines, but I may lower them if I have misjudged the level of mastery." In other words, the line for the A's may go down to 83%, but never rise higher than the contracted 85%.

Why this works: The students are empowered when they see that the outcome of their course grade is largely or entirely dependent on their work in the course, rather than on a comparison of their work with the work of others. Young people react very positively to fairness; a contract boosts confidence.

Potential problems: The professor will need to construct exams such that the level of mastery of the material is accurately reflected by the grade that students achieve on the test. In psychometrics, this is referred to as content validity.[12] To be realistic, some faculty have trouble doing this.

In particular, one has to watch for misjudgments of mastery in multiple choice exams of the type where the simplest arithmetic mistake will yield an incorrect answer.

The kind of contract we recommend is very scary to some department administrators, who may be insistent that each course have a predetermined median grade. Such worries, amusingly, reflect a lack of confidence in *faculty* members' ability to assess mastery levels.

2. BRING "REAL LIFE" INTO THE CLASSROOM

News, crises, and everyday life open the mind. Devote 5 minutes of each class to a discussion of science in current affairs, preferably using ideas from print media, or television or radio news reports. Every minute spent this way is worth it.

Newspapers (print or online editions) sadly carry little science; what they do carry is often health-related. As far as chemistry goes, the papers, or the web-based new media, rarely give chemical structures but sometimes name the molecules or drugs in their stories. Every morning the instructor might profitably scan the health and science stories in *The New York Times* or an equivalent resource. With Wikipedia, or WolframAlpha, or chemicalize.org, the structure of a molecule can be retrieved immediately. It should be shown in class, along with a screen shot or webpage of the newspaper or magazine story.

Our experience is that from a stream of such "short stories from the real world" comes appreciation of the relevance of what is taught. In chemistry, students begin to see that small differences in the structures of molecules may determine whether a substance will hurt or heal. Or both. For example, they may begin to understand that not all cholesterol is bad, or that crystal meth and the decongestant it's illegally made from in home laboratories differ by just one atom.

Why this works: Of course, motivation for learning is enhanced by the perception that the material studied is relevant. People are curious about how things work, and disasters arouse both fear and compassion towards our fellow human beings.

But the discussion of newsworthy topics may not be the most important part of this strategy. Ultimately, bringing real life into the classroom day in, day out builds a bond between teacher and student. Students begin to feel that the instructor has gone to the trouble of reading the paper or searching other media that very day (caring trumps content!) and of making a slide which demonstrates that s/he cares that students learn.

Potential Problems: The discussion of "What's in the news?" could eat up precious class time. Depending on the topic, student interest may be intense; if so, the professor can simply suggest continuing the discussion during the next class period, or outside of class during office hours.

3. "CHEAT SHEET" OR PROMPT TO LEARNING?

Here's a strategy on which the two of us disagree: Allow each student to bring into a test or final examination an 8½ by 11 page on which anything in the world can be written.

One of us feels strongly that as we move toward textbooks sold not as print copies but available for download, and as it becomes increasingly more difficult to forbid a student to use a computer or fancy calculator in an exam situation, we are moving toward open book exams anyway, like it or not. He asks "why not invent a new way of teaching for that eventual situation?"

The other co-author feels that this is *not* a good strategy because she has observed its negative consequences. In her experience, most students think that if they can bring in a "cheat sheet" to the test, they need not know anything because everything relevant can be written on the sheet. These students fail tests because they spend most of their study time looking for what to copy onto the sheet, information of which they have no conceptual understanding. Or they spend time during the exam regretting that they didn't put "the answer" on the sheet. She feels that professors should provide information (such as constants) that students need, but are not expected to memorize. She stresses to students that they can only think critically using information safely stored in their minds—information that they *own*. "Cheat sheets" are very attractive to many students, but she has seen very few who use this tool to their advantage rather than to their detriment.

Why this might work: The sheet serves as a security blanket for scared students, of course. But its true purpose is to make the student review the material, to make judgments about what is essential and what isn't, and to organize the material. The sheets can become a prime learning tool. With progress in the course, one of us has observed that students realize this, saying after an exam "I didn't even look at the sheet."

Potential problems: A number were mentioned above. Furthermore, even if everyone in academia used this strategy, the standardized professional entrance tests such as the MCATs, LSATs, and GREs of this world do not allow it.

4. TURN IT AROUND

Consider a typical classroom exposition of a principle: "You have A, you apply a way of analysis that allows you to conclude that B will happen." There's a natural tendency for students to view that particular analysis as the only approach to problems and exercises involving the principle. The teacher can combat that outlook, and reinforce the primacy and power of the principle, not the specific example, by immediately following up with a second example where the question is turned around: "You get B,' and applying the same ideas, you can figure out what A' led to that."

To be specific: Suppose a teacher in an introductory chemistry course has just gotten through discussing, say, the mass relationships in a combustion reaction: octane (C_8H_{18}) is burned with unlimited oxygen to give water and carbon dioxide. He or she then continues:

> Here we've seen how to figure out that if you burn 114 grams of octane with an unlimited amount of oxygen you will get 352 grams of carbon dioxide. But wait, the same ideas can be put to work in many more problems. Here are two:
>
> (a) I don't have an unlimited amount of oxygen (the air intake on my car is clogged), I have 200 grams of O_2. How much carbon dioxide would I get then from my 114 grams of octane? This is a so-called limiting reactant problem; seemingly different, tougher. Yet the same ideas are at work.
>
> (b) My Volvo runs 8,000 miles a year, at an average fuel consumption of 22 miles per gallon. How much CO_2 am I putting into the atmosphere each year?

The teacher elicits the solutions to these problems from the students, helping when they get stuck. Finally s/he says: "I've just used the basic ideas of mass relationships in chemistry in three different ways. Now, I want you to go home, and make up a fourth way to use the same equation to probe the idea of mass relationships."

A physics analogue would be to ask for the range of an artillery piece, given a firing angle, a mass of the projectile, and an initial velocity, assuming frictionless flight. Then the instructor could ask what an observed range tells you about the velocity and the other factors specified.

Why this works: Sure, repetition of a concept reinforces it. But there is more—understanding is applying productively a heuristic, theory, or model to a set of facts. There is nothing more convincing of a concept's value than the feeling that it can be used for not just the problem that occasioned it, but for many other problems. Also learning is made fun by one of the elements of humor—surprise. And turning things around has an element of surprise to it. Imagine—an environmental conclusion of vast importance could come from simple stoichiometry!

Potential problems: Repeating the same question type in different permutations may seem repetitive to the teacher; we think it is rarely so to the student.

5. USE HUMOR

When one of us asked a group of Louisiana State University students to explain the difference between studying and learning, most replied that studying involves forcing yourself to memorize uninteresting stuff (as they put it), whereas learning means gaining insight into stuff you actually care about. To one and all learning was most often fun, but studying was usually tedious.

Here then is one of our great problems. How can the teacher and the student break down the perceived barrier between studying and learning? How can we build into the travails of most study some of the psychological fun of learning—the joy, that tremendous self-enhancing sensation of understanding after not understanding, that empowering flow?

Judicious doses of humor help a lot. Few chemical stoichiometry problems or lists of names of the bones in a foot could be imagined to evoke raucous laughter. But lapsing into a fragment of "Dry Bones" (the thighbone is connected to the hipbone…), or playing Tom Lehrer's "Element Song," or "Blackalicious," "Chemical Calisthenics," or Diego Carrasco's "Química" breaks tedium, gives the feeling of fun. Such musical instances have been assembled.[13]

Humor is also a smile, or a surprise, or turning things around and looking from a different perspective. All of these things are part of what made the Marx brothers so good. Work in that direction, work to achieve surprise. Look in the course material for mistakes that lead to weird contradictions or unphysical results. These are the intellectual equivalent of pratfalls.

Why this works: Humorous situations are attention-grabbing, emotionally satisfying, and can create an environment that promotes long-term retention and learning.[14] Humor also reduces stress, allowing students to enjoy the learning experience. Humor humanizes the instructor, builds a bond.

Potential problem: If too much humor is used students may fail to take the course or the instructor seriously. Also, the sense of humor is very individual—what one person thinks is funny, another finds stale. Fun and hard work most certainly coexist, but students must not underestimate the serious effort and time commitment required for success.

6. DEMONSTRATE!

Not every subject lends itself to doing demonstrations. Chemistry certainly does. Mind you, demonstrations did not come easily to one of us, a theoretical chemist. But he took to it, and in fact learned how to turn white wine into red (and back again) from his coauthor.

The advice here is simple: Do still more demonstrations.

Demonstrations are somewhere between magic and science.[15] Somewhere between gripping theater and chemistry. Somewhere between circus and the Zen koan that bestirs the dormant knowledge in a student's mind.

We know no deeper silence in a classroom than that which accompanies the first seconds of a demonstration. Theater directors and nervous concert hall managers envy us those natural moments of rapt attention. The auditorium is hushed, awaiting change. The demonstrator does not fail to provide it, with color, flame, smoke, or explosion. There ensues catharsis for the lecturer, a catering to all the senses of the audience, and, sometimes the only thing the students remember from a course.

Yet in the hands of a good teacher, so many more and valuable things may flow from that demonstration:

- a repeated metaphor for the heart of chemistry, which after all is about substances and their transformations,
- the essential question, whose asking makes a teacher delirious if it arises in the minds of just a few: "What is happening?"
- a return to the real world; science works with flights of imagination interspersed with mundane reality. The process of teaching naturally stresses symbols—words, concepts, and theories. The demonstration touches the earth. It may be staged, but it is tangible.

Why this works: On a simple level, a demonstration is a shifting of gears, from lecture to action. It is an intellectual alarm clock—"time to wake up, something is going to happen!" As academic theater, it crosses for a moment the bridge from learning to entertainment. In the hands of a good demonstrator, it crosses back to content, and so enhances learning.

Potential problems: At times the link between demonstrations and what is being taught is weak—so few instructors explain the complex kinetics of a typical blue-yellow oscillating reaction. Moreover, a course overloaded with demonstrations could sacrifice learning for entertainment. But, perhaps in the lecture room it is as Daryle Singletary sings: "I ain't never had too much fun."[16]

So far, we have presented quite specific strategies. We now turn to some general observations about the education process, awareness of which can greatly enhance learning. These are directed toward both the learner and the teacher.

Three Transforming Motivators

1. LEARNING STYLES MATTER

A student's learning style impacts the way s/he prefers to take in information, process that information, and interact with others.[17] Some students prefer to memorize discrete facts and specific formulas and then apply them, whereas other students prefer to use broader concepts and organizing principles to derive the discrete facts and formulas themselves. Learning style can also refer to a person's preferred modality—visual, auditory, read/write, or kinesthetic,[18] or to a number of other characteristics, such as personality type, for example extrovert/introvert, sensing/intuitive, thinking/feeling, or judging/perceiving.[19]

It is important for students to become aware of their learning styles and for teachers to know that there are different ways to learn, that more roads than one lead to this Rome. Why impose *your* way (and get frustrated when people don't use it) when you can encourage your students to learn in their own, optimum ways? When students become aware of their learning preferences, they learn more efficiently by, for example, converting lecture notes or course manual or text into their preferred format.

But what if a student's learning style does not match an instructor's teaching style or, worse, the learning culture of an entire discipline? The rich fields of human thought and practice are not tilled in just one way. Becoming aware that a task *can* be done differently is the beginning of learning *how* to do it differently. In a multivalent, diverse world, we can learn to learn in diverse ways without necessarily giving up our preferred ways of coping with knowledge and process.

Still, there is a problem here, epitomized in the comments of a person (the daughter of one of us) who crossed the worlds of science and the arts:

> How to make Ulysses palatable to a materials scientist or stoichiometry bearable for a DJ? I do not think it is so easy to get non-quantitative people to learn quantitative stuff. It requires blood, sweat, and tears, just as I, formerly a primarily intellectual, non-physical person, have torn my hair out and endured many humiliations in order to become a fully realized, physically capable artist.[20]

Unbridgeable as it might seem, we have to try to cross the gulf. Some teaching strategies outlined above may serve—the use of science in the news, humor. One can look for chemistry in culture, culture in chemistry. The environmental and ecological is a natural bond—it's nice to worry about how much CO_2 cars put in the atmosphere, but approximately how much is it in fact?

A potential difficulty is that when students determine their preferred learning style, they may be tempted to think they can learn only in that way. A student who is a visual learner, when confronted with text that is devoid of any pictures or figures, may conclude that the material is impossible to learn. It is important to stress that learning styles can be learned; just being aware that something may be learned in a variety of ways helps. When students investigate a spectrum of strategies, consistent with the gamut of learning styles, they broaden their learning preferences and become better learners.

2. LEARN TO LEARN

Most students think that learning selected terms, definitions, and solutions to specific problems is the way to perform well in courses.[21] Few of them realize that learning is a process, and that there are various stages of learning.[22] Learning *how to learn*, through examples, really helps.

In 1956 Benjamin Bloom and colleagues identified levels of learning proceeding from rote memorization through comprehension, application, analysis, and synthesis, finally to evaluation.[23] Recently, this taxonomy has been revised[24] and verbs used to describe the levels. Additionally, the top two levels have been reversed. In the new taxonomy the levels proceed from remembering through understanding, applying, analyzing, and evaluating to creating. (See Figure 24-1 below.)

We have found that teaching students how to learn has transformed many of them from rote memorizers and regurgitators into independent, self-directed learners. Showing students how Bloom's Taxonomy is applied to "Goldilocks and the Three Bears" helps them understand the distinctions between the levels. (See Figure 24-2 below.)

Concept maps,[25,26] graphic representations of relationships and applications of concepts in a field, are a very useful tool for moving up the learning pyramid.

In addition to teaching students about Bloom's taxonomy, we have found that when students learn about metacognition (thinking about one's own thinking),[27,28] they transform their attitudes about learning, their methods of study, and their grades. Metacognition is a way of standing outside, of willed thinking about the acquisition of knowledge and understanding. That impartial and reflective outsiderness is also a useful quality for scientific research, or inquiry of any kind.

Is there a potential danger of talking too much about the meta-world, at the expense of applying what one has learned to the academic subject at hand? An introductory chemistry course is not a philosophy of education course. We may have a

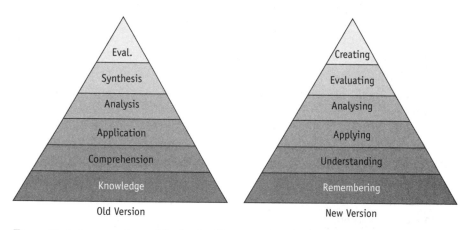

Old Version   New Version

Figure 24-1 A comparison of the levels of learning described by Benjamin Bloom and colleagues in 1956 and the revised version as introduced by his former students in 2001. The original version used nouns to describe the levels, whereas the newer version uses verbs. Additionally the newer version identifies creating as the highest level, whereas the original version places evaluation above synthesis. Both versions are currently widely used. Reproduced by permission of Rick Overbaugh.

~ *Bloom's Levels of Learning* ~

Applied to <u>Goldilocks and the Three Bears</u>

Creating	Write a story about Goldilocks and the Three Fish. How would it differ from Goldilocks and the Three Bears?
Evaluating	Judge whether Goldilocks was good or bad. Defend your opinion.
Analyzing	Compare this story to reality. What events could not really happen.
Applying	Demonstrate what Goldilocks would use if she came to your house.
Understanding	Explain why Goldilocks liked Baby Bear's chair the best.
Remembering	List the items used by Goldilocks while she was in the Bear's house.

Adapted from http://www.kyrene.k12.az.us/schools/brisas/sunda/litpack/BloomsCriticalThinking_files/v3_document.htm

Figure 24-2 Bloom's Taxonomy levels applied to learning tasks associated with the children's tale of "Goldilocks and the Three Bears."

disagreement between the authors here (again?!) One of us (guess who) can't get enough of metacognition (because she has seen countless students improve their test scores from below 50 to over 90 in a matter of weeks, just by using metacognitive learning strategies!), while the other one of us tires, wants to grapple with real teaching. We do agree that when students become fluent in the language of chemistry (or any subject) their metacognitive sophistication will increase to the level that they no longer have to consciously think about it.

Finally, a word about memorization, the first level in Bloom's taxonomy and the bane of teachers desperately trying to impart genuine understanding. Could memorization in fact be a learning style? There are courses ranging from history to chemistry to anatomy, where some memorization is required. Understanding is partly formed by finding a point where you stop asking questions, so that you don't get caught in an unproductive reductionist mode.[29,30] Consequently, instructors must take care to explicitly point out what material is simply to be memorized, and what material s/he expects students to understand and apply.

3. THE MENTOR/APPRENTICE BOND

The feeling that the teacher knows more than you—knows more ways to transform raw facts into understanding or how to actually make an object or molecule—can intimidate you, as a learner. You might think, "How could I possibly learn to do that?" But when respect for a teacher's mastery accompanies a second feeling, that the teacher cares deeply about transferring knowledge and understanding directly to you, then a mysterious psychological force is turned on—the mentor/apprentice relationship.

There is nothing about the mentor/apprentice linkage specific to learning science—it is a constant of human society. It is how the metalworkers of Benin passed on their expertise; it is how a 17th century student of a Raku master in Kyoto himself was transformed into a ceramic artist; it is how coaches motivate athletes.

The reason the relationship works so well as a learning/teaching strategy is, we believe, two-fold: First there is a simple motivating force: the student admires the mentor (admiration does not exclude resentment of a perceived taskmaster), the student *wants* to gain the mentor's ken. Second, learning is not a process which insists on perfect understanding at every step. That's a caricature of mathematical proof. At its best, learning in science is a nonlinear sequence of observing facts, then trying to explain them, and in the process gathering or being confronted with further facts and continuing to augment one's understanding. In the sequence we outlined, confidence (call it faith) that the mentor has wisdom and tools to impart can make the learner accept facts (on faith), secure in the psychological confidence that the mentor will explain, in time. To put it another way, the mentor/apprentice relationship can guide the learner through unavoidable boring or tough stages, toward mastery.

In some way the process of learning recapitulates the scientific method. And science is hardly just the building of theories.

Enabling Learning and Teaching

We have called the teaching process magical and mystical; so is learning. People have taught and learned for tens of thousands of years; the biological roots of learning are older still. There is no one way to teach or learn, yet we think there are some identifiable underlying psychological principles that enable good learning:

1. Empathy: The teacher must care, and God knows it is difficult to do so when there are four classes to teach, inadequate pay, distractions diverting attention from the students, and social problems as obstacles. But the students have carefully tuned emotional antennae that detect care. And a good number respond.
2. Active learning: Any teaching strategy that stimulates participatory activity on the part of the student—we have stressed a variety of such—will make learning so much easier.
3. Judicious interplay of groups and individuals: Learning is a solitary action, yet it can be enhanced by episodes of group activity. Such interplay is often observed in society, for example, in the way kids master any sport (dribbling practice in

soccer, a team scrimmage) or learn music through taking part in a marching band. And the group interplay at a meeting of professionals from any discipline demonstrates learning at its best!

4. Empowerment: Students love to feel capable.[31] We have seen countless students get hooked on studying and learning once they saw their abilities growing dramatically, through their own efforts. Anything we do that successfully empowers students will work, for they want to succeed.

We in the Academy expect students to acquire information, strategies, and critical thinking skills that allow them to learn from our teaching. There should be no less expectation that instructors think critically and seek out specific strategies to improve performance in the classroom or lecture hall. The suggestions we present here are not prescriptive; we just want to share with you some of the strategies we have improvised and developed over the years to facilitate learning for, rather than to deliver instruction to, the students we have taught. We hope that you will find them to be useful tools in your own teaching and learning.

Acknowledgements

We are grateful to many people for their thoughtful comments: Deena S. Weisberg, Brian Coppola, Stephanie McGuire, Robert Root-Bernstein, James Wandersee. A much abbreviated and different version of this article was published in *Science* 325 (2009): 1203.

Notes

1. http://sas.calpoly.edu/asc/ssl/notetaking.systems.html.
2. W. Pauk, *How to Study in College* (Boston: Houghton Mifflin, 2000).
3. http://www.thefulcrum.ca/node/580.
4. J. D. Bransford, A. L. Brown, and R. R. Cocking, eds., *How People Learn: Brain, Mind, Experience, and School* (Washington, D.C.: National Academy Press, 2000).
5. K. A. Kiewra, Investigating Note-Taking and Review: A Depth of Processing Alternative. *Educ. Psychol.* 20 (1985): 23.
6. A. King, Comparison of Self-Questioning, Summarizing, and Notetaking-Review As Strategies for Learning From Lectures. *Am. Educ. Res. J.* 29 (1992): 303.
7. M. G. Jones and L. Brader-Araje, The Impact of Constructivism on Education: Language, Discourse, and Meaning. *Am. Comm. J.* 5 (2002): 1.
8. D. Johnson and R. Johnson, *Learning Together and Alone. Cooperative, Competitive, and Individualistic Learning*, 4th ed. (Boston: Allyn and Bacon, 1994).
9. L. Springee, M. E. Stanne, and S. S. Donovan. Effects of Small-Group Learning on Undergraduates in Science, Mathematics, Engineering, and Technology: A Meta-Analysis. *Review of Educational Research* 69 (1999): 21; http://www.wcer.wisc.edu/archive/Cl1/CL/resource/scismet.htm
10. http://teaching.berkeley.edu/docs/study_groups.pdf; http://ezinearticles.com/?5-Tips-to-Form-a-Successful-Study-Group&id=2458407.

11. http://www.how-to-study.com/study-skills/en/studying/38/study-groups/; http://web.mit.edu/uaap/learning/study/groups/index.html; http://businessmajors.about.com/od/studentresources /a/Study_Groups.htm.

12. S. Haynes, D. C. S. Richard, and E. S. Kubany, Content Validity in Psychological Assessment: A Functional Approach to Concepts and Methods. *Psychol. Assess.* 7 (1995): 238.

13. S. Alvarez, Music of the Elements. *New J. Chem.* 32 (2008): 571.

14. M. K. Morrison, *Using Humor to Maximize Learning* (Lanham, MD: Rowman & Littlefield Education, 2008).

15. R. Hoffmann, preface to H. W. Roesky and K. Möckel, *Chemical Curiosities* (Weinheim: VCH, 1996), p. v.

16. Lyrics and music for "Too Much Fun" are by C. Wright and T. J. Knight.

17. S. Cassidy, Learning Styles: An Overview of Theories, Models, and Measures. *Educ. Psychol.* 24 (2004): 419.

18. www.vark-learn.com.

19. www.myersbriggs.org.

20. S. McGuire, personal communication, November 27, 2009.

21. S. Y. McGuire, Using the Scientific Method to Improve Mentoring. *The Learn. Assist. Rev.* 12 (2007): 33.

22. S. Y. McGuire, in *Survival Handbook for the New Chemistry Instructor*, ed. D. Bunce and C. Muzzi (New York: Pearson Prentice Hall, 2004), chap. 8.

23. B. S. Bloom, ed., *Taxonomy of Educational Objectives: The Classification of Educational Goals, Handbook I: Cognitive Domain* (Philadelphia: David McKay, 1956).

24. L. W. Anderson and D. R. Krathwohl, eds., *A Taxonomy for Learning, Teaching and Assessing: A Revision of Bloom's Taxonomy of Educational Objectives: Complete Edition* (London: Longman, 2001).

25. J. D. Novak, *Learning, Creating, and Using Knowledge: Concept Maps As Facilitative Tools in Schools and Corporations* (Mahwah, NJ: Lawrence Erlbaum Associates, 1998).

26. J. D. Novak and D. B. Gowin, *Learning How to Learn* (Cambridge: Cambridge University Press, 1984).

27. J. H. Flavell in *The Nature of Intelligence*, ed. L. B. Resnick (Mahwah, NJ: Erlbaum, 1973).

28. J. D. Bransford, A. L. Brown, and R. R. Cocking, eds., *How People Learn: Brain, Mind, Experience, and School* (Washington, D.C.: National Academy Press, 2000).

29. R. Hoffmann, *The Same and Not the Same* (New York: Columbia University Press, 1995).

30. R. Hoffmann and B. P. Coppola, Some Heretical Thoughts on What Our Students Are Telling Us. *J. Coll. Sci. Teach.* 25 (1996): 390; Chapter 23 in this book.

31. J. P. Raffini, *150 Ways to Improve Intrinsic Motivation* (Boston: Allyn and Bacon, 1995).

Part 5

ETHICS IN SCIENCE

25

Mind the Shade

ROALD HOFFMANN

Sentenced to create—be it molecules, or laws, or paintings you may love or hate—we give in, with feeling, make new substances, transform old ones. Still others in the economic chain sell them; I teach about them. Each of us has a role in the use of chemicals. That use does immense good. And just sometimes does harm to people or property. Even though molecules are molecules, not in and of themselves good or evil.

What is an individual chemist's ethical responsibility when this occurs? Well, each of us confronts ethical questions in the light of his or her traditions. Nothing is simple when goods collide. I don't want to preach; the only advice I would presume to give is: "Mind the shade."

Let me explain. Political campaign ads to the contrary, very little in this world is pure good or pure evil. Yet evil gets done. No, it is not the work of Satan; it is the work of pretty normal men and women, who are likely to be kind to their children and goldfish. And those who mean ill intuitively know that responsibility for exploitation or hurt had best be diffused, so that an individual in a necessarily long chain be little tempted to see the ethical consequences of the whole.

Also people intent on no good construct, subconsciously, for themselves (and their collaborators) a mind-set that transforms the act psychically, taking it outside some personal ethic. In the analysis of evildoing by real people, not comic-book characters, one finds incredible compartmentalization, and the fanning of dehumanizing prejudices. Why? To self-justify actions that—in another part of life, dealing with others—would clearly be counter to the ethics that everyone, even evildoers, carries around.

Given this tendency of evil to diffuse and transform itself, it is precisely those actions that are ethically gray or shaded, neither clearly good nor bad, which should be thought through in greatest depth. If there be a data point that indicates disagreement with a theory, or hints at side effects of a drug, shall I discard it before I tell my supervisor? To do so seems easy, so harmless, especially when little is certain. There will be other tests, right? But the cumulative effects of such selective shading may be disastrous.

Have I conveniently put the burden of ethical judgment on the chemist in industry rather than the academic? And am I constructing a guilt-ridden world where people are to be condemned, by others or themselves, for innocent creation that is used for evil purposes by others? Let me try to work through these concerns.

The mass production of a substance affects many people, and I think indeed needs to be subjected to greater ethical scrutiny than an individual action. But the initial discovery—even if it be just a playing out of curiosity—is not devoid of ethical content either. A common distancing strategy is to say, "Oh, I just made that, I couldn't imagine it would be misused." My personal response is, "Yes, that's true, you (the maker of the HCN in Zyklon B, the tobacco farmer) are not *legally* responsible. Your *moral* responsibility for misuse may be tiny, or it may be substantive. That needs to be negotiated in a dialogue with yourself, with the help of those who give you counsel. At the least, it should serve as a small mental red flag the next time you make a molecule."

I am aware that we do not need more sources of incapacitating guilt. And the situational complexity of ethical decisions is immense. The job that may be dangerous to oneself, producing something capable of misuse, may be the only way to feed a family.

Small things, small decisions. In the shaded areas, where nothing is clear. That's where real, tough ethical decisions are made. We should be grateful that we are presented with choices that only human beings can make.

26

Science and Ethics

A Marriage of Necessity and Choice for This Millennium

ROALD HOFFMANN

1

Think of the last two hundred years. Incredible things were given to us by other human beings, in art, music, and social structure—who will question the value of *War and Peace*, or a Cezanne painting, a Beethoven quartet. Or the end of slavery, the empowerment of women. One should not compare the incommensurate, but there is no question that among the greatest achievements of these years is the gain in our understanding of the world within us and around us, the outcome of science.

The achievements of science are of value to humanity in material and spiritual ways. Our own chemistry has so much to be proud of—the extension of life expectancy from forty years to seventy (in part of the world), birth control, synthetic fertilizers to feed twice as many people as could have been fed before, chemotherapy, synthetic fibers and plastics with their myriad of uses, a greater color palette for all. Science, coupled with technology, is democratizing in the deepest sense of the word—it makes available to a wider range of people the necessities and comforts that in a previous age were reserved for a privileged few.

The achievement is also spiritual. I am not talking only about the indirect benefit of being able to hear Bob Marley and the Wailers, or a raga, Sind Bhairavi, on a CD. Anywhere in the world. I speak of the direct spiritual value, of *knowledge* gained of how genetic information is transmitted, stars are born, or how the color of a cornflower comes about. A knowledge that may not be of material value, will not make millions, but still makes the human spirit soar.

2

Given this incredible gain in our knowledge, and the ever-so-clear material improvements in our life span and comfort (and that of much, not nearly all, of the world as well), it is clear as we stand at the beginning of the 21st century that (a) people are not any happier than they were, say, 100 years ago. And (b) that many do not praise the achievements of science and technology with enthusiasm, but question them. Or are suspicious of them.

How to deal with this? One could push the concerns aside, and on the first point say pessimistically that it is not in the nature of people to feel happy, that inherent in the human condition is to create dissatisfaction for ourselves and among us when happiness is to be felt. And one can counter the second observation, that people question or suspect science, by pointing to surveys of public attitudes that consistently place scientists high up, to be admired.

I think the ambivalence on both points—people pleased by technological progress and health, people finding ways to be unhappy; people admiring scientists, and at the same time suspicious of science—contain in themselves one starting point for understanding. It seems to me obvious (and I would urge chemists in their rationalist mode to accept it) that people are not machines; they are wonderful and vexing creatures that move to the work of creation (or survival) in ways that mix reason with emotion. Logic and science do not suffice to understand people or societies.

3

What are the concerns that bother people most about science, the source of their suspicion? I think the worries are environmental, ecological, and moral.

Within two centuries—the centuries of modern chemistry—science and technology have transformed the world. What we have added, mostly for the best of reasons, has modified qualitatively the great cycles of the planet. The amount of nitrogen fixed from the atmosphere by the Haber-Bosch process, that masterpiece of chemical ingenuity, is comparable to global biological nitrogen fixation. These changes have been wrought in the geological equivalent of the blink of any eye.

We see the effects of our intervention in big things, such as global warming, in small things, in ... why we wash an apple—reflect, please, how different the reason for that simple action we teach our children *is* from the time of our grandparents. There is a good reason why the original of Michelangelo's David was moved off the Piazza della Signoria in Florence. There are very good reasons why we should wake the environmentalist within all of us.

The moral concerns arise from a mix of assaults on basic human suppositions which are deeply moored in our culture. Often they are mind sets that have no logical basis (remember, I accept that the logic of emotions is different). The questioning feelings, a kind of innate queasiness that some order is crossed, often emerge out of a tension within us about the natural and the unnatural, a distinction that chemistry is often hell-bent on violating. The conviction of a Vietnamese farmer that he be able to replant next spring a staple crop from the seed of this season is so deep, that the intended violation of that time-honored cultural truth by an agricultural company is viewed as a moral transgression. Modern medicine's modifications of reproductive technology, all done for the best of reasons (the deep unhappiness of not being able to conceive cannot be touched by someone who does not experience it), raise a multitude of moral concerns.

I think that in the last fifty years our attitudes toward pollution and environmental violation have changed essentially. They have crossed some ill-defined borderline from being viewed as just a mess we have made, the clutter of our children's room, a giving in to entropy, to.... a word I will use in its rich theological sense... to a sin.

4

To push these concerns aside as unthinking, uninformed anti-intellectual opposition is to miss the point. Real, smart, normal, thinking and feeling people are concerned about where science is going, what it is doing to people's lives. How to deal with this?

What I see is a three-fold response in the millennium at whose door we stand. First, an acceptance of human nature. Second, an explicit introduction of ethical and ecological concerns in what we do as chemists. And third, a reawakening of one of the primary motives of science, to meliorate the human condition.

A word about each of these: As I've said, human beings are not logical machines. If they were, there would be no creation, no, not even scientific creation, for the facts, no matter how carefully obtained, are mute. Deduction and induction are both trivial—understanding builds slowly, out of wildly imagined stories that are continually tested against reality. An acceptance of the role of emotion and feeling in human reactions to science is essential. A superrationalist approach gets scientists nowhere in their interaction with people.

Second, the environmental and ecological consequences of our intervention in this world are very real. I have already mentioned our wholesale perturbation of the nitrogen cycle. The same goes for sulfur; and we all follow the carbon story with greatest concern. James Lovelock thinks that we are beyond sustainability in our relationship with Gaia. Sustainability smacks of equilibrium; the earth is an open system in the thermodynamic sense, and equilibrium may always be a local state, or never achieved. Still I think we must try. Technological fixes (e.g., a good way to fix carbon dioxide) we will attain, for sure. But I think that to gain the confidence of the world, we must go beyond that—we must show that we care about what bothers people. I think in this millennium science must change so that absolutely every one of its actions is accompanied by ethical and environmental impact assessments. On the part of scientists, and on the part of the institutions that support them.

Third, I think that we have to show people the bona fides of our concern, by directing our work for betterment of the human condition. All human creative activity—art and science—is driven by a mixture of selfishness and altruism. As optimistic as I am about human beings, I have a feeling that professionalized science has drifted toward selfishness and solipsism. Attempts by governments to direct research are viewed as encroachments (often for good reasons, admittedly). The new is patented too often, and food for the soul set out for sale to industry all too cheaply. I think we could do much better.

Can we change? Oh yes, I think science will change in the new millennium. It can change, you know, it did change—think about a time when there were no Asians, blacks, Jews, or women in the system. And, you know, people did science without computers. Science is not immutable, it is a socially and intellectually adaptable system.

5

Underlying all three of the points in my response are questions of ethics and the social responsibility of scientists. I think that as part of our motion to gain the trust of the world that we think we deserve, must be a perception that we aspire to be as ethical as human beings can be (no more), and that we are concerned about our responsibility to society as creators and experts.

Ethics, the pondering of what is right or wrong, of good and evil, and of human responsibility, enters into what we do in a number of ways. I divide, quite arbitrarily, ethical concerns into three categories, and then bring them together.

1. Interpersonal ethics, the consequences of interactions between individuals.
2. Ethical question that arise as individuals deal with societies. A very important subcategory here is the class of interactions between individuals in a microsociety, say chemistry, actions which build or damage the microsociety. Which is often a fragile community of spirit as well as professional substance. And, importantly, this is where the social responsibility of scientists enters.
3. Societal ethical concerns. In this category I see the transcendent questions of fairness between countries and cultures, between minority and majority groups, between richer and poorer societies.

Let me elaborate on each of these in detail, in the context of chemistry. It has been given to me personally to be involved in questions in each category in the last few years, and there is no way for me to avoid dealing with that involvement.

6

Interpersonal ethics is what we know best. Some things are easy—the matter of forming a queue as a bus loads, or not leaving off a paper a person who has contributed to the work. Whether we derive our ethics from some normative standard, or a utilitarian calculation, human actions have the capacity of causing physical or mental pain. The feeling of having been dealt with unfairly or unjustly come in that latter category, and is perceived as strongly as physical pain. And in science, where ideas are currency (curious, isn't it, in a science where universals are valued, not particulars—why should it matter who first wrote down $E = mc^2$?), perceived violations of intellectual property are experienced ever so strongly. You can see this at work in Corey's claim of having told Woodward the essence of the orbital argument for the course of electrocyclic reactions, a matter of which I have written elsewhere.

7

There are ethical concerns that arise between individuals and societies. The falsification of a measurement in an experiment or the report of a fictitious calculation falls here; no other individual may be directly hurt by such an action (some exceptions to be noted later), but part of the social contract between an individual scientist and his society is violated. Sometimes these actions may be illegal, sometimes that may not be legally actionable but strike at the heart of the precarious relationship that makes for a society of, say, chemistry.

Let me give two examples: Suppose groups of Professors A and B are making molecule X, group A submits a synthesis to a journal, which is sent to B for reviewing, and the review is delayed (for ostensibly good reasons, it is argued), while B submits a paper with its synthesis. And was it an editorial mistake to send A's paper to B, or was it not?

A second example is the taking of ideas or details from a proposal to a government granting agency on which one is a reviewer. These are society-damaging actions of an extreme nature; the second one—stealing ideas from a proposal—unimaginably destructive to the system, as young people contemplate what they should or should not put into a grant proposal.

In a very different vein, but still in this general category is the cooperation of scientists in the work of totalitarian governments. Some of you have seen the painful debate ensuing from sensationalist (and most people think ill-founded) claims of Nazi collaboration against a chemist/physicist of heroic stature, Peter Debye. I am in a minority of people who think that Debye's signing as head of the German Physical Society a letter expelling in one week after Kristallnacht 1938 all the remaining Jewish members of the Society was an unethical action, the outcome of unthinking opportunism. It was an action against chemistry.

I will come back to talk of the social responsibility of scientists.

8

In the third category I see the transcendent questions of fairness between societies, between nations and minority and majority groups, between richer and poorer societies. These are some of the most difficult ethical questions to deal with, and at the first sight it would seem that chemistry has little to do with these.

First some examples: Journals are our window into the world. I remember well my days at Moscow University in 1960, how carefully read each issue of *Science* or the *Journal of the American Chemical Society* was in Moscow in those days. Computers are supposed to have opened up the world to all. But what if not a single library in Rwanda can afford the electronic edition of that journal?

The recent developments in HIV therapies raised, as you have seen, important questions for some countries: Should these countries violate patent law and purchase low cost generics for their people?

Is there some reality in the argument that most large pharmaceutical companies would rather work on behavior-modifying drugs of the antidepressant or Viagra type than on malaria or infectious diseases?

I believe, with some economists, that market forces alone do not provide optimal conditions for considered human existence. One has to intervene. The dream of fairness and social justice will not vanish. Science has a great part in making it come true. I will return to this.

9

But let me first return to what I see as our social responsibility to our fellow human beings.

Molecules are molecules. Chemists and engineers make new ones, transform old ones. Still others in the economic chain sell them, and we all want them and use them. Each of us has a potential or real role in the use and misuse of chemicals.

We are sentenced by our nature to create. There is no way to avoid investigation of what is in or around us. If you don't find that molecule, someone else will—there is no way to hide a facile synthesis of an immunosuppresant that saves, or a potent narcotic that destroys. At the same time I believe that as we create, and in doing so testify that we are human and alive, we also have responsibility for thinking about the uses of our creation. Even of abuses by others. And, importantly, we have the responsibility of voicing our concerns.

What about the ethical neutrality of science? I think this seductive phrasing of the question leads one away from the essential ethical consideration, by distancing an object from the human being that created it. I believe, and there is some philosophical tradition that supports this, that in any action by a human being, the instrument of that action (a gun made or fired, a molecule synthesized and sold, yes, even a mathematical equation or a poem) must be accompanied by a moral judgment. The judgment is: "will the use of that instrument by me (or by others) hurt people, or not?" The invention or implementation of a tool without consideration of the consequences of its use is deeply incomplete.

Science is not ethically neutral.

10

But what if that moral consideration of use and abuse is made by an individual, in considered judgment? And then, as sometimes happens (with a host of narcotics, with chlorofluorocarbons, with Zyklon-B), the molecule is misused? What responsibility does the original creator have?

You will see the characters in my play, "Should've," voice different opinions about this. I would say that we have then that we have the makings of Greek tragedy. All of you know the terrible story of Oedipus, the ill-fated and good king of Thebes, as told in Sophocles' play. In an essential part of his story, Oedipus kills a man at a place where three roads meet. He kills him in self-defense, justifiably. Later he discovers that the man he killed was his father Laius, and Laius's wife, whom he married, and by whom he had four children, was his mother. Oedipus blinds himself, but he lives on.

I will tell you another story of responsibility. During the Civil War in the U.S., President Abraham Lincoln went off for retreats to the Soldier's Home on the outskirts of Washington. It was a place where soldiers came in, to heal. Or were brought to be buried in the cemetery nearby. The soldiers Lincoln had sent, for the best of reasons, to be killed. Lincoln saw them every day. He lived with that tragic responsibility. Later presidents...well, they have not emulated this great troubled man.

Scientists must live with the consequences of their actions. It is this which makes them actors in the glorious tragedy that is life and not comic heroes on a pedestal. It is this responsibility to humanity that makes scientists human.

11

I think the way science and technology will be done in this millennium will be different. The scientists and technologists of the future will have learned a lesson from the worries of people about the environment, concerns which at the deepest level are moral and ethical. Our largest companies are pretty close to being there, and have made sound environmental analysis—to factor into a business plan the consequences of a product to the consumer, to the workers who make it, and to the environment—a part of the way they work. Not that they don't try to get away with a lot—the story of the inexorable but contested road away from gasoline as a fuel for the automobile shows this well. Along that road lies the great (even if resisted) chemical achievement of the three-way catalyst, along that road lie two programs of discovery in progress, of an effective fuel cell, and the efficient solar cell system. But public and governmental pressure will prevail—don't let up.

I believe that individual scientists in the future will be much more aware of the ethical and environmental consequences of their curiosity-driven research. It has to be that way; as I said before the process of science without accompanying consideration of what the science is doing for or to human beings and the environment is deeply incomplete.

Marriages are not made in heaven; human beings, of choice and necessity, make them.

12

In the nineteenth century science and technology looked to people like a certain way to improve the world. Part of it was just talk. And part I think was real, a sense that reason, incarnate in science, would prevail. In curious ways, the ideology of Marxism partook of the optimistic nature of nineteenth-century science—if we could change Nature, we could perhaps change human institutions as well. This is not the place to talk of the failures of Marxism, a psychologically unperceptive, and in practice infinitely corruptible system. Curiously though, if Marxism is gone, something of what attracted people around the world to it, an underlying outrage at social injustice, remains. It is searching for an outlet.

Science meanwhile went its own way, became professional and institutionalized. The nineteenth-century narrative—of improving the world, somehow began to sound hypocritical in the rush to found companied, hype. Well, it is not that way for young people, who still enter our profession of transforming matter. Why do they do so? For the same reasons we did, attracted by the bangs, stinks, and colors. But they also enter it with unbridled optimism—you can see their palpable interest in green chemistry.

I think it's time to revive our dream. The UN's Millennium Development Goals are not a bad place to look at where we can begin, negotiated as they were out of the concerns of many nations and cultures.

Eradicate extreme hunger and poverty.
Achieve universal primary education.
Promote gender equality and empower women.
Reduce child mortality.
Improve maternal health.
Combat HIV/AIDS, malaria, and other diseases.
Ensure environmental stability.
Develop a global partnership for development.

There is a tremendous amount of chemistry in every one of them, even in places where you don't think, as in gender equality.

Let us take advantage of these goals. Chemistry is about change. We can, each in our small way, work to make the world a better place; we can ameliorate the human condition.

27

Honesty to the Singular Object

ROALD HOFFMANN

The theme before us is "Language, Lies, and Ethics." As a scientist and a writer I could think of the way language is used in science, how it differs from the language of poetry. I could examine the claims of science to approach truth, and how its standards of evidence differ from those of, say, the law. But let me take another tack, and begin by a look at storytelling in science, clearly a process couched in language. The moral implications of narrative will then take me to another place, to consider what ethical lessons, if any, might emerge from normative science.

Stories

Science tells some rollickin' good stories. So why are scientists so unappreciative of the necessity of storytelling for the success of their own enterprise? Why do they beatify Ockham's razor rather than the rococo inventiveness of their hypotheses?

Because they are afraid of "just so" stories. The Kiplingesque allusion points to one of science's historical antipathies—to the teleological. Countered by a human proclivity for exactly that, the teleological, in the telling of scientific stories. Is there also a suspicion of the particularity of language, when scientists are ideologically committed to infinitely paraphraseable universals?

Consider first the stories that emerge out of science. So many to choose from—the epics of continental drift, or the way one iron atom in hemoglobin communicates with another. Or amusing ones, like how the amount of vanilla claimed to be natural in French ice cream exceeds by a factor of ten the quantity of beans shipped from Madagascar. Which led to a cat and mouse game between the forgers of vanillin (the flavor principle here) and the scientific detectives who learned to distinguish between the natural and synthetic form of one and the same molecule.[1]

Or take a triumph of molecular biology, the working out of the chemistry and function of the ribosome. In Figure 27-1 is a schematic illustration—not an atom in sight in this representation—of this biomolecular "smart" factory. It is a complex

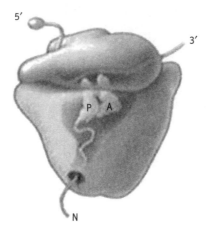

Figure 27-1 A representation of a ribosome, by Graham T. Johnson. Reproduced by permission from *Cell Biology* by Thomas R. Pollard and William C. Earnshaw (Philadelphia: Saunders, 2002), © Elsevier. The strand marked 5'–3' symbolizes the messenger RNA, the strand ending in N is the protein being synthesized.

of about 80 proteins and a few RNA molecules which takes a strand of messenger RNA (complementary to DNA) and initiates a process of linking, according to the RNA instructions, amino acids shuttled to site A by a transfer RNA to an already formed piece of the enzyme at another site P. And proofreading the enzyme coming off, at a rate of 20 amino acids a second.[2]

Shall I compare thee to a Rube Goldberg machine[3] (Figure 27-2)? (In England it would be Heath Robinson.) And is there a gaping trap in this simplistic mechanism of mechanistic visions? Oh, yes. The way we envisage the ribosome is mechanical, linear, and ... ephemeral. The representation, thrilling as it is, is transitory. Yet—and this is what some critics of scientific knowledge miss—this most unfaithful representation doesn't hinder us from designing real, functional antibiotics that throw a wrench into the workings of microbial ribosomes.

The ribosome story allows me to shift to something much more interesting. This is the utility, nay necessity, of storytelling *for* practicing science.

Why should storytelling be essential for science? Well, every time the simple is proffered, human beings fall for it. So admiration for the symmetrical molecules, exemplified by the ones shown in Figure 27-3, or for a simple mechanism of a chemical reaction, the aesthetic imperative in physics (if an equation is beautiful, it must be right) seems.... natural. And related to our falling for political ads, of any persuasion. But what if honest investigation of the real world reveal *complexity*, bound to be discovered in any biological or cultural entity that has been subject to inherently complexifying evolution? Even in a molecule. Take a look at hemoglobin, the oxygen carrier in our blood (Figure 27-4). This is already a much simplified

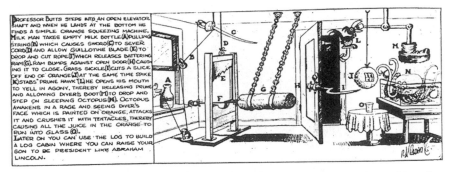

Figure 27-2 A cartoon by Rube Goldberg, one of the series of inventions by Prof. Lucifer Gorgonzola Butts. Reproduced by permission of Rube Goldberg, Inc.

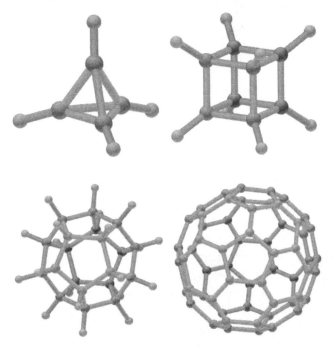

Figure 27-3 Some lovely symmetrical molecules, beautifully simple, simply beautiful, and…devilishly hard to make (except for buckminsterfullerene, the last to be synthesized).

representation, omitting the vast majority of the more than 9000 atoms in this $C_{2954}H_{4516}N_{780}O_{806}S_{12}Fe_4$ molecule. Where, and how, does one then find pleasure in such contorted complexity?

Storytelling seems to be ingrained in our psyche. I would claim that with our gift of spoken and written language, this is the way we wrest pleasure, psychologically, from a messy world. Scientists are no exception. We tell stories for they first satisfy, then keep one going. Stories "domesticate unexpectedness," to use Jerome Bruner's phrase.[4,5]

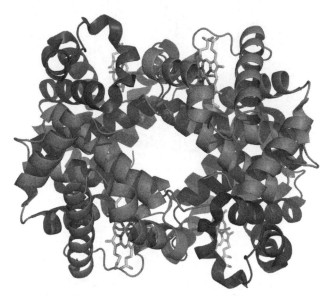

Figure 27-4 A schematic, "ribbon" drawing of the structure of hemoglobin. There are four subunits in the molecule, roughly identical in pairs. The ribbon traces the backbone of the biopolymer; note the helices. The oxygen is held at the iron atoms that center the four platelet shapes nestled in the folds of the protein.

Ethics Growing Out of Science

Almost every story has a moral, explicit or not. As Hayden White asks: "When it is a matter of recounting the concourse of real events, what other 'ending' could a given sequence of such events have than a 'moralizing' ending?"[6] Emily Grosholz notes perceptively that "...when we hear a story, we evaluate the agents and the action. Maybe this is because of the irrevocability of human action (it only happens once, so it better be good), and intentionality (we always do things for an end or reason)...The very choice of beginning and ending confers meaning."[7]

In the endings I have seen, more often in scientific seminars than papers, there is a curious mixture of celebration of the human achievement, with serendipity co-opted to serve design. And there's a double resacralization of the mundane, of what has just been demystified. In a spiritual process that I find refreshing, awe is expressed at what happens in nature or what human beings (OK, usually the author) can do. And the mysteries as yet to be resolved are articulated. With complete faith that they will be resolved.

This has led me to think not only about the morality emerging out of endings, but also of the potential of the process of science for constructing ethics. Here I follow, half a century later Jacob Bronowski's path. My *landsman* in more ways than one was much less tentative than I will be when he said "... the practice of science compels the practitioner to form for himself a fundamental set of universal values."[8]

Can ethics grow out of science? The very question may seem ludicrous to two communities of scientists: those who really believe that science is ethically neutral, and those who believe that scientists are inherently ethical. So let me first contend with these, as provocatively as I can.

To claim that science is ethically neutral ("I just worry about the technology of cloning. Someone else can decide if it's good for people.") puts scientists squarely in the company of anti-gun-control activists ("Guns don't kill; people do."). By contrast, I believe, and there is some philosophical tradition that supports this, that in any action by a human being, the instrument of that action (a gun, a molecule synthesized, yes, even a mathematical equation or a poem) must be accompanied by a moral judgment. The judgment is: "will the use of that instrument by me (or by others) hurt people, or not?" The invention or implementation of a tool without consideration of the consequences of its use is deeply incomplete.

As for the claim that scientists are born with ethics—well, that's just as likely as their being born with aesthetics or logic. That the latter is not true, you learn from reading the "peer review" referees' comments on your paper. We scientists are smart people who have opted to engage in a remarkable social system for garnering reliable knowledge, that knowledge being of great practical and spiritual value. The critical components of that Western European social invention, science, are (a) normal, curious people, some of whom like mathematics; (b) people not afraid of getting their hands dirty—experimenters; (c) an open system for dissemination of what one finds, and a communal urge to do so; and (d) a method that encompasses frequent dipping back and forth between approximations to reality (gauged by our occasionally misleading senses and our tools) and flights of imaginative fancy in hypothesis formation and theory building.

So...is there is something in the practice of science that can enhance the ethics brought to it by scientists, or that possibly can engender an ethical outlook? As the above makes clear, I do not come to this because I think scientists are "better" than other people—far from it. Nor do I dare presume that a relatively late social invention, science, could provide a broad rationale for a human quality as fundamental as ethics. (Or is ethics itself a social invention? If so, it is older than science. But not as old as curiosity.)

When goods collide, where do we get our criteria for deciding among them? From the usual sources, like them or not: our socialization at home and in our schools, i.e., from our parents and teachers. Perhaps from our genes, though not as much as E. O. Wilson would like us to believe. From churches and religions. From reading—novels are especially strong moral instruments. Not a tad diminished by deconstruction. By the time science enters a young person's moral consciousness, he or she is usually a pretty well-defined moral human being. For some, ethics may be a set of rigid schemata to be applied to any situation. Yet the web of life has a way of generating new quandaries; one's personal sense of what's right, how to act in difficult times, evolves even as it is moored in the past.

Two of the components I gave of science, publishing and the nervous motion twixt theory and reality, depend on texts, talks, and conversations. These generate narrative. And, even forgetting the moralizing endings, such acts of communication inevitably confront scientists with ethical choices—to be faced, evaded, negotiated. Let me expand on this.

An important part of the system of science is publication, with the potential of replication. How reproducible scientific findings are, and whether the reality of reproducibility is essential to belief, is a matter of contention.[9] It took several years for public questioning to surface of the all too novel measurements, a multitude of them, of Hendrick Schöen in solid state physics.[10]

Could it be that the primary emotional motivation for a scientist who does not falsify a synthesis or measurement is simply fear, rather than the psychoethical drive to report facts honestly? Perhaps, though I find definite positive value in fear in making us behave righteously. To a point. And fear of damnation, big and small, is certainly important in Christian ethics. It may be painful for most of us to see others, never ourselves, "... do the right deed for the wrong reason," as T. S. Eliot says in *Murder in the Cathedral*.[11] But I accept the way we are: The "habit of truth," as Bronowski called it, is formed in many ways.

Ethics is like a limb that needs exercise to function. The importance of publication is that it provides exposure to potential testing. Time and time again. Fraud in science is ultimately unimportant. There is much prurient interest in it, for sure. With the same origins as our fascination with the sexual misdeeds of our ministers. Priests of the truth have a longer way to fall. But fraud is unimportant because the psychopathology of its perpetrators is such that their sense of fear of being proven wrong is somehow abrogated, and they never forge the dull, only the interesting. Thus, the normal workings of the system ensure that others—out to prove the makers of the startlingly new wrong, not right—will repeat the experiment.

So the system works, but is the individual scientist, motivated by loss of reputation if proven deceptive, likely to become more ethical? A cynical viewpoint is that they will learn to sanitize, embroider and manicure just enough to get away with what they can. And pile on the hype. A more charitable viewpoint is that we learn that data are not only not to be trusted, but that they are mute, and inherently conservative. That a human being must interpret them—yes, tell a story about them. And that it is OK (within a self-correcting system such as science) to risk an imaginative, ornate hypothesis which does an end run around Ockham's razor.

Something salutary takes place in the writing of an experimental part of a scientific paper. I have trouble in picking one of my own to show you, for, sad-to-say, I'm just a theoretician. But here's a piece of a mixed experiment/theory paper in which I am a coauthor, ergo in part responsible:

Crossover Experiments. *In a 25-mL reaction flask was placed 0.050g (0.094 mmol) each of $Cp_2^*Th(^{12}CH_3)_2$ and $Cp_2^*Th(^{13}CH_3)_2$. The vessel was* *evacuated, and then 10 mL of Et$_2$O was condensed into the flask at -78°C. The suspension was stirred at this temperature until all of the material had dissolved and a colorless solution was obtained. The flask was then backfilled with 1 atm of CO and the solution stirred vigorously. After 4 h at -78°C, the solution was allowed to warm to room temperature whereupon a colorless solid ([Cp$_2$*Th (μ-O$_2$C$_2$-(CH$_3$)$_2$)]$_2$) precipitated. Next, 2 mL of degassed 1 M H$_2$SO$_4$ was added to the reaction mixture via syringe under a flush of argon. After the resulting suspension was stirred for 15 min, the mixture was centrifuged to remove a colorless, flocculent solid. The Et$_2$O layer was then separated from the aqueous phase. The aqueous phase was next washed with four 2-mL portions of Et$_2$O, and the washings were combined and then dried over MgSO$_4$.[12]*

Basta! You see a report of what was done, almost in iambic pentameter. Not the average run, but the best that was done, to be sure. It's there, this experimental part of a longer paper, for historical reasons: as evidence that it was done, that it can be done, with details reproducible by anyone (well, maybe). But why give the evidence? Isn't there trust in the community, aren't we all gentlemen? Or were, now that 35% of Ph.D.s in chemistry in the U.S. are women...

In citing another's experimental (or theoretical) work, there's a similar wrestling match on. To cite is an act of trust. Which can also be viewed as an act of mistrust, for by citing someone else without questioning the result, one is protected should it be faulty. To say that the mistrust complicit in the statement of conditions of an experiment, or citing someone's work, negates the trust overtly expressed by using the work, and that that's all there is to science, is to miss the fertile tilled orchard of science—the creation of molecules as well as frameworks of understanding.

This is the essential tension of which Thomas Kuhn wrote, between trusting and not trusting.[13] I think writing an experimental part of a paper, or reading it in someone else's text, not once but many times (I have written five hundred such, not untypical) is an ethically productive action. In which both subconsciously and overtly the issues of trust and mistrust are negotiated by chemists. The important word here is "negotiation": the web of habitual description and citation subconsciously (and explicitly) forces the creator to confront the other. It is an inherently social web, built out of real and imagined interactions with other human beings. In it are the makings of a gift economy. And of empathy.

First Time Narratives

I see two other places where in an interesting way ethics emerges from normative science. The first is the responsibility taught by first time narratives, first time representations. I remember for instance, when Fred Hawthorne—now at UCLA—came one 1961 day to Harvard, when I was a graduate student there, and told us how

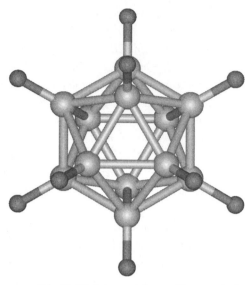

Figure 27-5 The structure of the $B_{12}H_{12}^{2-}$ ion synthesized by Pitochelli and Hawthorne, Ref. 14.

he had made $B_{12}H_{12}^{2-}$, a molecule shaped like an icosahedron (see Figure 27-5).[14] Nothing like it had been seen before; he described its properties with evident and appropriate excitement. Hawthorne knew instinctively that there was a story to be told of $B_{12}H_{12}^{2-}$, that it sufficed to tell it straight. *Das Ding an Sich* was indeed beautiful enough; it was sacred even as it came to be in his profane hands. In another day, another time, Fred would have said that it was given to him by the grace of God. In 1961 he called it serendipity.

There was no more questions of Hawthorne making up a fib around $B_{12}H_{12}^{2-}$ than of Haydn writing a dissonant section in one of his piano trios.

Representation as Furniture

Much of what we do in science is to represent reality. Those representations, whether in language or not, are murky mirrors. But, as Emily Grosholz says, "representation is also generative: we say more than we know we are saying, and we induce order by our orderings, and good representations, as intelligible things, add to the furniture of the world. So representations are both more and less than what they represent. By misrepresenting, they also allow us to know, and to create."[15]

Speaking of language and veracity, Oliver Sacks tells an interesting story in a film he made in the *Mind Traveler* series. In Eureka, CA he met a family of deaf Mexican farm laborers. Among the five children, the three older brothers, who did not sign, were suspicious of their younger siblings, who were learning sign language. Because they would learn to lie, the older ones said.[16]

When you see something for the first time, you don't know what it is. When you describe it for the first time, language will fail you. You grope for meaning. But there is no lie. Were we given more such moments!

Honesty to the Singular Object

A second experience is one shared by poets and scientists. Something is seen, felt, then described. Now not for the first time, but for the umpteenth. So love has fled, and it hurts to remember what was good. It *has* happened to others, though that thought seems not to comfort at all. A poem needs to be written—one is in the Luberon, in winter, one walks out in the morning into the vineyard, it's sad to face that beauty alone. But then there's a grape cluster, like no other grape cluster. It must be described:

RAISINS FOR BEING
They left small bunches
on the vine, green late-
comers; the farmers

knew the day to pick,
sugar rising in the
berries, rain offshore. But

four sunny days broke
the pattern; the vines free
of their luscious burden

filled out the stragglers.
And then I came, just
before pruning,

and walked out in
the morning frost, the sun
clearing the Luberon,

and a thousand droplets,
on a grape cluster,
muscat pavé, told me

that I had a latecomer's
right, to live life out
reflecting, free albeit

tethered, at an angle
to the sun, sweet to you.[17]

I describe, and I am not sad any longer. For a while.

Elsewhere there is a molecule I see in a journal (Figure 27-6).[18] I talk about it to one of my graduate students, Pradeep Gutta. It has at its center a ring, with two tins and two nitrogens in it. But as you see, the environments of the two tins are strikingly different. Why? Could it be because of their different substituents, the chemical shrubbery hanging off the tin atoms? No, for the molecule with all substituents identical is calculated to have exactly the same geometry. " 'Tis a puzzlement," as Yul Brynner said. And could one exchange the environments of the two tins, as shown in Figure 27-7? We calculate the way the electrons move in this molecule, their orbitals, orbits writ large. And we reason out a reason, because ... that's our métier. There's a story to be told, I tell it, as well as I can.

The language I use to tell my story is that of science, which is not the language of poetry, at least not much of the time. There is no premium on ambiguity in science. But that a word mean two things and sound like three other words, that ... is the stuff of poetry. What science and poetry share, even though they parted company, it seems centuries ago, is an honesty to the singular, determinate object. We tend to think science is after universals, the infinitely paraphraseable, to use again Gunther Stent's idea.[19] But science is not one thing, and maybe chemistry is different—we build shape, motion, and reaction on specific, variably persistent group of atoms. Trends matter, general theories less. And individual molecules, examined up close, most of all.

Craving understanding, we circle around the object of our affections. In love with the particularity, the "thingness" of this powder, just this shade of turquoise, we study it. Here is what William Blake said:

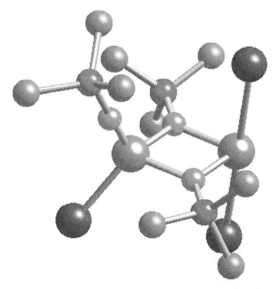

Figure 27-6 The structure of a molecule made by Chitsaz et al., Ref. 18 (Large gray spheres = tin).

Figure 27-7 A drawing at left of the same molecule as in Fig. 8, now pruned to its geometrical essentials. The sequence of molecules, left to right, indicates a hypothetical way that the environments of the two tins could interconvert—a set of steps in a molecular ballet.

> *He who would do good to another, must do it in Minute Particulars: General*
> *Good is the plea of the scoundrel hypocrite & flatterer:*
> *For Art & Science cannot exist but in minutely organized Particulars.*[20]

And A. R. Ammons, the American poet for whom art and science were not separated, in a section of his *Hymn*:

> *And I know if I find you I will have to stay with the earth*
> *inspecting with thin tools and ground eyes*
> *trusting the microvilli sporangia and simplest*
> *coelenterates*
> *and praying for a nerve cell*
> *with all the soul of my chemical reactions*
> *and going right on down where the eye sees only traces*
>
> *You are everywhere partial and entire*
> *You are on the inside of everything and on the outside*
>
> *I walk down the path down the hill where the sweetgum*
> *has begun to ooze spring sap at the cut*
> *and I see how the bark cracks and winds like no other bark*
> *chasmal to my ant-soul running up and down*
> *and if I find you I must go out deep into your*
> *far resolutions*
> *and if I find you I must stay here with the separate leaves*[21]

Sixty-five million of the seventy million compounds known are white crystalline solids. I give you four vials—all white powders: one is sugar, another salt, the third penicillin, the fourth tetradotoxin—the poison of the fugu or pufferfish. Will you play Russian roulette with these? Your body knows the difference. The difference, and its definition by the fallible powers of our mind and hands, is as beautiful as it is essential. The description of difference is one task the scientist does as well as it can be done.

Does the ethical bent inherent in the precision of language sought by scientists and poets make scientists and poets better human beings? No, no more than it improves those who professionally lead the considered life. The ethical impulse is strong, inherently human. It can be suppressed, most alarmingly by crowds and power, to use Canetti's phrase. And, remarkably enough, it can be suppressed by the flush of first creation: I'm thinking of the susceptibility to this of the saints—Sakharov and Bethe in science, Lowell and Sexton in poetry. Ethical thinking can be awakened, it needs to be reawakened, by consideration of whether a molecule can harm, by advances in reproductive technology, and…just what one can invent in a historical play, and whether a poem hurts a lover. Even a soap opera can teach ethics. We should be grateful for these little (or big) prods to ponder ethical choice.

The First (Fruitful) Intersection of Science and Ethics

I want to make a final point that returns to our cultural roots. The tree in the Garden of Eden of our primeval religious narrative was the Tree of Knowledge of Good and Evil. I take the *etz hadaat tov vera* as…the tree of ethics (a word not in ancient Hebrew), and the first link between science, narrative, and ethics. Let me bypass the question of why a just God would put ethics out of reach. He did. Continuing in my unrespectful/respectful *midrash* (which is similar to that of Zygmunt Bauman, Jean-Pierre Wils, and more recently of Leon Kass[22]), is not Adam and Eve's transgression implicit in the tale, serpent or no serpent? Without it there would be no narrative, no story of humankind. We'd still be between the four rivers, right?

Even before eating of the fruit of the tree—and the rabbis discuss whether it was wheat, grapes, or fig, with no apple in sight—Eve makes a decision: "When the woman saw that the tree was good for eating and a delight for the eyes, and that the tree was desirable as a source of wisdom, she took of its fruit and ate."[23]

The science in my mildly sacrilegious *midrash* is manifold. It sparkles in the knowledge that the tree conveys. Of what? Of oppositions and polarities. Of matter particulate and continuous, opposites attracting each other or repelling, of analysis and synthesis. Of what is to be hidden and what is to be revealed, of the same and not the same (those four vials). And the verse speaks, directly, of experiment. For this is what Eve hazarded, isn't it? She saw, and thought, and acted. She acted on beauty, for wisdom. Kant would approve. Eve did what had to be done, not to end but to begin a story. In which curious human beings have the choice between good and evil.

Acknowledgments

Friends helped me a lot in responding this topic, among them Sylvie Coyaud, Margery Arent Safir, and Emily Grosholz. I am grateful to Jennifer Cleland for her assistance with some research, and to M. M. Balakrishnarajan, Pradeep Gutta, and Beate Flemmig for some drawings.

Notes

1. R. Hoffmann, Fraudulent Molecules. *American Scientist* 85 (1997): 314.
2. D. N. Wilson and K. H. Nierhaus, The Ribosome Through the Looking Glass. *Angew. Chem. Int. Ed. Engl.* 42 (2003): 3464.
3. P. C. Marzio, *Rube Goldberg: His Life and Work* (New York: Harper & Row, 1973).
4. J. Bruner, *Making Stories* (New York: Farrar, Straus, and Giroux, 2002), p. 90.
5. R. Hoffmann, Narrative. *American Scientist* 88 (2000): 310; R. Hoffmann, Why Buy That Theory? *American Scientist* 91 (2002): 9.
6. H. White, *The Content of the Form* (Baltimore: Johns Hopkins, 1987) p. 23.
7. E. Grosholz, Personal communication.
8. J. Bronowski, *Science and Human Values* (New York: Harper & Row, 1965), p. xiii.
9. R. G. Bergman, Irreproducibility in the Scientific Literature: How Often Do Scientists Tell the Truth and Nothing But the Truth? *Perspectives* 8.2 (1989): 2.
10. L. Cassuto, "Big Trouble in the World of 'Big Physics.'" Online. Internet. September 16, 2003. Available: salon.com/tech/feature/2002/09/16/physics/print.html
11. T. S. Eliot, *Murder in the Cathedral, Part I* (New York: Harcourt, Brace & World, 1963), p. 44.
12. K. Tatsumi, A. Nakamura, P. Hofmann, R. Hoffmann, K. G. Moloy, and T. J. Marks, Double Carbonylation of Actinide Bis(cyclopentadienyl) Complexes: Experimental and Theoretical Aspects. *J. Am. Chem. Soc.* 108 (1986): 4467.
13. T. S. Kuhn, *The Essential Tension: Selected Studies in Scientific Tradition and Change* (Chicago: University of Chicago Press, 1977).
14. A. R. Pitochelli and M. F. Hawthorne, The Isolation of the Icosahedral $B_{12}H_{12}^{-2}$ Ion. *J. Am. Chem. Soc.* 82 (1960): 3228.
15. E. Grosholz, Personal communication.
16. O. Sacks, Personal communication.
17. R. Hoffmann, "Raisins for Being," in *Soliton* (Kirksville, MO: Truman State University Press, 2002), p. 28.
18. S. Chitsaz, B. Neumüller, and K. Dehnicke, Synthese und Kristallstruktur des gemischt-valenten Komplexes $[Sn_2I_3(NPPh_3)_3]$. *Zeitschrift für Anorganische und Allgemeine Chemie* 626 (2000): 813; P. Gutta and R. Hoffmann, Unusual Geometries and Questions of Oxidation State in Potential Sn(III) Chemistry. *Inorganic Chemistry* 42 (2003): 8161.
19. G. S. Stent, Prematurity and Uniqueness in Scientific Discovery. *Scientific American* 227 (1972): 84.
20. W. Blake, *Jerusalem: The Emanation of the Giant Albion* (Princeton: William Blake Trust/Princeton University Press, 1991), p. 219.
21. A. R. Ammons, "Hymn," in *The Selected Poems* (New York: Norton, 1986), p. 9.
22. J.-P. Wils, Pleasure and Punishment: The Temptation of Knowledge. *Future* (2003): 74; Z. Bauman, What Prospects of Morality in Times of Uncertainty? *Theory, Culture & Society* 15.1 (1998): 11. See also L. Kass, *The Beginning of Wisdom: Reading Genesis* (New York: Free Press, 2003).
23. Genesis 3:6. The translation is from *The Torah* (Philadelphia: The Jewish Publication Society, 1962).

28

The Material and Spiritual Rationales Are Inseparable

ROALD HOFFMANN

There are sound spiritual reasons for the ecological and environmentalist perspective—for minimizing pollution and harm to ourselves, to future generations, to the earth. Are these consistent with the material reality and aspirations of chemistry and chemical industry? One would like to think they are. But what of the realities? I want to take a hard, personal look at this fundamental tension. And also search for what is special about Green or Sustainable Chemistry,[1] facing up to the obstacles confronting the field. And, while reaching for a measure of transformation, a multifaceted Green Index, to come back to a moral perspective on our creative activities.

The Problem

Chemists and chemical engineers are prone to believe that the general public does not recognize the contributions that chemistry has made to our health and our standard of living. And we often cringe at the perception that others blame us (and the great industries that employ us) for fouling our own nest, the infinity of ways we have found of affecting adversely our bodies and the earth by producing on the megaton scale the unnatural.

Each of these adverse opinions can be productively discussed—both with the people whose adversarial or anguished arguments chemists react to, and with the chemists' exaggerated and defensive response to them.

The facts remain that the industries that transform matter (to which chemistry is central) have flourished to an extent that is staggering. They've played an essential material role in prolonging life, and while not making people any happier, they have provided spiritual value. The value I'm thinking of is not in creating the materials for CDs and books, ancillary tools to spiritual satisfaction, but in providing partial, yet unprecedented knowledge of the world. And the transformative industries are also responsible for an immense quantity of hazardous waste. The scale of their fecund creative enterprise is such that the major cycles of the world are perturbed. More than half the N and S atoms in our bodies have seen the inside of a chemical factory. And C, O, and H atoms too, through agriculture, food preparation, and sewage treatment.

Environmental and ecological concerns have evolved, from simple revulsion at smell and touch of the discarded, to a moral concern. With all the ambiguity of what constitutes sin in the Judeo-Christian tradition, I have no doubt that in the last fifty years the line dividing the ugly, unwanted and uncouth from the sinful and immoral has been crossed, in thinking of ecological violations. One can get people to ponder the creative ambiguities of natural/unnatural and find their way to weighing unpalatable ecological choices. But I think one will not change the underlying moral nature of their environmental concerns.

The intensity of the spiritual concern is new. But not its imperative. It has always been there, in our roots, in the story of Genesis. I will return to this at the end of the paper.

The Economic Imperative for Green Chemistry

It is clear that a chemical process working on cheaper feedstocks, or using a more efficient catalyst, or an alternative process with a lesser waste stream that must be treated, will be adopted by industry. Examples are available in the production of genetically engineered insulin, or vanillin from cellulose rather than petroleum origin. Individual companies may have resident brakes of an institutional resistance to change, so that innovation is likely to come from rivals. The excesses of monopolistic resistance—acquiring intellectual property, but then suppressing patently better processes, are the realm of much anecdotal speculation; the reality will always be hard to document.

Is a transition to considerations of atom economy[2] or the choice of solvents or reagents from a green perspective now, or will it be, an economic consideration for a company? Of course it is.

But no one has any illusions, except when they choose to close their mouths because jobs are at stake, on the inherent goodness of any business. If I were to make a new catalogue of the primary sources of moral corruption, I would put profit and money first, closely followed by politics (that aspect of a necessary human activity in which one acts on unthinking conviction), and crowds and power (the loss of individual moral perspective in a mass situation).

There are positive forces pushing companies the green way. An enterprise must have a product that is useful, that fills a demand or niche. It must also advertise and market that product. In communicating to the public, an image of benefit and responsibility and moral probity is central. So positive images, be they of a portly

Quaker or a barely sweating bicyclist nearing the top of a hill, abound. The public's moral concern with ecology and the environment has been used by industries, in advertising alternative biological agents, ice cream flavoring, or a hybrid vehicle.

At the same time, I fear that if no money has to be spent on testing drug safety, or on finding alternative chemistries, or disposing of a certain concentration of mercury, it will not be spent. The cases of abuse, petty to criminal, are legion. I don't want to retell here the thalidomide story, the Spanish cooking oil crime, or the patent medicine that poisoned so many in prohibition times. Let me rest in a middle ground, the response of the U.S. automobile manufacturers to regulatory demands on automobile emissions by the nice conservative people of California. Incredible lobbying followed, with claims that (a) the automobile was unimportant as a source of air pollution, (b) emissions of CO, unburned fuel, and nitrogen oxides just couldn't be reduced to the degree demanded, or if they could, (c) it would be prohibitively expensive. Once the California guidelines were adopted, the automotive industry went ahead to do what the creativity of its employees promised—to produce an economic three-way catalyst that reduced all emissions by factors of 10 to 40 (a remarkable achievement), meliorated California photochemical smog, and quite incidentally, saved the lives of thousands of ill people intent on committing suicide by running their car engines in an enclosed space.

One also has to pay tribute here to the push of our trade unions toward safer working conditions in the chemical industry. That we don't read of deaths in megaton industrial hydrogen and sulfuric acid production is something to marvel at. Negotiations and the threat of strikes are part of the process; one has to admit that the role of the unions has weakened over the last decades as a shrinking world and the lowering of trade barriers have allowed companies to shift dangerous and polluting manufacturing to developing countries. The unions' overall positive role is not damaged by aberrations such as the baseball players' "union," which has consistently given priority to the pay of its members rather than their health.

I think green chemistry is indeed part of the economic planning of major transformative industries. But it is only so because of continued regulatory pressure by municipalities, states, and the federal government.

Which brings me to regulation. Attitudes toward regulation depend first of all on politics, and then on whether one is at the receiving end of regulatory practice, or a citizen consumer of the products regulated. People are inconsistent here, because they are people. My own prejudices should be made clear—I'm a Democrat and a liberal; I think there is insufficient regulatory activity to safeguard American health and well-being.

Even with years of Republican administrations cutting regulatory agency budgets and purview, there has been sufficient legal activity to make compliance with regulatory codes a substantial part of the financial planning of companies. This, in turn, assures that practices which are effectively Green Chemistry are given a place in industrial planning at the very highest level. Not willingly, I repeat, but still a place. What unwilling compliance does to attitudes in industry to innovation in Green Chemistry, to the kind of innovations where the immediate economic benefit is not obvious, I don't know. But it's OK, let's follow the push of the economical and regulatory incentives, they make people think. Push has a way of turning into self-propulsion.

Maybe there's a bigger problem for Green Chemistry in academia.

Fundamental Chemistry, Mainly in Academia

Though much important fundamental research gets done in industry, the spiritual center of chemistry in our country remains in the research universities. Where people are taught, values are formed. The Catholic Church knew that well in medieval Europe, Islamic madrasas try it today. Thinking more positively, in the process of education, our teachers enable us to find within ourselves what we did not know we had the power to do. No wonder that we admire them, emulate them. Or, in the ultimate profession of the new, science, in an Oedipal fit, we seek to forget them, if not slay them.

The way to atom and energy economy, to benign solvents and waste stream processing has a harder road ahead in academia than in industry. Here's a sampling of the negative responses to be overcome: (a) "it's boring tuning of known syntheses, or the umpteenth synthesis of something ordinary;" (b) "the principles are old hat, it's mostly chemical engineering" (of course said by chemists who've never faced up to scaling up a process); (c) "it's not a road to the National Academy of Sciences."

The overcoming of so much prejudice in a profession (university teaching and research) that is second only to the building trades in its conservatism, is not going to be easy. In industry there was the stick of the bottom line and the law. These don't seem to matter for professors.

But don't lose hope. I think the components in a strategic response are, in my opinion (a) money, (b) sticks and carrots, but since professors don't respond to rules they way industry has to, I suppose it has to be mostly carrots. Ergo more money; (c) role models; (d) textbooks; (e) and intellectual positioning, which I will try to provide.

It is a time of constant dollar funding, fortunately also a time of stabilization (in the U.S., not China) of the number of scientists. It seems worse than it was, so people are bitching. And while they are incensed about the tendency toward more and more mission-oriented research, they'll go, of necessity for where the money is. I would like to see NSF (yes, basic research), and NIH (yes, the rubric being preventative medicine, broadly interpreted) and Dept. of Commerce (as with catalytic converters, the U.S. has something to sell here) programs directed overtly toward Green Chemistry.

Industry, despite much publicity, supports universities in a tiny way, getting the Ph.D., its most valuable asset, essentially free of charge. Industrial enterprises are also hard pressed to do their own research, internally—sadly, this is where they foolishly economize in tough times. Can U.S. industry find places where Green Chemistry related research institutes can serve a common need for fundamental knowledge: I can think of supercritical fluid properties, catalysis, and a common institute to evaluate environmental impact (more on this later).

More money: The initially DOD and then NSF-funded series of Materials Science and Engineering Research Centers around the country have been both productive and provided visibility for a science in formation. I think a series of similar centers in Green Chemistry would really take off. My bet is that most universities with strong agricultural and engineering faculties would find that they already do an awful lot of Green Chemistry, and all that is needed is a thematic thread. At first such themes seem just artificial, a funding maneuver. But human beings have a way of transforming pro forma connections into true collaboration and innovation. And universities are hopeless addicts to centers.

Role models: As I said, universities are awfully conservative, even as they provide a social setting for incredible individual innovation. Green Chemistry is inherently applied. Its creativity is expressed by working within constraints—not quite organic and inorganic chemistry's way of being original. There the first synthesis matters, and maybe one doesn't care if an esoteric or dangerous solvent is used. Perhaps it even adds a tinge of vicarious excitement.

If students are to be attracted to Green Chemistry, they need models. We need proponents of this way of creation, and even some heroic figures. Who, in the way of heroes, may or may not be really heroic. To avoid getting into trouble, let me move to a look at what I see as a parallel case of successful ascendancy by an applied discipline into the conservative academic chemical mainstream. This is polymer chemistry. Perhaps we have something to learn from that field.

So here was this large group of talented applied organic chemists, who had learned to exercise fine control in the synthesis of nearly infinite linear systems. And, boy, did they add value. Polymer chemists were also eminently employable. But there were almost no synthetic polymer chemists at the major universities, and in organic textbooks polymer chemistry was either a bit of color or a tacked-on chapter.

That was 25 years ago. Today every Chemistry Dept. wants polymer chemists, and graduate students are very interested in the field. There is no time here to go into the complex of reasons for the transformation, but part of the story is the recognition by some very good organic and organometallic chemists, in the U.S. Robert Grubbs and Charles Casey prominently among them, that there were fascinating chemical problems in polymer chemistry. They received recognition for their work because they already had more than an apprenticeship, but real status within the traditional community. And they played the important role of a legitimizing bridge.

Polymer chemistry had the advantage of working from a solid industrial base; people wanted polymer chemists, even if the major universities were not producing them. A similar economic base moved the field of analytical chemistry, suffering from lack of academic prestige in the second half of the 20th century. Role models, eminent converts, heroes and heroines are needed in Green Chemistry.

Textbooks may be an easier part of the academic enterprise to penetrate. The current trend in texts is for real world examples (those boxes with too expensive photographs), and an inherent context of justification in societal terms. No complaint, let's take what we get, no matter how we get it. Just dissolve the edges of those boxes, and let them "penetrate" the main body of the text.

An Intellectual Place for Green Chemistry

Science is a remarkably successful Western European social invention, a system for gaining reliable knowledge from the collective effort of curious people, some of them, but hardly all, mathematically inclined. Essential to its functioning is experiment, the willingness to get one's hands dirty. This alternates with flights of organizing theoretical fancy. Crucial to science is a healthy skepticism about the reliability of knowledge acquired by others, if not oneself. And a requirement, turned into compulsion, to share with others what one has found. Also, this is important for what follows, a desire to improve this world, be it just a little.

And chemistry? A two-fold art, craft, business, and science of, from one perspective, substances and their transformations. Also, at the same time now, of persistent groupings of atoms, molecules, and their transformations.

Is there a place for Green Chemistry in the intellectual scheme of things? Of course. But I mean is there a niche, a special feature? One other than just economic viability, one that would provide an underlying ideology, pull in young people at the intellectual level, and give satisfaction to the older practitioners?

I see two ideas: the meliorative and the ludic. And an approach, intensely particular.

Fancy words these. So let me flesh them out, the methodology first.

Intensely particular: Grandiose claims about the value of science and technology changing the world for the better are met today with infinitely more skepticism than when such claims were voiced in the nineteenth century. Indeed chemistry has been a major democratizing force, contributing strongly to an unprecedented extension of the life span, to control of women over fertility, to access to color and information and art to all. Yet we have also seen two World Wars, in one of which my father and three of four grandparents were killed, and the suicidal terrorism of today. All helped along by technology, if not science. It's somewhere between silly and insensitive to chastise thinking and feeling people for failing to see what science has wrought for them; the time has arrived to come to terms with the persistent, justified ambivalence of human beings about science and technology. And, incidentally, with the true change of which chemistry is emblematic.

Young people today, worldwide, are much more cynical than they once were about the positive claims of science, as they are of claims to the general good. Self-centered, professionally minded to an excess, materialistic, I still think that young people respond to particular, individual cases of service to humanity, especially their own. As small an action as giving blood, or hearing detailed accounts of Doctors Without Borders service in Chechnya, produces a moral high.

In the case of Green Chemistry, a revision of the Monsanto logo to "food, health, and hope" is met by our young people with apathy. But a description in class of the Green Chemistry Presidential Award winners' work, in all detail, will move them. Direct involvement in such a synthesis, or a senior undergraduate level exercise in redesigning an aspirin synthesis to run entirely in aqueous solvent—these will be valued.

A part of science, with its heavy burden of a reductionist core and emphasis on the reproducible, tends to devalue a particularist approach to knowing the world. The latter is best represented by a poem. About a weed, and no other weed. This particular weed. Or, in history, the exploration of medieval life through a selection of the Inquisition records on the events in Montaillou, a small French town that "succumbed" to the Cathar heresy. The particular need not be lost in chemistry—as everyone trying to repeat a synthesis that is in the literature knows. It matters in whose hands chemistry is done.

So, no slogans, please, just case studies. And not just masterpieces. Drafts and failures are incredibly instructive.

Meliorative: After I got through saying that nineteenth century slogans of the inherent goodness of science won't work, I want to still argue for a reemphasis of that function of chemistry which claims as one of our aims the improvement of the human condition. But I would like to do it modestly, with a recognition of our limitations, and emphasizing the personal and particular.

The modest derives from the history of the 20th Century (not that the 21st has opened auspiciously), and the environmental and ecological disasters (always understandable after the fact, always the product of mistakes. By others. Right?) brought about by our reaching for...a better world. I'm still an optimist (about most everything except myself), even after having lived through the Holocaust. So, without dreaming, I've used the word "meliorate" rather than the definite "ameliorate" (and here followed Peter Medawar.) Our work in chemistry should aspire to help humanity. Not to save it—no magic bullet, no technological fix. Just little people, following their curiosity, but still trying in small, specific ways to do something for others. While also improving their own lot—we are good at combining altruism with selfishness.

Something has been lost in the education of chemist. It's not the hagiographies of Louis Pasteur, Marie Curie and George Washington Carver, though they played a part for sure in getting *me* into science. I think Green Chemistry has a wonderful future, in well-told stories of specific pollution abatement, of preventative ecological detective work. I think that through the particular, the ecological, good young people can be handed a rationale for entering the long apprenticeship of chemistry.

All experience with introducing Green Chemistry in the laboratory or lecture, all that I have heard of, is extremely positive. Ears perk up; there is this tremendous desire by young people to do not only well, but to do good.

Ludic: A fancy word for having fun playing games. I could have brandished *Homo Ludens*, and cited Hesse's classic *The Glass Bead Game*. And dated myself with both. It's clear there's a game-playing aspect to science, and especially chemistry. Designing a synthesis, *and* making it work when the design fails, have this enticing puzzle-solving aspect. As does the detective work of a sporting structure determination.

Green Chemistry has a good chance of playing to this ludic aspect of science. In the existing industrial processes, people made things work, ingeniously so, subject, of course, to economic constraints. But maybe not the additional environmental and ecological limitations that must be faced today. It's a new game! The rules are not only "use the 50 top (cheapest) chemicals with a high yield," but also economize on atoms, minimize waste, have products and waste streams biodegradable.

Yes, the first synthesis is most fun. But we already know that syntheses have to be modified with scaling. And there is cachet in higher yield, with simpler reaction conditions and greater control. So there is in making a sustainable synthesis. The game is harder, it takes a cleverer solution. No cheating is allowed, by using some magic but dangerous solvent!

Here's this clever new game, with neat rules. And while these rules are not rigid, neither are they arbitrary. They're based on something very real, something urgent for the planet—our approximation, always tentative, of what will keep the earth and us whole.

Let's talk about some of the rules of this grand game.

Towards a Multifaceted Green Index[3]

We often say that action A is environmentally more benign than action B. Or want to say so. A could be having tea in washable ceramic mug, B using a disposable polystyrene cup. A could be the Monsanto isocyanate synthesis, an innovation, B the traditional one (isocyanates are produced worldwide to the extent of about 5 million tons a year. And not for fun.)

Often there is a gut feeling at work—anything must be better than phosgene as a reagent, right? Phosgene, a most toxic gas, a chemical weapon, is in fact widely used in the production of isocyanates. Well, it matters in whose hands the chemistry proceeds—some of us (not this theoretician) have learned to handle megakilos of dangerous chemicals, such as hydrogen, sulfuric acid, or phosgene, safely. You would have heard about any major disasters involving phosgene, if there had been any.[4]

Still more often, in the complex reality of this world, the actions of humans are simply difficult to compare in terms of their environmental cost. We are left unsure, if not paralyzed.

It would be good to have a measure of sustainability or greenness. We (Ignacio Permanyer Ugartemendia and I) think there could be one—we will suggest a research program, an ambitious and difficult one, to creating a reasonable and compassionate measure, a multifaceted gauge of transformation.

The multifaceted descriptor signifies for us a commitment to reality and a deliberate denial of one of humanity's banes—to see everything reduced to a single number or an ordinal ranking. Amartya Sen has warned of this: "The passion for aggregation makes good sense in many contexts, but it can be futile or pointless in others."[5] We will forward the idea of a tripartite view, devising first of all a reasonable physically-based transformation index, and coupling it with views of risk perception (as distinct from risk assessment), and of a bringing out of the social costs of the transformation posited.

The image we have is one founded in the present or foreseeable world of the computer. Imagine that we want to assess the consequences of using ceramic cups or Styrofoam ones for a year. In our dream world, the choice would be written up in reasonable simple language. A computer program, no more sophisticated than what is already out there working for Amazon would process the question, and return with some more questions (What is the usage envisioned, i.e., how many cups a year?) The program would then analyze the problem, presenting its results in the shape of the object envisaged—a mug and a plastic cup in this case. On its front face it would project the entropy-based transformation index we will describe. A twist of the cup (sensing controls at work) and up on another side of the cup would flash up a list of risk perceptions. Another rotation of the cup and there would come a sequence of images playing out social consequences.

We want the medium to be the message—that cup works in different contexts. To paraphrase Magritte, sometimes a cup is just a cup. Sometimes it is a small but much magnified (so many cups) manufactured object, with dispersed resource and pollution costs. It has connotations of comfort or even social justice (not only the rich should be able to have brightly colored cups). It would be instructive to see these. If one wanted to see them.

A Measure of the Artifactual

Some time ago, I found that I could begin a discussion of natural/unnatural by talking about cotton, rayon, and nylon. People consider cotton as natural, nylon as synthetic. Most don't know what rayon is, but generally dub it as synthetic. A gentle exposition makes clear the semi-synthetic nature of rayon, with a most natural material, wood pulp, at its source. The cellulose is modified by some neat chemistry, and then regenerated, some fraction of it remaining chemically modified (Figure 28-1). One can find good statistics on the chemically intensive agriculture of commercial cotton, with use of fertilizers, herbicides, and pesticides in excess of most crops, as well as genetically modified varieties. And most people accept the plant origins of petroleum, the "feedstock" from which nylon's monomers derive.

From that point, people can be induced to reflect, gently and thoughtfully, on the fact that of the three plant *seeds* that eventually led through a sequence of human interventions to cotton, rayon, and nylon, that only the seed that eventually became nylon could honestly be called natural. It alone was never touched by a human being.

A trivial insight, perhaps. But it serves as a lead-in to a discussion of other human motions at the natural/artifactual border. Still the feeling persists, even after such a discussion on origins, that nylon is more *transformed* than the others. Maybe it is, maybe it isn't. I am not a great fan of quantification (as surprising as that may seem to be for a theoretical chemist), but I would like to have some "transformation index," built up of pieces which I can understand, which could tell me the "transformation distance" of cotton, rayon, and nylon from their plant origins.

I am led, and not because I love thermodynamics (it is a beautiful structure, I do like to teach it) to a measure that focuses on the most fundamental of physical

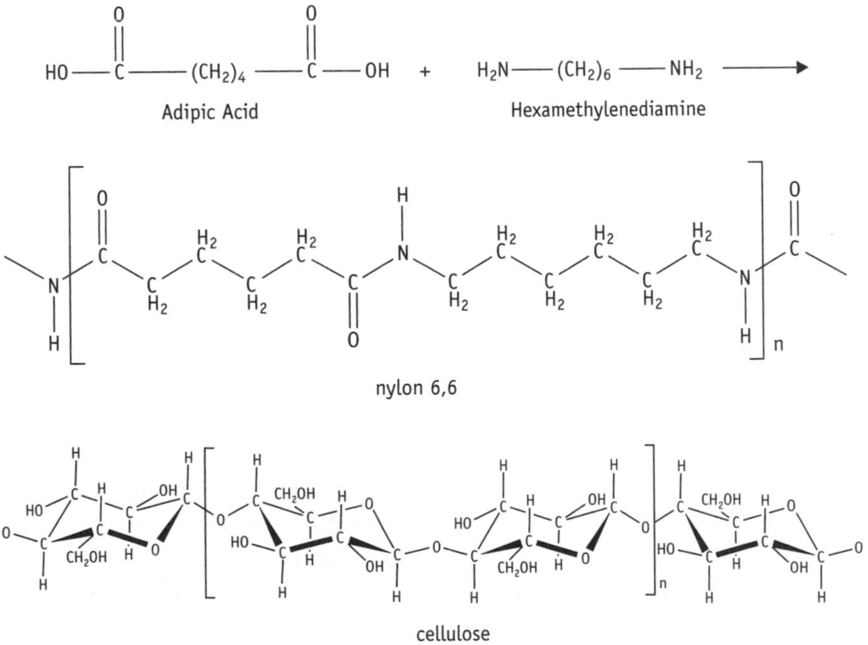

Figure 28-1 Top two lines: Nylon, a synthetic copolymer. Bottom: Cellulose, the basic structure in cotton and rayon.

quantities, energy and entropy. Cotton, rayon, nylon are organized parts of our environment—as are we, our poems, our houses, our laws. The second law of thermodynamics tells us that in every process there is an increase in the entropy of the universe. The more efficient the process, the less entropy is dissipated. So why not take the sum of the entropy changes, alternatively the useless heat added to the environment, as a measure of transformation?

There exist already closely related and better reasoned (and sometimes controversial) alternative analyses to what I propose—those of "exergy" and "emergy."[6,7,8] Exergy, for instance, is also a thermodynamically based quantity, the useful energy. An exergy analysis focuses on what energy remains available in a process; the transformation distance measure I suggest on what is lost.

Two more things need to be said: First, contemporary industrial practice, of course seeking lower cost, already has effectively implemented the tools for such an analysis. At least for the piece of transformation it chooses to consider—the manufacture of a product. Second, one has to face head on the problem that greenness will not, cannot be measured by a single number. The influence of transformative processes on society has to be taken into account, as do the perceptions of people on any accompanying risks. I will return to these.

Toward Measuring the Costs of Transformation

First, some assumptions that might enter the construction of an index:

(a) Chemical reactions are transformations of matter.
(b) Some will proceed spontaneously, others will require an input of energy.
(c) At the beginning of every sequence of transformations and every energy source are natural materials.
(d) In the way human beings are, subverting their own efforts to categorize things, living organisms, bred by humans, are often used in transformations. So I think agriculture must be viewed as a transformative activity.
(e) Each transformation requires *materials* (reactants and auxiliaries, such as solvents and catalysts), *energy*, a *locus* for the transformation (a chemical plant or factory, with tools). And *labor*.

It would seem that the simplest index one could use would be a cost one. Labor, place, energy have to be paid for. I'm not an economist, but I can see that each category I've mentioned in (e) above has problems in using a price measure—all energy forms are subsidized (security costs in Middle East to keep U.S. gasoline prices low; solar energy research). Labor costs vary widely. One will have to use dollar costs at some point in the calculation, but let me go ahead and discuss the implementation and problems for an index such as I would suggest.

The general idea is that the measure of transformation be the sum of the total energy "wasted," not recoverable, in the overall process. To put it technically, for any one individual step in the overall process:

Transformation index = unrecoverable heat added to universe = (entropy change of universe in that step) x (temperature of surroundings) – energy stored as chemical energy of products

One reason I suggest this measure is that for many industrial chemical reactions the heating protocol is accessible. So one knows how much heat is put into the universe. And the chemical energy stored is also known. Though one makes an effort to make use of the heat produced in chemical reactions, in practice I believe that most of it is unrecovered, dissipated. The transformation index measures the fundamental inefficiency of the transformation of matter. The lower, the better.

There already are in industrial practice many components of such an analysis. So we see use of essential considerations of atom economy, and of the concept of an E factor, the ratio of the mass of the masses of waste produced to useful products (introduced by R. A. Sheldon[9]) in a process. The energy needs of a process are routinely analyzed in terms of their mass and energy requirements at source, and their flow thereafter. No company could turn a profit without such analyses.

I still think there is a place for a global index of transformation, long term and long range. A global entropic view—the heat dissipated in the sum of all components, even those hidden, in a cradle to grave chemical process—would allow us to get a comparative measure of seemingly incommensurable processes.

It's Not Going to Be Simple

There are problems, for sure, in implementing a dissipated energy or entropy index of transformation.

Equations that chemists are used to will have to be modified, for

$$A + B \rightarrow C + D$$

by itself doesn't tell you much about the greenness of the process. Yes, I know about equilibrium, but almost all useful transformations are desired perturbations of equilibria, so I use the unidirectional arrow.

Equations will look more like

A + B + energy input + labor + other costs + equipment + medium (solvent)
→ C + D + reusable solvent + reusable equipment + wastes from making
medium and equipment reusable + heat dissipated

or, in another form, call it (to use a common term) "cradle to gate":

$$natural\ substances \rightarrow C + D + wastes + heat\ lost$$

or in a lifetime, or grand cycle, or "cradle to grave," analysis one will have to add

$$C + D \rightarrow substances\ modified\ but\ reusable + wastes + heat\ dissipated.$$

Current industrial practice is superb at the "gate-in" to "gate-out" analysis. We and industry need help in quantifying, or just becoming aware of the "cradle to gate-in" and "gate-out to grave" pieces of human transformation.

As obvious as these considerations are, there is still little that is unambiguous in the above equations. Take energy input. We have a clear picture of what it takes to make a chemical reaction run, say 100 kJ/gram of C produced. The 100 kJ comes in as electricity. In a given country we can make an estimate of the origins of that energy—petroleum, coal, gas, water power, nuclear. We can analyze for each type of energy the heat added to the universe. Roughly. Much thought already has gone into such calculations; they would be the first thing that a hypothetical institute (it's not a job for one person, or even one group) devoted to the implementation of a transformation index would focus on.

There are unknowns that make the process difficult—for instance, to this day, despite the expense of over $10 billion at Yucca Flat, we have no good figure for the costs of long-term nuclear waste disposal. And government subsidies, which I have already mentioned, confuse the analysis of energy costs.

How to Include Labor and Ingenuity?

The most troublesome components in constructing a transformation distance will be "other costs" and labor. By "other costs" I mean the expense of creating the environment where the labor of transformation can be done. Governments have called it "overhead," and its range is wide, a multiplier of tangible costs in the range of 2. The "other costs" include the social setting of educating people with expertise, and the marketing, public relations and personnel costs that allow a modern industrial enterprise to provide the setting for A + B → C + D.

And it all can't proceed without ideas—the inventiveness of the underlying science, the ingenuity of taking that knowledge into practice. How to relate labor and intellectual property to the entropy added to the universe?

In the cost perspective, the market cost of labor is relatively easy to calculate. From an entropy or heat loss viewpoint it isn't. Shall we take a mix of workers, give them a working lifetime, look at that fraction of time worked, and calculate back from their diet in kJ to a contribution to the entropy of the universe? I think this somewhere between inhuman and silly, and certainly does not recognize the imaginative component of work.

Economists have been ingenious at valuing immaterial goods. It's worth to explore various possibilities for giving an entropic cost to the seemingly unmeasurable.

Less Tangible, but No Less Important Considerations in Any Measure of Greenness

Supposing a believable transformation distance index could be devised, with much work. It will not suffice as a green index, I believe, for it does not address human perceptions of risk, nor intangible costs to society. Let me explain why I think so.

A distinction has to be made between risk *assessment* and risk *perception*. Perception is individual (though it may be firmly rooted in group mentality) and subjective. To me, as a writer, subjective is hardly a bad word. Assessment tries to be more objective and more general. It must do so, but there is a hidden tension, from which so much anguish and misunderstanding ensues—in trying to be objective, assessment risks being felt as arrogant and inhuman. In writing about this elsewhere,[10] I have urged scientists to avoid scientistic stances based on their perceived assessment, especially in times of crisis.

Control over a risk factor, the empowerment to do something about it, both figure importantly in risk perception. Nuclear power, arguably the least polluting source of energy, has an almost insurmountable risk perception factor operating against it, because of (a) its association with nuclear weapons, (b) its intangibility, (c) its distance from our control, and (d) our inability to put a cost factor on long-term disposal of nuclear wastes.

Consequence to society requires a crystal ball that is less cloudy than what we carry around in our head. The heirs of Prometheus, despite his name, are better at tinkering (call it technological fixes) than foresight. And yet we have to try.

Consider, for instance, global warming, largely due (I think the consensus is in place) to anthropogenic CO_2 production. The transformation index might take any fuel to CO_2 and potable water, but we can't stop there, given what we know. A typical car puts into the atmosphere more than its weight of CO_2 each year; the long-term consequences on the environment need to be described.

Of course, people have worried about the carbon cycle. One hope has been biomass production of fuel. My colleague David Pimentel (together with Tad W. Patzek) several years ago made an analysis of the energy costs, in the spirit of what I have proposed, but much more professional. Their best estimate is that it's not worth it; the energy cost of production, from a variety of crops, is greater than the energy made available. Burning biomass to heat homes is apparently economical. I believe that the physical or chemical conversion of light into electricity, not biofuels will be the central component of a future life cycle of the world.

As described, I would propose that a green index should contain three components, the first a number, the other parts descriptive. The first piece is the transformation index I'll discuss. Summed over history and extrapolated to the future, it will ideally be a measure of "transformation distance" from the natural origins, based on heat dissipated, entropy added to the universe. The second component of a green index, call it "perception," is its most subjective piece; a description of dread potential, of lack of control, lack of empowerment in a transformation. Psychologists, journalists, historians would be enlisted in the task of compiling this part.

The last part of a three-tiered index, call it "societal cost," would contain some estimate of the effects on society of the transformation, aside from their assessed inefficiency. Subjective for sure, and always subject to revision.

Essential Change

Nothing is simple in what faces us here, but imagine the reaction of eighteenth-century chemists to Lavoisier's insistence that nothing is gained, nothing is lost in a chemical reaction. They might have argued (some did) "don't bother us about mass balance; all we care about is that phosphorus burns." Worrying about mass conservation is precisely what led to modern chemistry. A concern about what exactly heat and work meant gave us the marvelous structure of thermodynamics.

I believe there is a new chemical revolution in the making. Its components: analyzing the atom and energy costs of all reactions (not just the one we desperately want) for sure. And, more difficult for scientists, listening to the perception of human beings, and looking all the while, as best as we can, for environmental and ecological consequences. The outcome of such a revolution will be the creation of real sustainable chemistry, a chemistry in balance with the world.

It's complicated, and to me seems a life of work for a group of dedicated, ingenious people. Not a back of the envelope calculation, and no, molecular orbitals will not help much. I wish I were younger; I would take on the challenge. It's worth trying to do.

The Two Adams

There need not be a no-man's land between environmentalism and science. I would like to advocate a common ground shaped, interestingly enough, in the Judeo-Christian religious tradition.

Everyone who has read Genesis will have noted that there seem to be two Creations conflated into one. So in Genesis 1 it says (King James version here):

> And God said, Let us make man in our image, after our likeness: and let them
> have dominion over the fish of the sea, and over the fowl of the air, and over the

cattle, and over all the earth, and over every living thing that moveth upon the earth. So God created man in his own image, in the image of God he created he him; male and female created he them. And God blessed them, and God said unto them, Be fruitful and multiply, and replenish the earth, and subdue it:

In Genesis 2 God forms man again!

When the Lord God made earth and heaven—when no shrub of the field was yet on earth and no grasses of the field had yet sprouted, because the Lord God had not sent rain upon the earth and there was no man to till the soil, but a flow would well up from below the ground and water the whole surface of the earth—the Lord God formed man from the dust of the earth. He blew into his nostrils the breath of life, and man became a living being ... The Lord God took the man and placed him in the garden of Eden, to till it and tend it.
　　　　　　　　　　　　　　　　— Jewish Publication Society translation[11]

This man is alone, and needs a helpmeet.

The story of the two Adams has puzzled many. Literary scholars have seen in the two Adams an imperfect merger of two texts. Rabbi Joseph Soloveitchik, an important Jewish thinker of the 20th Century, sees something deeper—two sides of humanity. In a beautiful piece of writing, he characterizes the two Adams:

Adam the first who was fashioned in the image of God was blessed with great drive for creative activity and immeasurable resources for the realization of this goal, the most outstanding of which is his intelligence, the human mind, capable of confronting the outside world and inquiring into its complex workings ... God, in imparting the blessing to Adam the first and giving him the mandate to sub-due nature, directed Adam's attention to the functional and practical aspects of his intellect through which man is able to gain control of nature.[12]

Adam the first, Soloveitchik says, wants to be human, to discover his identity. And he acquires *"dignity through glory, through his majestic posture vis-à-vis his environment."* Solveitchik juxtaposes the personality of Adam the first with that of the modern sci-entist, the transformer of nature who does not necessarily understand it.

Soloveitchik describes Adam the second as a very different creature. Alone, so alone at first, he is led to think, to ponder the questions of Why? What? Who? To answer these questions Adam the second

does not create a world of his own. Instead, he wants to understand the living, "given" world into which he has been cast. Therefore he does not mathematize phenomena or conceptualize things. He encounters the universe in all its col-orfulness, splendor, and grandeur, and studies it with the naïveté, awe, and admiration of the child who seeks the unusual and wonderful in every ordinary thing and event. While Adam the first is dynamic and creative, transforming sensory data into thought constructs, Adam the second is receptive and beholds the world in its original dimensions. He looks for the image of God not in the mathematical formula or the natural relational law but in every beam of light, in every bud and blossom, in the morning breeze and the stillness of a starlit evening.[12]

Soloveitchik continues:

Adam the first He told to exercise mastery and to "fill the earth and subdue it," Adam the second, to serve. He was placed in the Garden of Eden "to cultivate it and to keep it."[12]

Too many of the critics of the alleged Judeo-Christian responsibility for the environmental crisis we must face have chosen to castigate Adam 1. Some scientists, in a fit of nineteenth-century confidence, have chosen to emulate the same Adam. It is time to see both Adams, both Eves, irrevocably choosing to live, not without anguish, but to live, on the only earth given to them. To us.

Notes

1. S. Boeschen, D. Lenoir, and M. Scheringer, Sustainable Chemistry: Starting Points and Prospects. *Naturwissenschaften* 90 (2003): 93.
2. B. M. Trost, Atom Economy—A Challenge for Organic Synthesis: Homogeneous Catalysis Leads the Way. *Angew. Chem. Int. Ed. Engl.* 34 (1995): 259.
3. The construction of an index of transformation is a work in progress, in collaboration with Ignacio Permanyer Ugartemendia.
4. Nothing in this world is completely safe. On March 6, 2000, a phosgene gas leak from a Thai plastics factory killed 1 person and injured 814 others. A laboratory accident involving inadvertent phosgene release in Fuzhou, China, on June 16, 2004, killed 1 person and injured more than 260 others. A phosgene-containing pipe rupture on September 8, 1994, in Yeochon, Korea, resulted in multiple injuries and 3 deaths. http://emedicine.medscape.com/article/832454-overview
5. A. K. Sen, J. Mullbauer, R. Kanbur, K. Hart, and B. Williams, *The Standard of Living, The Tanner Lectures 1985* (Cambridge: Cambridge University Press, 1987), p. 33.
6. I. Dincer and M. A. Rosen, *Exergy: Energy, Environment, and Sustainable Development* (New York: Elsevier, 2007).
7. J. L. Hau and B. R. Bakshi, Expanding Exergy Analysis to Account for Ecosystem Products and Services. *Environmental Science and Technology* 38 (2004): 3768, and references therein.
8. H. T. Odum, *Environmental Accounting: Energy and Environmental Decision Making* (New York: Wiley, 1996).
9. R. A. Sheldon, E Factors, Green Chemistry and Catalysis: An Odyssey. *Chem. Commun.* 2008: 3352.
10. R. Hoffmann, *The Same and Not the Same* (New York: Columbia University Press, 1995).
11. The Torah (Philadelphia: The Jewish Publication Society of America, 1962).
12. J. B. Soloveitchik, *The Lonely Man of Faith* (New York: Doubleday, 1965).

Sources

Most of the articles in this collection have been previously published. Some have been slightly revised by Roald Hoffmann and the editors to eliminate redundancies and to bring the scientific content and references up to date. A few articles are the texts of unpublished lectures. The sources for the articles are listed below. All are used with permission from the publishers, which is gratefully acknowledged. Unless otherwise noted, Roald Hoffmann is the sole author of that chapter.

Chapter 1   "Trying to Understand, Making Bonds," unpublished lecture at 2007 ACS Meeting Symposium for Roald Hoffmann's 70th birthday.

Chapter 2   "Why Buy That Theory?" *American Scientist*, **91**, 9–11 (2003).

Chapter 3   "What Might Philosophy of Science Look Like If Chemists Built It?" *Synthèse*, **155**(3), 321–336 (2007).

Chapter 4   "Qualitative Thinking in the Age of Modern Computational Chemistry, or What Lionel Salem Knows," *Journal of Molecular Structure (Theochem)*, **424**, 1–6 (1998).

Chapter 5   "Narrative," *American Scientist*, **88**(4), 310–313 (2000).

Chapter 6   "Unstable," *American Scientist*, **75**, 619 (1987).

Chapter 7   "Nearly Circular Reasoning," *American Scientist*, **76**, 182 (1988).

Chapter 8   "Ockham's Razor and Chemistry," R. Hoffmann, V. I. Minkin, and B. K. Carpenter, *Bulletin de la Société Chimique de France*, **133**, 117–130 (1996); Reprinted in *Hyle*, **3**, 3–28 (1997).

Chapter 9   "Learning from Molecules in Distress," R. Hoffmann and Henning Hopf, *Angewandte Chemie*, **47**, 4474–4481 (2008).

Chapter 10   "Why Think Up New Molecules?" *American Scientist*, **96**, 372–374 (2008).

Chapter 11   "How Should Chemists Think?" *Scientific American*, **268**, 66–73 (February 1993).

Chapter 12   "Protean," R. Hoffmann and P. Laszlo, *Angewandte Chemie, International Edition*, **40**(6), 1033–1036 (2001).

Chapter 13   "Under the Surface of the Chemical Article," *Angewandte Chemie, International Edition,* **27,** 1593–1602 (1988).

Chapter 14   "Representation in Chemistry," R. Hoffmann and P. Laszlo, *Angewandte Chemie, International Edition,* **30,** 1–16 (1991).

Chapter 15   "The Say of Things," R. Hoffmann and P. Laszlo, *Social Research,* **65**(3), 653–693 (1998).

Chapter 16   "How Symbolic and Iconic Languages Bridge the Two Worlds of the Chemist: A Case Study from Contemporary Bioorganic Chemistry," E.R. Grosholz and R. Hoffmann, in Nalini Bhushan and Stuart Rosenfeld, *Of Minds and Molecules: New Philosophical Perspectives on Chemistry,* Oxford University Press (2000).

Chapter 17   "How Nice to Be an Outsider," *Canadian Review of Comparative Literature,* **26** 163–169 (1999).

Chapter 18   "The Metaphor, Unchained," *American Scientist* **94**(5), 406–7 (2006).

Chapter 19   "Art in Science?" *Q (A Journal of Art),* 62–65 (May 1990).

Chapter 20   "Science and Crafts," in *The Nature of Craft and the Penland Experience,* ed. Jean McLaughlin, New York: Lark Books, 58–64 (2004). We are grateful to the Penland School of Crafts for permission to reprint this chapter.

Chapter 21   "Molecular Beauty," *Journal of Aesthetics and Art Criticism,* **48**(3), 191–204 (1990).

Chapter 22   "Teach to Search," *Journal of Chemical Education,* 73(9), A202–A209 (1996).

Chapter 23   "Some Heretical Thoughts on What Our Students Are Telling Us," R. Hoffmann and B. P. Coppola, *Journal of College Science Teaching,* **XXV,** 390–394 (1996).

Chapter 24   "Specific Learning and Teaching Strategies That Work, and Why They Do So," R. Hoffmann and Saundra Y. McGuire, a condensed version was published with the title, "Learning and Teaching Strategies" in *American Scientist,* 98, 378–382 (2010).

Chapter 25   "Mind the Shade," *Chemical and Engineering News,* **75,** 3 (Nov. 10, 1997).

Chapter 26   "Science and Ethics: A Marriage of Necessity and Choice for this Millennium," unpublished lecture at IUPAC meeting 2007.

Chapter 27   "Honesty to the Singular Object," an essay for Sprache, Lügen, und Moral edited Margery Arent Safir, Suhrkamp Insel, Frankfurt, 2009 (published in German), pp. 84-110.

Chapter 28   "The Material and Spiritual Rationales Are Inseparable," an unpublished lecture at the Gordon Conference in Green Chemistry, 2004.

Index